Encyclopedia of Seismic Waves

Encyclopedia of Seismic Waves

Edited by **Agnes Nolan**

R CALLISTO REFERENCE

New York

Published by Callisto Reference,
106 Park Avenue, Suite 200,
New York, NY 10016, USA
www.callistoreference.com

Encyclopedia of Seismic Waves
Edited by Agnes Nolan

International Standard Book Number: 978-1-63239-295-4 (Hardback)

Printed in the United States of America.

Contents

Preface   VII

Chapter 1   **Seismic Wave Interactions Between the Atmosphere - Ocean
             - Cryosphere System and the Geosphere in Polar Regions**   1
             Masaki Kanao, Alessia Maggi, Yoshiaki Ishihara,
             Masa-yuki Yamamoto, Kazunari Nawa, Akira Yamada,
             Terry Wilson, Tetsuto Himeno, Genchi Toyokuni,
             Seiji Tsuboi, Yoko Tono and Kent Anderson

Chapter 2   **Electric and Electromagnetic Signals Under, On, and Above
             the Ground Surface at the Arrival of Seismic Waves**   21
             Akihiro Takeuchi, Kan Okubo and Nobunao Takeuchi

Chapter 3   **Electric Displacement by Earthquakes**   45
             Antonio Lira and Jorge A. Heraud

Chapter 4   **Wavelet Spectrogram Analysis of
             Surface Wave Technique for Dynamic Soil
             Properties Measurement on Soft Marine Clay Site**   59
             Sri Atmaja P. Rosyidi and Mohd. Raihan Taha

Chapter 5   **Quasi-Axisymmetric Finite-Difference Method
             for Realistic Modeling of Regional and Global
             Seismic Wavefield – Review and Application**   85
             Genti Toyokuni, Hiroshi Takenaka and Masaki Kanao

Chapter 6   **The Latest Mathematical Models of Earthquake Ground Motion**   113
             Snezana Gjorgji Stamatovska

Chapter 7   **Coupling Modeling and Migration for Seismic Imaging**   133
             Hervé Chauris and Daniela Donno

Chapter 8   **Using a Poroelastic Theory to Reconstruct
             Subsurface Properties: Numerical Investigation**   151
             Louis De Barros, Bastien Dupuy, Gareth S. O'Brien,
             Jean Virieux and Stéphane Garambois

Chapter 9   **Wave Propagation from a**
             Line Source Embedded in a Fault Zone   173
             Yoshio Murai

Chapter 10   **Effects of Random Heterogeneity**
             on Seismic Reflection Images   191
             Jun Matsushima

Chapter 11   **Seismic Modeling of Complex Geological Structures**   213
             Behzad Alaei

Chapter 12   **Modelling Seismic Wave**
             Propagation for Geophysical Imaging   237
             Jean Virieux, Vincent Etienne, Victor Cruz-Atienza,
             Romain Brossier, Emmanuel Chaljub, Olivier Coutant,
             Stéphane Garambois, Diego Mercerat, Vincent Prieux,
             Stéphane Operto, Alessandra Ribodetti and Josué Tago

Chapter 13   **Shear Wave Velocity Models Beneath**
             Antarctic Margins Inverted by Genetic
             Algorithm for Teleseismic Receiver Functions   289
             Masaki Kanao and Takuo Shibutani

Chapter 14   **2.5-D Time-Domain Finite-Difference**
             Modelling of Teleseismic Body Waves   305
             Hiroshi Takenaka and Taro Okamoto

             Permissions

             List of Contributors

Preface

Detailed information regarding the field of seismic waves has been presented in this insightful book. Seismic wave research not only enables us to assess earthquakes and tsunamis but it also acts as an indicator to determine Earth's features and composition. It has also led to the discovery of Mohorovicic discontinuity. With the development of theoretical understanding of physics behind seismic waves, numerical and physical modelings have remarkably advanced. It has also contributed towards some advanced applications like utilization of seismic stimulation to enhance the productivity of oil wells and utilization of artificially-induced shocks to examine subsurface of Earth. This book illustrates developments in seismic wave analysis from a theoretical perspective. It also discusses numerical simulations, research applications, data acquisition and interpretation. It includes contributions of scientists and researchers from all over the world.

This book has been the outcome of endless efforts put in by authors and researchers on various issues and topics within the field. The book is a comprehensive collection of significant researches that are addressed in a variety of chapters. It will surely enhance the knowledge of the field among readers across the globe.

It is indeed an immense pleasure to thank our researchers and authors for their efforts to submit their piece of writing before the deadlines. Finally in the end, I would like to thank my family and colleagues who have been a great source of inspiration and support.

Editor

Seismic Wave Interactions Between the Atmosphere - Ocean - Cryosphere System and the Geosphere in Polar Regions

Masaki Kanao et al.[*]

National Institute of Polar Research, Tokyo
Japan

1. Introduction

At the time of the International Geophysical Year (IGY; 1957-1958), it was generally understood by a majority of seismologists that no extreme earthquakes occurred in polar regions, particularly around Antarctica. Despite the Antarctic being classified as an aseismic region, several significant earthquakes do occur both on the continent and in the surrounding oceans. Since IGY, an increasing number of seismic stations have been installed in the polar regions, and operate as part of the global network. The density of both permanent stations and temporary deployments has improved over time, and has recently permitted detailed studies of local seismicity (Kaminuma, 2000; Reading, 2002; 2006; Kanao et al., 2006).

Several kinds of natural seismic signals connected to the atmosphere - ocean - cryosphere system can be detected in polar regions (Fig. 1). Ice-related seismic motions for small magnitude events are generally named 'ice-quakes' (or 'ice-shocks') and can be generated by glacially related dynamics (Tsuboi et al., 2000; Anandakrishnan et al., 2003; Kanao and Kaminuma, 2006). Such cryoseismic sources include the movements of ice sheets, sea-ice, oceanic tide-cracks, oceanic gravity waves, icebergs and the calving fronts of ice caps. At times, it can be hard to distinguish between the waveforms generated by local tectonic earthquakes and those of ice-related phenomena. Cryoseismic sources are likely to be influenced by environmental conditions, and the study of their temporal variation may provide indirect evidence of climate change..

In the Arctic, particularly in Greenland, the largest outlet glaciers draining the northern hemisphere's major ice cap have suffered rapid and dramatic changes during the last

[*] Alessia Maggi[2], Yoshiaki Ishihara[3], Masa-yuki Yamamoto[4], Kazunari Nawa[5], Akira Yamada[6], Terry Wilson[7], Tetsuto Himeno[1], Genchi Toyokuni[1], Seiji Tsuboi[8], Yoko Tono[8] and Kent Anderson[9]
[1]*National Institute of Polar Research, Tokyo,* [2]*Institut de Physique du Globe de Strasbourg, CNRS and University of Strasbourg,* [3]*National Astronomical Observatory, National Institutes of Natural Sciences, Iwate,* [4]*Kochi University of Technology, Kochi,* [5]*National Institute of Advanced Industrial Science and Technology, Tsukuba,* [6]*Geodynamics Research Center, Ehime University, Ehime,* [7]*Ohio State University,* [8]*Japan Agency for Marine-Earth Science and Technology, Yokohama,* [9]*Incorporated Research Institutions for Seismology, Washington, DC,* [2]*France;* [1,3,4,5,6,8]*Japan;* [7,9]*USA*

decade. They have lost kilometers of ice mass at their calving fronts, thinned by 15% or more in their lower reaches, accelerated by factors of 1.5 (Howat et al., 2005; Rignot and Kanagaratnam, 2006), and generated increasing numbers of large glacial earthquakes (Ekström et al., 2003; 2006). These significant changes, which have occurred as the climate has warmed and surface melting on Greenland has increased (Steffen et al., 2004), highlight the importance of dynamic processes operating within the polar ice sheet and at its outlet glaciers.

In this chapter, several features of seismic wave propagation in polar regions are illustrated, through discussion of travel time anomalies, wave amplitudes and the frequency content of power spectral densities (PSD). Characteristic seismic signals are classified into one of three main categories according to their origin: ice-related phenomena, oceanic loading effects and atmospheric perturbations. The physical interaction mechanisms between the atmosphere - ocean - cryosphere system and the geosphere (solid Earth) in polar regions are analyzed, and their possible use as climate change indicators is discussed.

Fig. 1. Photographs of atmosphere - ocean - cryosphere and geosphere environments in polar regions, particularly around the Lützow-Holm Bay (LHB) region, Antarctica. Left , from top to bottom: a glacier, tide-cracks, iceberg and sea ice. Middle: Syowa Station (SYO) in the Ongul Islands, surrounded by sea-ice and separated from the continent by the Ongul Channel. Right, from to to bottom: glacier and mountains, the 'spring river' between sea-ice and the coast, ice fall at the edge of a small glacier. All photos are provided by M. Kanao and NIPR

2. Seismic signals from the ocean

All seismic stations deployed on the Earth's surface record ubiquitous signals at periods between 4 and 25 s are commonly referred to as "microseisms". In the absence of earthquakes, microseismic waves are the strongest amplitude signals worldwide. Microseisms are considered to be dominated by Rayleigh waves that arise from gravity waves in the ocean that are forced by surface winds. The period ranges of microseisms are dictated by the physics of gravity wave generation, and are constrained by the speed and extent of Earth's surface winds (Aster et al., 2008; Aster, 2009; Bromiriski, 2009).

The microseism spectrum has a bimodal composition, caused by the existence of two distinct physical mechanisms that transfer ocean wave energy to seismic waves in geosphere

(Fig. 2). The first spectral peak between approximately 12 and 30 s, commonly called "primary" or single-frequency microseism (SFM), arises from the transfer of ocean gravity wave (swell) energy to seismic waves as oceanic waves shoal and break in the shallow waters. The highest amplitude and longest period swells are created by large and intense storms that generate strong sustained winds over a large area. Swell propagates dispersively across ocean basins, which results in longer period swell arriving at the coast before the shorter period swell. This period-dependent delay is readily measured in data recorded by seismic stations, ocean buoys, and seismographs, such as those deployed recently on a giant Antarctic iceberg (MacAyeal et al., 2009).

The second, more prominent, microseism peak between approximately 4 and 10 s, commonly called secondary or double-frequency microseism (DFM), arises from nonlinear interaction of interfering ocean wave components that produce a pressure pulse at double their frequency. This pressure pulse propagates with little or no attenuation to the sea floor where it generates seismic waves. The DFM is thought to be generated both near the coasts, where coastal swell reflection can provide the requisite opposing wave components, and in the deep southern ocean.

An example of PSD and the origin for microseisms

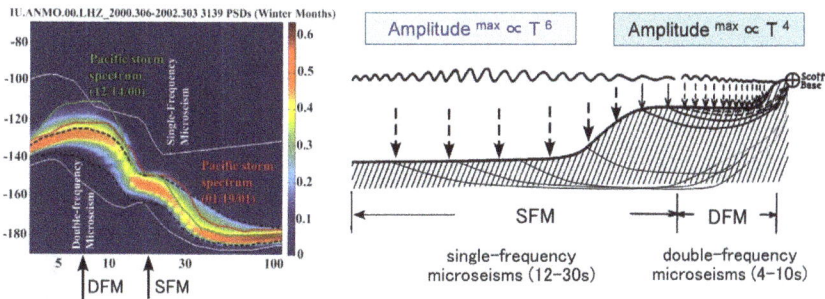

Fig. 2. Left: An example of a typical power spectral density plot (PSD) (modified after Aster, 2009). Right: An illustration of the origin of microseisms (modified after Hatherton, 1960)

3. Microseisms in polar regions

On a global scale, microseism amplitudes are generally highest during local winter, because nearby oceans are stormier in winter than in summer (Stutzmann et al., 2009). In polar regions, particularly from the evidence of Antarctic stations, the opposite observation is made: microseism amplitude is attenuated during local winter for both primary and secondary microseisms (Hatherton, 1960). The observation is explained by the presence of the sea-ice extent impeding both the direct ocean-to-continent coupling that generates the SFM and the coastal reflection which is an important component in the generation of the DFM (Grob et al., 2011).

In order to illustrate the variability of microseismic amplitude over time we have calculated power spectral densities (PSD) for data from the broadband seismometer (STS-1) at Syowa Station (SYO; 39E, 69S, Lützow-Holm Bay, East Antarctica). During time period shown, continuous STS-1 waveform data with 20 Hz sampling were automatically transmitted from

SYO to the National Institute of Polar Research (NIPR) by an Intersat telecommunication system (Aoyama and Kanao, 2010; Iwano and Kanao, 2009).

Figure 3 shows PSDs of the vertical broadband seismometer at SYO for a typical austral summer day in January, 2010, over the period band 0.1-100s. The DFM is clearly visible, as are several high-frequency dispersed signals that may be caused by variations of the ice environment in the vicinity of the station.

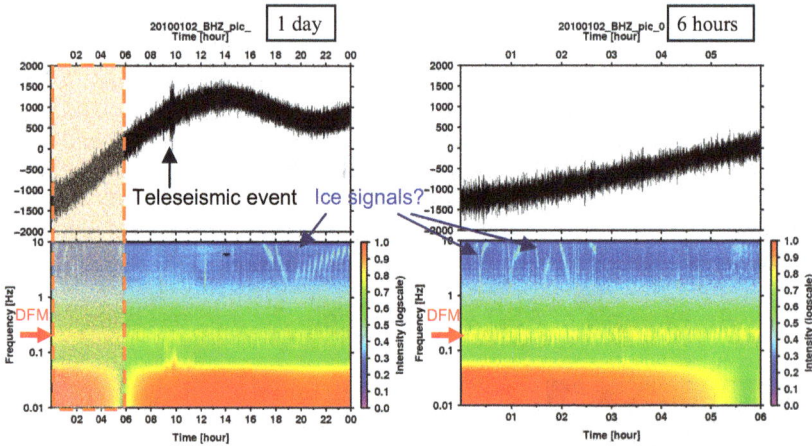

Fig. 3. The power spectral densities (PSD) of the vertical broadband seismometer (STS-1V) at SYO for January 02, 2010. Left: one day of data. Right: 6 hours of data, corresponding to the shadowed area of the left figure.

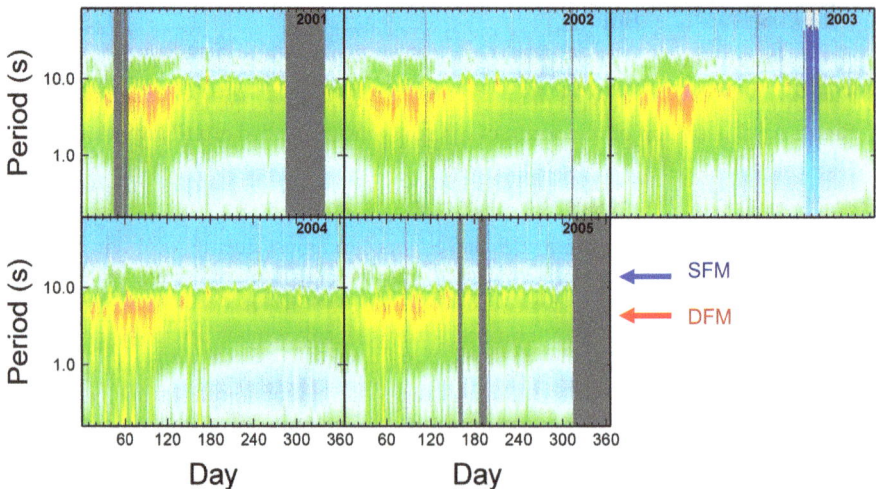

Fig. 4. Power spectral densities (PSD) of the broadband seismometer (STS-1V) at Syowa Station (SYO), Antarctica, for a period in 2001-2005. Signals corresponding to SFM and DFM are indicated by blue and red arrows (modified after Grob et al, 2011)

Figure 4 shows similar PSDs calculated for 5 consecutive years (2001-2005) over the period band 0.1-80 s. The DFM can be identified all year round, though with distinctly lower amplitude during the local winters (april-october). The relatively high degree of inter-annual variability presumably reflects the large influence of extratropical cyclonic storms that commonly affect both the northern and southern oceans. On the contrary, the SFM is observed only under excellent storm conditions during the austral local winter. The strength of both DFM and SFM are strongly related to the seasons, but presumably also to local ice conditions. For example, summers with lower amplitude microseisms at SYO correspond to residual sea-ice extension area near the Enderby Land coast (Grob et al, 2011).

In contrast, one-day PSD images for broadband seismograph at SYO clearly represent the continuous DFM; which was detectable in any time slots when storms or blizzards visited the station (Fig. 3). The DFM could probably be generated from the near southern oceans, including the vicinity of Lützow-Holm Bay.

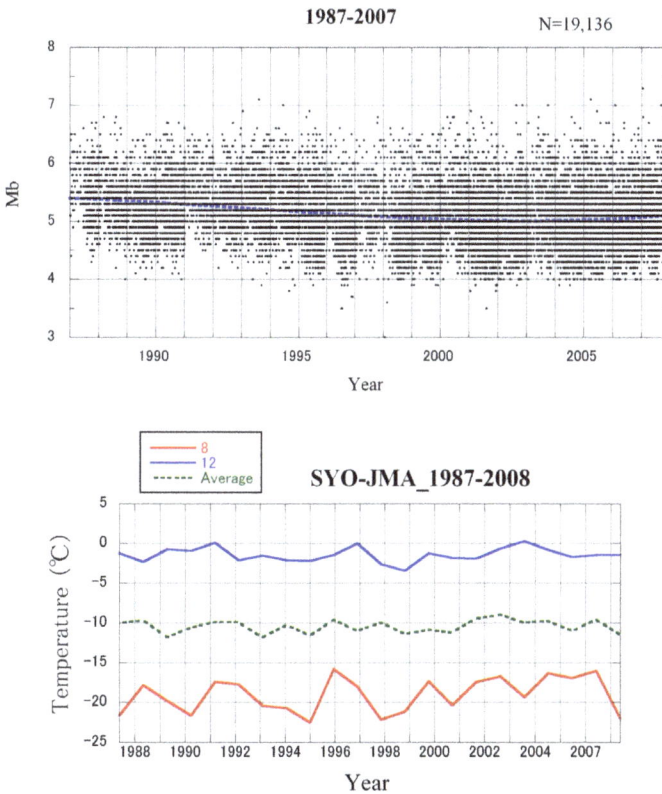

Fig. 5. Top: Teleseismic detectability at Syowa Station (SYO), Antarctica in 1987-2007 (modified after Kanao, 2010) as a function of Mb. Broken blue line indicates the smoothed average of Mb values for the whole period. Bottom: Temperature variations at SYO for the same years. Blue solid line: August (austral winter); red solid line: December (austral summer); green broken line: averaged values for a whole season

An important parameter for station operators is the teleseismic detectability, i.e. the capability of a station to detect a seismic event occurring at teleseismic distances (over 30° away). This parameter is strongly correlated to the noise level at the station. Temporal variations in teleseismic detectability at SYO were investigated for the period from 1987 to 2007 by Kanao, 2010. Figure 5 shows the body-wave magnitudes (Mb) for the events detected at SYO over the past two decades. The magnitudes of detected events range from a maximum of 6.5-7.0 to a minimum of 4.0-4.5. The average detected magnitude has decreased slightly over time, as the station quality has improved. During the austral summer, the station shows less teleseismic detectability (i.e. the station detects fewer low magnitude events) than during the austral winter, because of high noise level in the local summer due to both natural factors and human activity in the vicinity of the station.

The magnitude variations in teleseismic detectability imply strong relationship with the surrounding environment, such as meteorological events, sea-ice thickness and its spreading area (Ushio, 2003), and more particularly the amplitude of microseisms which is strongest during the austral summers.

4. Microbaroms and ice signals on infrasound

Infrasound is defined as sub-audible sound, i.e pressure waves with frequencies ranging from the cut-off frequency of sound for a 15°C isothermal atmosphere (3.21 mHz) to the lowest frequency of the human audible band (20 Hz). This frequency range is a new horizon for the remote sensing of the Earth's atmospheric physical environment. For example, the Sumatra-Andaman great earthquake of December 26th 2004 not only produced a tsunami that was recorded as far as Antarctica (Nawa et al., 2007), but also produced recordable infrasound waves in the atmosphere (Iyemori et al., 2005). Another example is given by the infrasound and seismic recording of the shock waves generated by a meteorite that overflew Japan (Ishihara et al, 2004).

Over the last few decades, in order to monitor the nuclear tests, a global infrasound network was constructed by the Comprehensive Nuclear-Test-Ban Treaty Organization (CTBTO; Fig. 6; Butler and Anderson, 2008). One objective of CTBTO is to estimate the detection and location capabilities of this network at regional and global distances, another is to explore ways to improve these capabilities and enhance the understanding of wave propagation through the atmosphere of the observed events. At this time, the CTBTO has sixty infrasound stations, each containing at least four sensors (arrayed stations), that can detect a several-kiloton TNT level explosion at a range of ~1000 km. Although the full capability of the global infrasonic network is yet to be established, it has been found to be adequate for monitoring nuclear tests, but too sparse for analyzing natural infrasound phenomena in detail.

In 2008, a Chaparral type sensor was installed on a rock outcrop at SYO station in East Antarctica as an International Polar Year (IPY 2007-2008) project (Ishihara et al., 2009). From the analysis of data recorded during the last two winter seasons in 2008-2010, we found continuous background infrasound noise (Fig. 7) that seem to correspond to co-oscillation of the DFM and SFM as observed by the seismometer at SYO. Time variations similar to those observed on seismic spectra are also observed in infrasound data, and also seem to correspond to storms that occur with intervals of a few days. These observations indicate physical interaction between the atmosphere – ocean system and the solid earth (geosphere) in the microseism/microbarom frequency ranges.

Comprehensive Nuclear Test-Ban Treaty Organization (CTBTO)

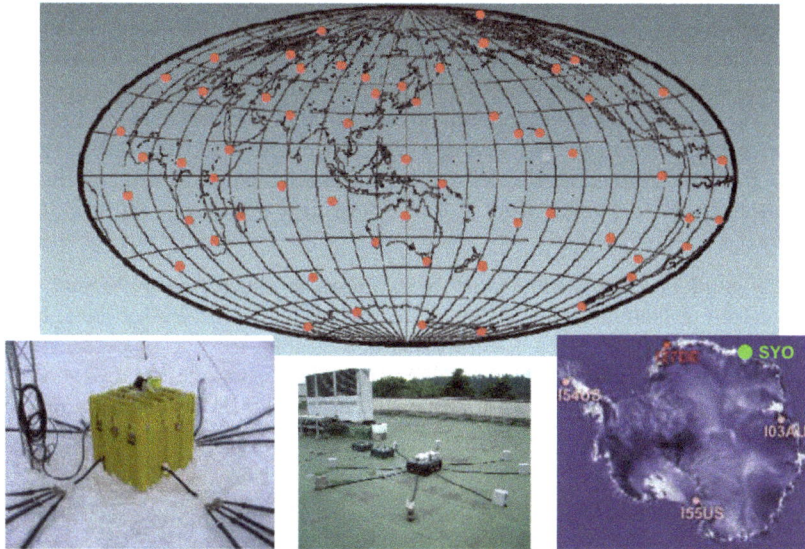

Fig. 6. Top: A global distribution map of the Comprehensive Nuclear Test-Ban Treaty Organization (CTBTO) infrasound network. Bottom, from left to right: infrasound station photo at 155US in West Antarctica; an observation test running at Tohoku University, Japan; infrasound station distribution in Antarctica

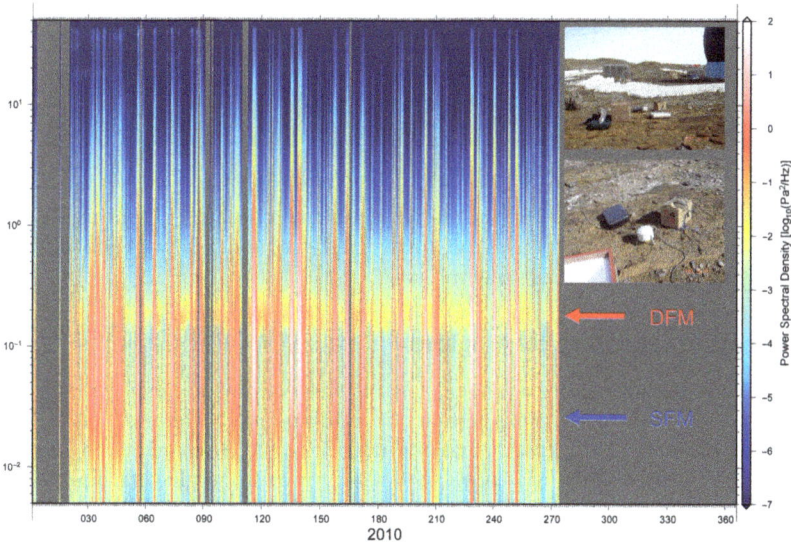

Fig. 7. Power spectral densities (PSD) of the infrasound signals at Syowa Station (SYO), Antarctica, during the period in January – October, 2010. Signals corresponding to SFM and DFM are indicated by blue and red arrows respectively

Figure 8 represents the PSD of infrasound signals at SYO during April 2009. The DFM is indicated by a red arrow, and varies significantly both in amplitude and frequency content over the month. These variations appear to correspond to the atmosphere variations tied to changes in weather conditions, as well as the spreading area and the thickness of sea ice in the surrounding bay. Also visible are the repeating signals with harmonic over-tones at a few tens of Hz (labeled 'ice sheet motion?' in Fig. 8), that may be related to ice dynamics caused by various environmental changes in cryosphere near the station (i.e. sea-ice movement, tide-crack opening shocks, ice-berg tremors, basal sliding of the ice-sheet, calving of glaciers etc.). Energy from storms sometimes extends to the low frequencies, below 0.1Hz.

A theoretical modeling approach would be required to determine the actual sources of several kinds of infrasound signals in the polar region. It would be also useful to compare these signals with other data, such as broadband seismograms that share sensitivity over part of the infrasound frequency range. The array alignment of the infrasound stations, moreover, could provide robust information about the arrival direction and epicentral distance from the infrasound sources.

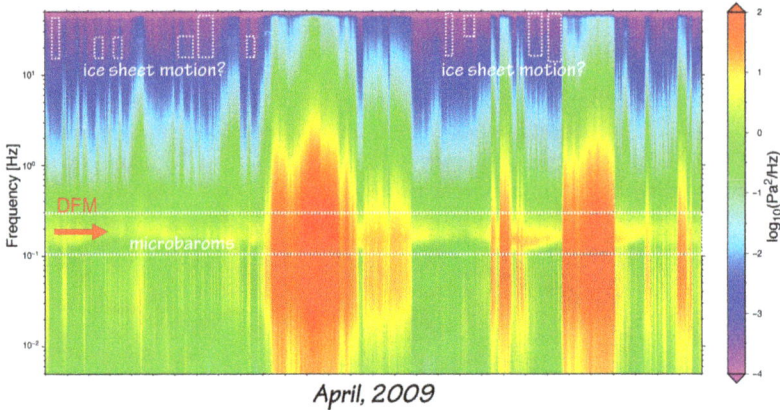

Fig. 8. Power spectral densities (PSD) of the infrasound signals at SYO during the period April 2009. The DFM is indicated by a red arrow. High frequency signals above 10Hz are also identified, and labeled as 'ice sheet motion?'

5. Glacial earthquakes in polar regions

Over the past few decades, more and more seismic observations in the polar regions by both temporary seismic networks and permanent stations have detected local seismicity. Bannister and Kennett (2002) found that the majority of the seismicity in the McMurdo Station area was located along the coast, particularly near large glaciers. They suggested a few generation mechanisms for these events, distinguishable by their focal mechanism and depth: basal sliding of the continental ice sheet, movement of ice streams associated with several scales of glaciers, movement of sea-ice, and tectonic earthquakes. Müller and Eckstaller (2003) deployed a local seismic network around the Neumayer Station, and determined hypocenters of local tectonic events, located along the coast and the mid area of the surrounding bay. Seismic signals involving ice-related phenomena are called «ice-quakes (ice-shocks for smaller ones), and are most frequently reported in association with glacially related mass movements of ice-sheets, or with sea-ice, tide-cracks and icebergs in the other polar areas (Wiens et al., 1990; Wiens, 2007; Anandakrishnan and Alley, 1997; Kanao and Kaminuma, 2006). The so-called «ice-micity» detected around the Bransfield Strait and Drake Passage by a local network of hydrophone arrays in 2006-2007 illustrate the dynamic behavior of sea ice in the Bransfield and Antarctic Peninsula regions(Dziak et al., 2009).

Local seismicity around the Lützow-Holm Bay (LHB) from 1987 to 2003 was reported by Kanao and Kaminuma (2006) (Fig. 9). The seventeen events were only detected by local seismic network deployed around the LHB, except for the September 1996 Mb=4.6 earthquake in the southern Indian Ocean. Almost all the hypocenters were located along the coast, apart from a few on the northern edge of the continental shelf. Several of these events could be large ice-quakes associated with the sea-ice dynamics around the LHB or in the southern ocean.

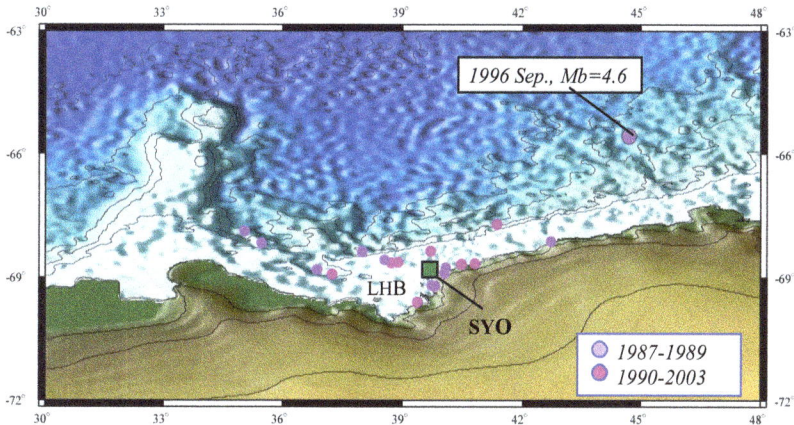

Fig. 9. Local seismicity around the LHB region from 1987 to 2003 (modified after Kanao and Kaminuma, 2006). These events, except for the September 1996 Mb=4.6 earthquake in the Indian Ocean, were detected by the local seismic network deployed at the LHB

Sea-ice dynamics and icebergs also affect seismic signals. A large volume of sea ice was discharged from LHB during the 1997 austral winter, and clearly imaged by the NOAA satellite (Ushio, 2003). The broadband seismographs at SYO recorded distinct waveforms associated with the discharge events (Fig. 10). The long-duration sea-ice tremors had very distinct spectral characteristics that distinguished them clearly from ordinary teleseismic and/or local tectonic events. Several sequences of harmonic over-toned signals, presumably associated with the merging of multiple ice volumes, appeared on the PSDs. The PSDs also showed surge events that seem more closely related to the break-up process of the sea-ice mass. Both kinds of cryoseismic waves occurred continuously for few hours, and repeated themselves several times within a few days during late July, 1997. Identification of the exact sources that produced these characteristic signals has not yet been completed, and theoretical modeling will most likely be required to explain the physical processes. Similar cryoseismic phenomena were also reported around the Ross Sea region (MacAyeal et al., 2009), the marginal sea of the Antarctic Peninsula (Bohnenstiehl et al., 2005; Dziak et al., 2009), as well as the continental margin of Dronning Maud Land (Muller and Eckstaller, 2003). In particular, iceberg-originated harmonic tremor emanating from tabular icebergs was observed by both seismo-acoustic and local broadband seismic signals (MacAyeal et al., 2009). The tremor signals consisted of extended episodes of stick-slip ice-quakes generated when the ice-cliff edges of two tabular icebergs rubbed together during glancing, «strike-slip» iceberg collisions. Source mechanisms of such harmonic tremors might provide useful information for the study of iceberg behavior, and a possible method for remotely monitoring iceberg activity.

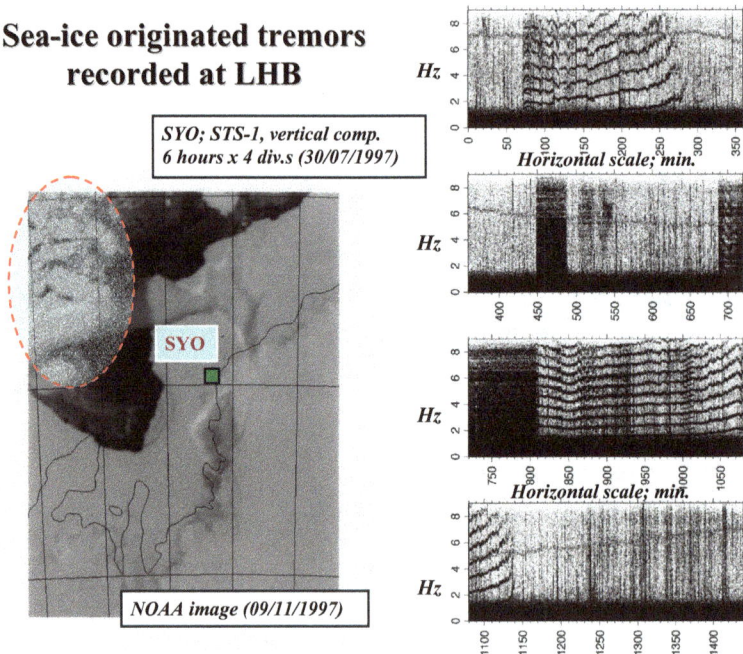

Fig. 10. A large volume of sea-ice discharge from the LHB occurred during the 1997 austral winter. Left : NOAA image in September 11, 1997. The broken red circle with light-blue shading highlights the estimated residuals of the discharged sea-ice volume. Right: PSDs of the broadband seismographs at SYO in July 30, 1997, in four successive time-periods of 6 hours each. Several characteristic signals with harmonic spectra were identified on the seismograms

Several kinds of natural signals were recorded by a seismic experiment with 161 temporary stations on the continental ice sheet (Mizuho Plateau in the LHB region) during the 2002 austral summer (Miyamachi et al., 2003). The experiment recorded chiefly the artificial waveforms originated by seven large explosions, but also detected tectonic earthquakes and ice related phenomena. The recorded signals have been classified into teleseismic events, local events assumed to be ice-quakes and the unidentified events (X-phases, Yamada et al., 2004). These ice-related phenomena are expected to be sensitive to climate change (Kanao et al., 2007). The recordings display variations in frequency content and arrival-times along the seismic profile consistent with documented abrupt variations of the sub-ice topography.

The features of the X-phases were clearly different from those of the small local ice-quakes (Fig. 11). A possibility for the origin of the X-phases may be regional intra-plate earthquakes. Such regional events around Antarctica from 1900 to 1999 are compiled by Reading (2002). East Antarctica from 90°E to 180°E, particularly the areas of Wilkes Land, the Transantarctic Mountains, and the Ross Sea, was the region showing the highest seismicity in the Antarctic. From the comparison with the arrival data at SYO, the maximum amplitudes of seismic phases appear to arrive at SYO with the delay of several seconds. Therefore, the X-phases could possibly travel to the seismic observation line and then SYO from the relatively active, intraplate seismogenic region in Wilkes Land – Ross Sea area. The estimated origin of the unidentified X-phases might be an intraplate earthquake or possibly a large ice-quake (glacial earthquake) around East Antarctica.

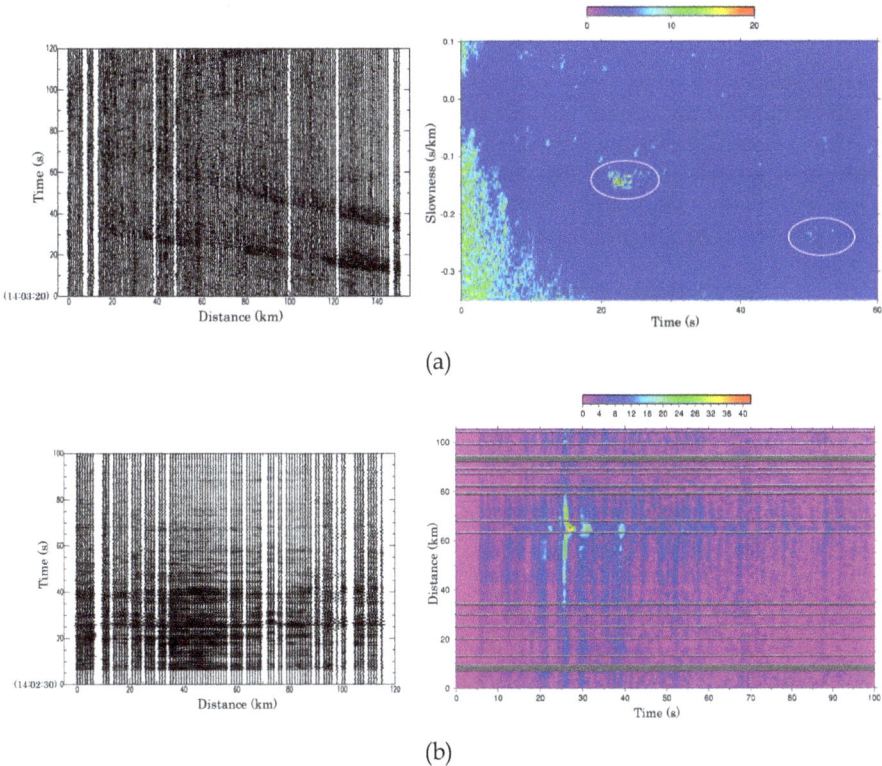

(a)

(b)

Fig. 11. Seismic signals recorded at a linear profile of stations deployed on ice sheet of the Mizuho Plateau (modified after Yamada et al., 2004). (a) Left: Record section showing seismic waves of ice-quakes. Vertical axis starts from Jan. 14, 2002, 14:03:20 (UTC). Right: Contour map of phase-weighted stacking (PWS) applied to the ice-quakes (two circled area). (b) Left: Record section showing seismic waves of «X-phases». Vertical axis starts from Jan. 27, 2002, 14:02:30 (UTC). Right: Contour map showing envelope amplitudes of band-pass filtered traces (1.0-2.0 Hz)

However, it should also be pointed out that several small to middle magnitude natural seismic events could not be located accurately, since they have ambiguous arrivals in the waveforms recorded by the present global network, particularly around Antarctica. In spite of the development of local networks in last two decades, we can hardly distinguish a difference between waveforms generated by local tectonic earthquakes and those of large ice-related phenomena.

In addition to the short-period cryoseismic signals mentioned above, a new class of seismic events associated with melting of large ice cap was discovered recently (Ekström et al., 2003 and 2006; Nettles et al., 2008; Fig. 12). These large events were called "glacial earthquakes", generated long-period (T>25 s) surface waves equivalent in strength to those radiated by standard magnitude five earthquakes, and were observable worldwide. The glacial earthquakes radiated little high-frequency energy, which explains why they were not detected or located by traditional earthquake-monitoring systems. These events are two magnitude units larger than previously reported seismic phenomena associated with glaciers, a size difference corresponding to a factor of 1,000 in a seismic moment.

The long-period surface waves generated by glacial earthquakes are incompatible with standard earthquake models for tectonic stress release, but the amplitude and phase of the radiated waves can be explained by a landslide source model (Kawakatsu, 1989). Over the fourteen-year period between 1993 and 2006, more than 200 glacial earthquakes were detected worldwide. More than 95% of these have occurred on Greenland, with the remaining events in Alaska and Antarctica (Dahl-Jensen et al., 2010).

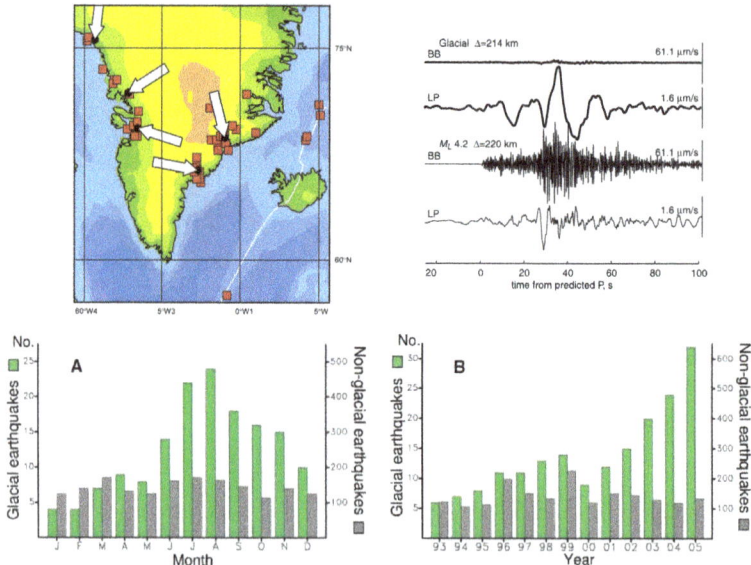

Fig. 12. (upper left) A distribution of the glacial earthquakes around Greenland. (upper right) An example of the comparison between a glacial earthquake and a tectonic crustal earthquake (after Ekström et al., 2003). (lower) Number of glacial and non-glacial earthquakes as a function of month (A) or year (B) (after Ekström et al., 2006)

6. Greenland ice sheet dynamics

Glacial earthquakes have been observed along the edges of Greenland with strong seasonality and increasing frequency from the beginning of this century (Ekstrom et al, 2003), by continuously monitoring data from the Global Seismographic Network (GSN). These glacial earthquakes in the magnitude range 4.6-5.1 may be modeled as a large glacial ice mass sliding downhill several meters on its basal surface over the duration of 30-60 s. Greenland glacial earthquakes were closely associated with major outlet glaciers of the ice sheet. Ekstrom et al. (2006) reported on the temporal patterns of the occurrence of events, finding a clear seasonal signal and a significant increase in the frequency of glacial earthquakes on Greenland after 2002. These patterns are positively correlated with seasonal hydrologic variations, recent observations of significantly increased flow speeds, calving-front retreat, and thinning at many outlet glaciers.

The last four decades of seismicity in Greenland and surrounding regions, including tectonic and volcanic events together with glacial earthquakes, have been investigated by Kanao et al. (2010). Statistically estimated seismic activity using data compiled by the International Seismological Center (ISC) indicates a slight increase in magnitude-dependency b-values from 0.7 to 0.8 from 1968 to 2007 (Fig. 13). This seems to indicate that the total seismicity in this area, including glacial earthquakes, has increased in magnitude over the last four decades. Before attributing this evidence to global warming, the other

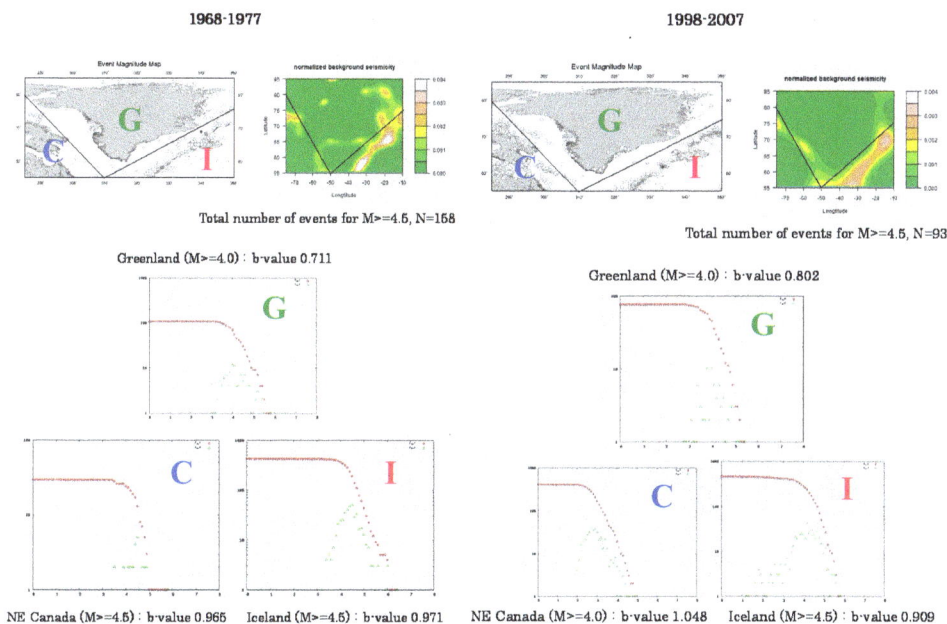

Fig. 13. Background seismicity and Magnitude–dependent 'b'-values for Greenland and the neighboring areas, on the basis of the statistic ETAS model using the hypocentral data collected at the International Seismological Centre (ISC). G: Greenland block; I: Iceland block; C: northern Canadian block. (left) 1968-1977; (right) 1998-2007

possibility for this increase are to be considered, such as of improvement of sensitivity of instrument over time and stations densities in the Arctic region.

The detection, enumeration, and characterization of smaller glacial earthquakes has for a long time been limited by the propagation distance to globally distributed seismic stations of the Federation of Digital Seismograph Networks (FDSN). Although glacial earthquakes have been successfully observed at stations within Greenland in recent years (Larsen et al, 2006), the station coverage was too sparse for detailed studies. In order to define the fine structure and detailed mechanisms of glacial earthquakes within the Greenland ice sheet, a broadband, real-time seismic network needed to be installed throughout onshore Greenland and around its perimeter (the Greenland Ice Sheet Monitoring Network; GLISN; Anderson et al., 2010; Kanao et al., 2008). The 2007-2008 IPY was an opportunity to initiate the new program by international collaboration.

In Greenland, long-term seismic monitoring of the ice sheet will be used to establish a baseline for the seismic activity in Arctic polar region. Deviations from the baseline would be useful indicators of dynamic changes that could signal, for example, new mechanisms of dynamic collapse of the ice sheet. At least as importantly, the seismic data obtained by the GLISN network can provide, along with the monitoring capability, new constraints on the dynamics of ice sheet behavior and its potential role in the sea-level rise during the coming decades.

Greenland Ice Sheet Monitoring Network (GLISN)

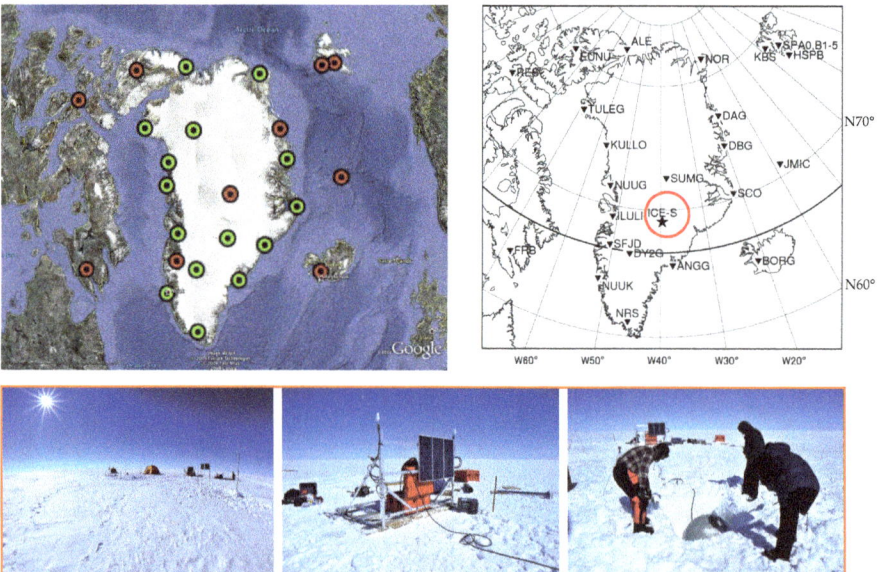

Fig. 14. (upper left) Location map of the broadband seismic stations deployed by the GLISN project. The solid red circles denote the existing FDSN stations and the solid green circles indicate the GLISN sites. (upper right) Location of the Ice-S station by open red circle. (lower) Installation of the Ice-S station on June 2011, taken by G. Toyokuni

7. Monitoring the atmosphere-ocean-cryosphere-geosphere system, the contribution of IPY and CTBTO

In the previous sections of this chapter, we have shown that seismic data contain significant information regarding wave activity, ice-dynamics and weather-related phenomena (e.g. storms). Microseism measurements in polar regions are a useful proxy for characterizing ocean wave climate and global storm intensity in the high latitudes, complementing other estimates such as marine surveys and satellite images. Individual stations respond most strongly to wave activities at regional shorelines, and the sensitivity of specific stations to ocean wave climate is controlled by factors such as storm tracks and coastal bathymetry. Continuous digital records from the Global Seismographic Network (GSN), the Federation of Digital Seismographic Network (FDSN) and their precursor networks extend back more than 40 years, and hence open up the possibility of using seismic data to investigate climate change. The new permanent network in Greenland (GLISN) significantly increases coverage of the surrounding Arctic region.

A program containing several field-campaigns was launched during the International Polar Year (IPY 2007-2008) and complements the networks of permanent stations at the high latitudes of the polar regions. In Antarctica, the most ambitious seismological field campaign conducted for the IPY was the «GAmburtsev Mountain SEISmic experiment» or GAMSEIS, an internationally coordinated deployment of more than 50 broadband seismographs over the crest of the Gambursev Mountains (Dome-A – Dome-F area) in East Antarctica. The aim of this experiment was to provide detailed information on the crustal thickness and mantle structure of the region and find key constraints on the origin of the Gamburtsev Mountains (Wiens, 2007; Hansen et al., 2010; An et al., 2010). GAMSEIS and many other seismological deployments, including a French deployment between Concordia and Vostok (CASE-IPY) and a Japanese deployment around the Lützow-Holm Bay (JARE-GARNET), were coordinated under a larger program called the 'Polar Earth Observing Network (POLENET; http://www.polenet.org/; Fig. 15)' whose aim was to establish a geophysical network to cover the whole Antarctic continent as well as Greenland for the duration of the IPY.

The seismic data obtained by the combined POLENET network are being used to clarify the heterogeneous structure of the Earth, particularly in the Antarctic region, by studying the crust and upper mantle and the Earth's deep interior, including features such as the Core-Mantle-Boundary (CMB), the lowermost mantle layer (D" zone) and the inner core. In addition to conventional seismological targets (e.g. crust and lithosphere structure, inner core structure), the IPY seismic stations could be used to help monitor geographical variations in climate indicators, over the span of 2-3 years. All data from IPY experiments will be distributed to the scientific community.

Together with the seismic networks, the infrasound stations in the Antarctic contribute to both CTBTO and the Pan-Antarctic Observations System (PAntOS) under the Scientific Committee on Antarctic Research (SCAR). The combination of seismic, infrasound and hydro-acoustic observations is required to understand in more detail the atmosphere-ocean-cryosphere-geosphere system and its variations. We are hopeful that the large quantity of data of these three types accumulated over the past decades by the CTBTO will one day be distributed to the scientific community.

POLENET in IPY 2007-2008

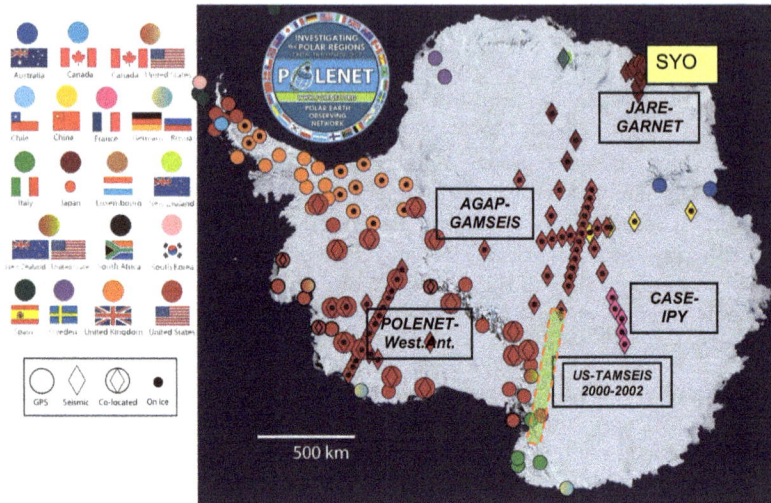

Fig. 15. Distribution map of seismic and the other geophysical stations deployed during the IPY 2007-2008. The major project names are labeled as; JARE-GARNET, AGAP-GAMSEIS, CASE-IPY, POLENET-West.Ant. and US-TAMSEIS, respectively. All stations in Antarctic continent contributed to POLENET program

8. Conclusion

We have described several features of seismic waves, and how they are related to the atmosphere-ocean-cryosphere system. Microseisms and microbaroms from the southern ocean are clearly recorded by both broadband seismographs and infrasound sensors deployed at Antarctic stations, and are modulated by the presence of sea-ice. Microseism measurements were a useful proxy for characterizing ocean wave climate and global storm intensity, complementing other estimates by ocean buoys or satellite measurements. Using the infrasound data at SYO, we have detected long duration signals with harmonic over-tones that may be related to the ice dynamics near the station.

Most of the community agrees that the polar regions play a critical role in the Earth's system. The Greenland ice sheet and its response to climate change potentially have a great impact upon mankind, both through long-term sea-level rise and through modulation of fresh water input to the oceans. Monitoring the dynamic response of the Greenland ice cap and the Antarctic ice sheet, would be important components of a long-term effort to observe climate change on a global scale. Future directions in global monitoring targets will emerge from multidisciplinary projects combining the data of several global networks.

There are still a lot to be learned about the physical mechanisms of interaction between the atmosphere-ocean-cryosphere system and the geosphere in the polar regions. Continuous observation by a sufficiently large number of high quality stations, as well as theoretical work, will probably be necessary to make progress in this field. Given the high cost and technical difficulties of continuous observation in the polar regions, such work would require strong international collaboration beyond the end of the International Polar Year.

9. Acknowledgements

We would like to express our sincere appreciation of our many collaborators in both polar regions. We thank all the members of IPY Antarctic projects of the AGAP/GAMSEIS (Prof. Douglas Wiens of Washington University, Prof. Andy Nyblade of Penn. State University, and other members), the POLENET (Prof. Terry Wilson of the Ohio State University, and other members), as well as the Japanese Antarctic Research Expeditions (JARE; Prof. Kazuyuki Shiraishi of NIPR and many other members). We also thank the Arctic ice sheet monitoring program of GLISN (Prof. Trine Dahl-Jensen of Geological Survey of Denmark and Greenland, and other members). Infrasound observation at SYO was supported by the Ministry of Education, Science, Sports and Culture, Grant-in-Aid for Young Scientists (B) 19740265, 2007 (P.I. as Yoshiaki Ishihara). The authors would like to express their sincere thankfulness for reviewers and the publisher for their useful comments and supports in publication management for the special issue on "Seismic Waves". The production of this paper was supported by an NIPR publication subsidy.

10. References

An, M., Wiens, D., Zhao, Y., Feng, M., Nyblade, A., Kanao, M., Maggi, A. & Lévêque, J.J. (2010). Lithospheric S-velocity structure of Antarctica inverted from surface waves, AGU Fall 2010 Meeting, 13-17 December, T21D-2188, San Francisco, California, USA.

Anandakrishnan, S. & Alley, R. B. (1997). Tidal forcing of basal seismicity of ice stream C, West Antarctica, observed far inland, J. Geophys. Res., Vol. 102, pp. 15,183-15,196.

Anandakrishnan, S., Voigt, D. E., Alley, R. B. & King, M. A. (2003). Ice stream D flow speed is strongly modulated by the tide beneath the Ross Ice Shelf, Geophys. Res. Lett., Vol. 30, doi:10.1029/2002GL016329.

Anderson, K. R., Beaudoin, B. C., Butler, R., Clinton, J. F., Dahl-Jensen, T., Ekstrom, G., Giardini, D., Govoni, A., Hanka, W., Kanao, M., Larsen, T., Lasocki, S., McCormack, D. A., Mykkeltveit, S., Nettles, M., Agostinetti, N. P., Tsuboi, S. & Voss, P. (2010). The Greenland Ice Sheet Monitoring Network (GLISN), AGU Fall 2010 Meeting, 13-17 December, C06-957393, San Francisco, California, USA.

Aoyama, Y. & Kanao, M. (2010). Seismological Bulletin of Syowa Station, Antarctica, 2008, JARE Data Report, Vol. 317 (Seismology 44), pp. 1-85.

Aster, R., McNamara, D. & Bromirski, P. (2008). Multidecadal climate-induced variability in microseisms, Seismol. Res. Lett., Vol. 79, No.2, pp. 194–202, doi:10.1785/gssrl.79.2.194.

Aster, R. (2009). Studying Earth's ocean wave climate using microseisms, IRIS Annual Report, pp. 8-9.

Bannister, S. & Kennett, B. L. N. (2002). Seismic Activity in the Transantarctic Mountains - Results from a Broadband Array Deployment, Terra Antarctica, Vol. 9, pp. 41-46.

Bohnenstiehl, D. R., Dziak, R. P., Parlk, M. & Matsumoto, H. (2005). Seismicity of the polar seas: The potential for hydroacoustic monitoring of tectonic and volcanic processes, the 12th Seoul Inter. Sympo. on Polar Sci., pp. 11-14, May 17-19, Ansan, Korea.

Bromirski, P. D. (2009). Earth vibrations, Science, Vol. 324, pp. 1026-1027.

Butler, R. & Anderson, K. (2008). Global Seismographic Network (GSN), IRIS Report, pp. 6-7.

Dahl-Jensen, T., Larsen, T. B., Voss, P. H. & the GLISN group. (2010). Greenland ice sheet monitoring network (GLISN):a seismological approach, GEUS "Report of Survey Activities" (ROSA) for 2009, pp. 55-58.

Dziak, R. P., Parlk, M., Lee, W. S., Matsumoto, H., Bohnenstiehl, D. R., & Haxel, J. H. (2009). Tectono-magmatic activity and ice dynamics in the Bransfield Strait back-arc basin, Antarctica, the 16th Inter. Sympo. on Polar Sci., pp. 59-68, June 10-12, Incheon, Korea.

Ekström, G., Nettles, M. & Abers, G. A. (2003). Glacial earthquakes, Science, Vol. 302, pp. 622-624.

Ekström, G., Nettles, M. & Tsai, V. C. (2006). Seasonality and increasing frequency of Greenland glacial earthquakes, Science, Vol. 311, pp. 1756-1758.

Grob, M., Maggi, A. & E. Stutzmann, E. (2011). Observations of the seasonality of the Antarctic microseismic signal, and its association to sea ice variability, Geophys. Res. Lett., Vol. 38, L11302, doi:10.1029/2011GL047525.

Hansen, S. E., Nyblade, A. A., Heeszel, D. S., Wiens, D. A., Shore, P. & Kanao, M. (2010). Crustal Structure of the Gamburtsev Mountains, East Antarctica, from S-wave Receiver Functions and Rayleigh Wave Phase Velocities, Earth Planet. Sci. Lett., 10.1016/j.epsl.2010.10.022.

Hatherton, T. (1960). Microseisms at Scott Base, Geophys. Jour., Vol. 3, pp. 381-405.

Howat, I. M., Joughin, I., Tulaczyk, S. & Gogineni, S. (2005). Rapid retreat and acceleration of Helheim Glacier, East Greenland, Geophys. Res. Lett., Vol. 32, L22502, doi:10.1029/2005GL024737.

Ishihara, Y., Furumoto, M., Sakai, S. & Tsukuda, S. (2004). The 2003 Kanto large bolide's trajectory determined from shockwaves recorded by a seismic network and images taken by a video camera, Geophys. Res. Lett., Vol. 31, L14702.

Ishihara, Y., Yamamoto, M. & Kanao, M. (2009). Current Status of Infrasound Pilot Observation at Japanese Islands and SYOWA Antarctica, and Development of New Infrasound Sensor using Optical Sensing Method, AGU Fall 2009 Meeting, 14-18 December, A13D-0244, San Francisco, California, USA.

Isse, T. & Nakanishi, I. (2001). Inner-Core anisotropy beneath Australia and differential rotation, Geophysical Journal International, Vol. 151, pp. 255-263.

Iwano, S. & Kanao, M. (2009). Seismological Bulletin of Syowa Station, Antarctica, 2007, JARE Data Report, Vol. 313 (Seismology 43), pp. 1-101.

Iyemori, T., Nose, M., Han, D. S., Gao, Y., Hashizume, M., Choosakul, N., Shinagawa, H., Tanaka, Y., Utsugi, M., Saito, A., McCreadie, H., Odagi, Y. & Yang, F. (2005). Geomagnetic pulsations caused by the Sumatra earthquake on December 26, 2004, Geophys. Res. Lett, Vol. 32, L20807, doi:10.1029/2005GL024083.

Kaminuma, K. (2000). A revaluation of the seismicity in the Antarcitc, Polar Geosci., Vol. 13, pp. 145-157.

Kanao, M. (2010). Detection Capability of Teleseismic Events Recorded at Syowa Station, Antarctica - 1987~2007-, Antarct. Rec., Vol. 54, No. 1, pp. 11-31.

Kanao, M., Himeno, T., Tsuboi, S., Dahl-Jensen, T. & Anderson, K. R. (2010). Glacial earthquake activities around the Greenland and surrounding regions, The 2nd International Symposium on the Arctic Research (ISAR-2), G4-P3, Tokyo, Japan.

Kanao, M., Tsuboi, S., Butler, R., Larsen, T. & Anderson, K. (2008). Planning of the Greenland Ice Sheet Monitoring Network (GLISN) for observing global warming, Drastic Change under the Global Warming, The 1st International Symposium on the Arctic Research (ISAR-1), pp176-179, Tokyo, Japan.

Kanao, M., Yamada, A., Yamashita, M. & Kaminuma, K. (2007). Characteristic Seismic Signals Associated with Ice Sheet & Glacier Dymanics, Eastern Dronning Maud Land, East Antarctica, In Antarctica: A Keystone in a Changing World, edited by A.K. Cooper and C.R. Raymond et al., USGS OF-2007-1047, pp. 182-186.

Kanao, M., & Kaminuma, K. (2006). Seismic activity associated with surface environmental changes of the Earth system, East Antarctica, In: Antarctica: Contributions to global earth sciences, Futterer, D. K., Damaske, D., Kleinschmidt, G., Miller, H. & Tessensohn, F. (Eds.), Springer-Verlag, Berlin Heidelberg New York, pp. 361-368.

Kanao, M., Nogi, Y. & Tsuboi, S. (2006). Spacial distribution and time variation in seismicity around Antarctic Plate - Indian Ocean, Polar Geosci., Vol. 19, pp. 202-223.

Kawakatsu, H. (1989). Centroid single force inversion of seismic waves generated by landslides, J. Geophys. Res., Vol. 94, pp. 12,363-12,374.

Larsen, T. B., Dahl-Jensen, T., Voss, P., Jørgensen, T. M., Gregersen, S. & Rasmussen, H. P. (2006). Earthquake seismology in Greenland – improved data with multiple applications, Geological Survey of Denmark and Greenland Bulletin, Vol. 10, pp. 57–60.

MacAyeal, D., Okal, E., Aster, R. & Bassis, J. (2009). Seismic Observations of Glaciogenic Ocean Waves on Icebergs and Ice Shelves, J. Glaciology, Vol. 55, pp. 193-206.

Miyamachi, H., Toda, S., Matsushima, T., Takada, M., Watanabe, A., Yamashita, M. & Kanao, M. (2003). Seismic refraction and wide-angle reflection exploration by JARE-43 on Mizuho Plateau, East Antarctica, Polar Geosci., Vol. 16, pp. 1-21.

Muller, C. & Eckstaller, A. (2003). Local seismicity detected by the Neumayer seismological network, Dronning Maud Land, Antarctica: tectonic earthquakes and ice-related phenomena. IX Intern. Sympo. Antarc. Earth Sci. Programme and Abstracts, pp. 236.

Nawa, K., Suda, N., Satake, K., Sato, T., Doi, K., Kanao, M. & Shibuya, K. (2007). Loading and gravitational effects of the 2004 Indian Ocean tsunami at Syowa Station, Antarctica, Bull. Seis. Soc. Am., Vol. 97, S271-278, doi:10.1785/0120050625.

Nettles, M., Larsen, T. B., Elósegui, P., Hamilton, G. S., Stearns, L. A., Ahlstrøm, A. P., Davis, J, L., Andersen, M. L., de Juan, J., Khan, S. A., Stenseng, L., Ekström, G. & Forsberg, R. (2008). Step-wise changes in glacier flow speed coincide with calving and glacial earthquakes at Helheim Glacier, Greenland. Geophys. Res. Lett., Vol. 35, L24503, doi:10.1029/2008GL036127.

Reading, A. M. (2002). Antarctic seismicity and neotectonics, In: Antarctica at the close of a millennium, Gamble, J. A. (Ed.), Wellington, The Royal Soc. of New Zealand Bull., Vol. 35, pp. 479-484.

Reading, A. M. (2006). On seismic Strain-Release within the Antarctic Plate, In: Antarctica: Contributions to global earth sciences, Futter, D. K., Damaske, D., Kleinschmidt, G., Miller, H. & Tessensohn, F. (Eds.), Springer-Verlag, New York, pp. 351-356.

Rignot, E. & Kanagaratnam, P. (2006). Changes in the velocity structure of the Greenland ice sheet, Science, Vol. 311, pp. 986-990.

Steffen, K., Nghiem, S. V., Huff, R. & Neumann, G. (2004). The melt anomaly of 2002 on the Greenland Ice Sheet from active and passive microwave satellite observations, Geophys. Res. Lett., 31, L20402, doi: 10.1029/2004GL020444.

Stutzmann, E., Schimmel, M., Patau, G. & Maggi, A. (2009). Global climate imprint on seismic noise, Geochem. Geophys.Geosyst., Vol. 10, Q11004, doi:10.1029/2009GC002619.

Tsuboi, S., Kikuchi, M., Yamanaka, Y. & Kanao, M. (2000). The March 25, 1998 Antarctic Earthquake: caused by postglacial rebound, Earth Planets Space, Vol. 52, pp. 133-136.

Ushio, S. (2003). Frequent sea-ice breakup in the Lützow-Holmbukta, Antarctica, based on analysis of sea ice condition from 1980 to 2003, Antarct. Rec., Vol. 47, pp. 338-348.

Usui, Y., Hiramatsu, Y., Furumoto, M. & Kanao, M. (2008). Evidence of seismic anisotropy and a lower temperature condition in the D″ layer beneath Pacific Antarctic Ridge in the Antarctic Ocean, Physics of the Earth and Planetary Interiors, doi:10.1016/j.pepi.2008.04.006.

Wiens, D., Anandakrishnan, S., Nyblade, A. & Aleqabi, G. (1990). Remote detection and monitoring of glacial slip from Whillans Ice Streeam using seismic rayleigh waves recorded by the TAMSEIS array, EOS Trans. AGU, Vol. 87, pp. 52.

Wiens, D. A. (2007). Broadband Seismology in Antarctica: Recent Progress and plans for the International Polar Year, Proceedings of International Symposium –Asian Collaboration in IPY 2007-2008-, March 1, Tokyo, Japan, pp. 21-24.

Yamada, A., Kanao, M. & Yamashita, M. (2004). Features of seismic waves recorded by seismic exploration in 2002: Responses from valley structure of the bedrock beneath Mizuho Plateau, Polar Geosci., Vol. 17, pp. 139-155.

Electric and Electromagnetic Signals Under, On, and Above the Ground Surface at the Arrival of Seismic Waves

Akihiro Takeuchi[1], Kan Okubo[2] and Nobunao Takeuchi[3]
[1]Earthquake Prediction Research Center, Institute of Oceanic Research and Development,
Tokai University 3-20-1 Orido, Shimizu-ku, Shizuoka 424-8610,
[2]Division of Information and Communications Systems Engineering,
Tokyo Metropolitan University 6-6 Asahigaoka, Hino 191-0065,
[3]Research Center for Predictions of Earthquakes and Volcanic Eruptions,
Graduate School of Sciences, Tohoku University
6-6 Aza-aoba, Aramaki, Aoba-ku, Sendai 980-8578,
Japan

1. Introduction

Numerous reports describe electromagnetic phenomena that occur before major earthquakes: anomalous electrotelluric potential changes (Varotsos, 2005), anomalous geomagnetic fields (Fraser-Smith et al., 1990), anomalous transmission of electromagnetic waves (Hayakawa et al., 2010), electron content perturbations in the ionosphere (Oyama et al., 2008), and infrared thermal anomaly on the ground surface detected from space (Saradjian and Akhoondzadeh, 2011). Researchers, expecting that these electromagnetic phenomena are useful for short-term prediction of earthquakes, have developed observation networks of various types throughout the world (Eftaxias et al., 2004; Uyeda et al., 2009). Although many researchers have explored such seismo-electromagnetic precursors, most reports are based on retrospective analyses. Scientific proof of the precursors apparently remains elusive. Earthquake prediction using these phenomena cannot be realized easily at this stage. To make steady progress in the study of seismo-electromagnetic precursors, our group has held that it is primarily important to prove the existence of phenomena at the occurrence of earthquakes (co-faulting signals) and at the arrival of seismic waves (co-seismic signals). Secondarily, it is important that these phenomena be evaluated quantitatively. Therefore, this chapter treats only co-seismic signals in that context.

Our group has used observation sites in northeastern Japan. Figure 1 depicts the locations and operation periods of our observation sites, which were set up on the ground, except at (H) Hosokura site, established *in* the ground. We have observed electric signals using reference electrodes buried in the ground and using condenser-type electrodes insulated from the ground, as described later. We explain co-seismic electric signals detected in (H) Hosokura site in section 2. Then, we describe those detected at other ground surface observation sites in section 3. Thereafter, in section 4, we discuss co-seismic electric/electromagnetic signals above the ground surface, along with signals detected under and on the ground surface, based on observed data. Finally, we summarize our studies and seismo-electromagnetic phenomena in section 5.

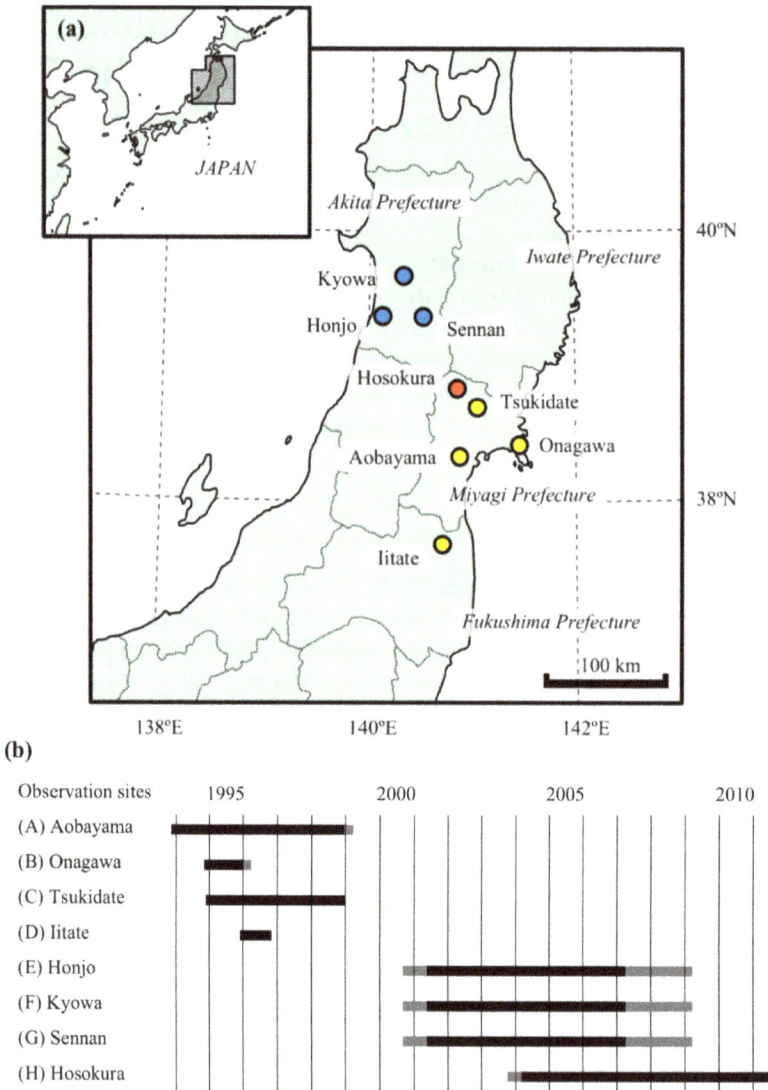

Fig. 1. (a) Locations of observation sites in northeastern Japan. Four yellow circles represent sites on the ground surface in the early stage. Three blue circles show sites on the ground surface in the late stage. The red circle denotes the underground site in the current stage. (b) Operation periods of the sites. Black zones show the period of normal operation. Gray zones show the period of construction, test operation, or bad condition because of superannuation. Only (H) Hosokura site is currently operational

2. Under the ground surface

2.1 Observation site and system

Our underground observation site, (H) Hosokura site (N38°48′, E140°53′), is located in the middle of the main gallery in the Hosokura mine in Miyagi Prefecture, Japan (Figure 1). It intersects sulfide veins in the propylite and green-tuff bedrock (Figure 2a). These rocks are types of hydrothermally altered andesite and tuff, respectively, which are widely distributed throughout northeastern Japan. A shaft near the room connects to the ground surface and galleries including the main one. The lower galleries are flooded. The mine operation has already ceased. However, maintenance continues. The underground water that gushes out is pumped out to maintain the water level at 10 m below the main gallery. It is then chemically treated before discharge. Although the entrances of the room and the main gallery are usually closed, air can leak through openings in the doors. The surface entrance of the shaft is open, and air can pass freely in and out. The observation room is located about 1.5 km from the entrance of the main gallery and about 70 m below the ground surface. The room area is ca. 15 × 12 m²; its height is ca. 2.5 m (Figure 2b). The air temperature in this room is kept at 23–27 °C year-round using a heater. The air is dusty. The walls, ceiling, and floor are dry, but the main gallery walls are wet.

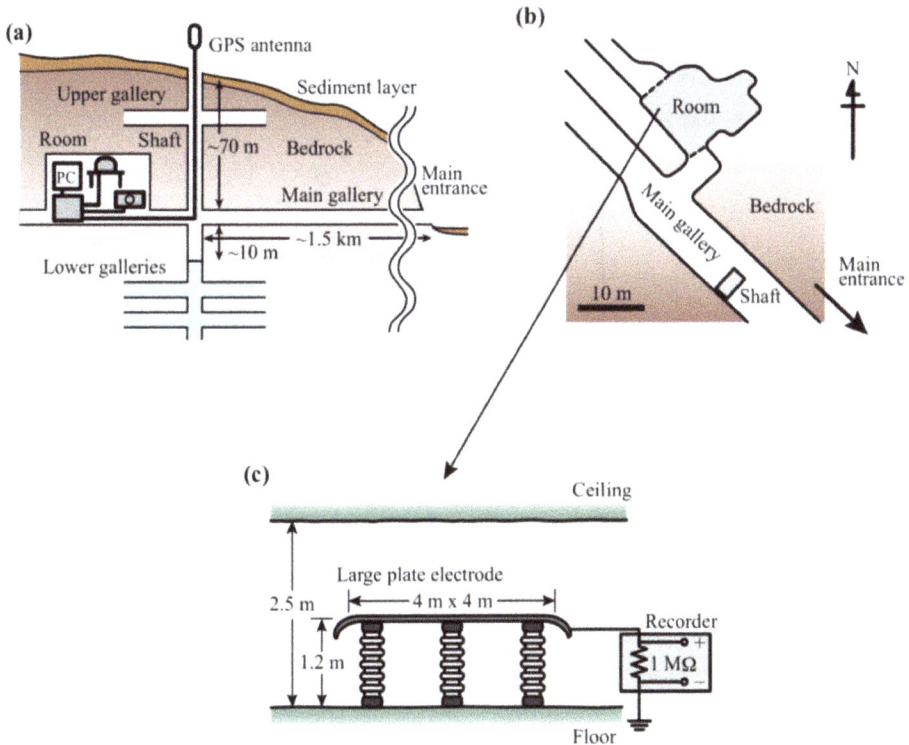

Fig. 2. (a) Side view of (H) Hosokura site in the Hosokura mine. (b) Top view. (c) Observation system in the room. Modified from Okubo et al. (2006) and Takeuchi et al. (2009)

We started observations here with a large plate electrode and a seismometer in 2003 (Okubo et al., 2006a). This large plate-aluminum electrode, with an area of 4 × 4 m^2, and is placed in the middle of the room (Figure 2c) on 15 1.2-m-long ceramic insulators that are terminated at both ends by metal caps. The plate electrode is connected to the floor through a data recorder (DR-1021; TOA-DDK Corp.) with 1-MΩ input impedances on each channel. This electrode functions as a condenser-type sensor with a time constant, probably of the order of ~10^{-4}–10^{-3} s. The seismometer, which is fixed on the floor near the electrode, detects the up-down acceleration of seismic waves. It is also connected to the data recorder.

2.2. Observed signals

Figure 3 presents an example of the observed signals at the arrival of seismic waves. The source earthquake, with magnitude of 4.6, occurred on 27 February 2004. The epicenter (N38°08′, E141°06′) was about 75 km east of this observation site, with focal depth of ca. 70 km. When the S-waves arrived and vertically displaced the room floor, the potential turned rapidly more negative and then reverted gradually to the former background level (Okubo et al., 2006a). The potential superimposed on the background is the potential difference across the internal resistance of the recorder, caused by a transient current, rather than the potential difference between the plate electrode and the floor. Considering the negative-ward shift of the potential and the overall setup of the electrode system, we infer expect the generation of an electric field with its upward electric lines of force. Similar signals were sometimes detected until this electrode system was changed later (Takeuchi et al., 2009).

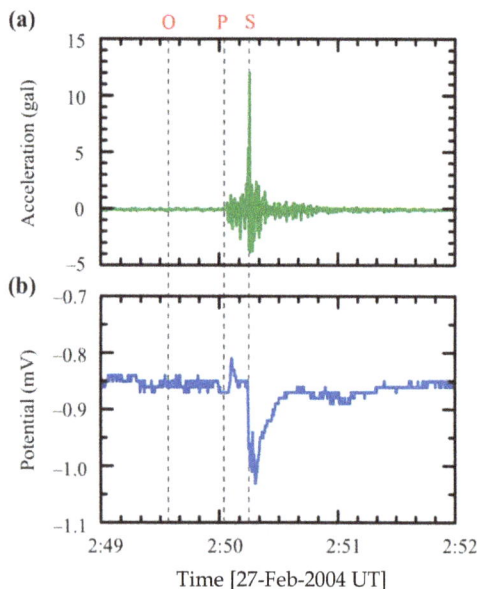

Fig. 3. Examples of electric signals detected at the arrival of seismic waves. (a) Ground up-down acceleration. (b) Electric potential detected by the large plate electrode. "O", "P", and "S" respectively signify the origin time of the earthquake, the time of the P-wave arrival, and the time of S-wave arrival. Modified from Okubo et al. (2006)

As possible mechanisms of the co-seismic signal depicted in Figure 3, we can present four candidates (Okubo et al., 2006a; Takeuchi et al., 2010):

1. **Piezoelectric effect:** Andesite bedrock surrounding the room slightly involves quartz. Therefore, seismic waves cause electric polarizations at each quartz grain/vein, thereby generating an electric field in the room if the summation of the polarizations effectively induces electrification of the floor and walls.
2. **Interfacial electrokinetic effect:** The lower galleries are flooded and the main gallery is wet. Therefore, the bedrock below the room floor will include pore water. When seismic waves push the wet bedrock, the pore water will flow and cause streaming potential because of the interfacial electrokinetic effect. This potential will generate an electric field in the room.
3. **Radon gas:** Radon is widely involved in rocks and soils of various types. When seismic waves arrive, radon gas will emanate from the surrounding bedrock into the room. Radioactive decay of radon gas can increase the air ion concentration. This increment will temporally increase the permittivity of the air and the capacitance of the large plate electrode, which will cause an apparent electromotive force between the electrode and floor if an electric field always exists there.
4. **Positive hole:** In general, igneous rocks subjected to non-uniform loading can activate positive holes (Details are discussed later). When seismic waves arrive, the andesite bedrock will activate positive holes. Positive electrification of the room floor will generate an electric field in the room.

All candidates may occur simultaneously. Their contribution ratios are determined by complex conditions (e.g., the arrival direction of seismic waves, the inhomogeneous bedrock, the wet/dry condition, etc.) and may differ each time. We treat the second candidate again in the next section. In the next subsection, we describe our specific examination of the fourth candidate and verify the positive electrification of the andesite bedrock using laboratory experiments (Takeuchi et al., 2010).

2.3 Laboratory experiments

Figure 4 presents the experimental setup. The rock sample was a block ($4.5 \times 4.0 \times 9.7$ cm^3) quarried from andesite bedrock surrounding the underground observation room. It was air-dried, as was the room floor. The block was placed into an aluminum enclosure, acting as a Faraday cage. The lower volume of the block was loaded uniaxially using a vise equipped with a load cell (CMM1-2T and CSD-819C; Minebea Co., Ltd.). The lower volume was grounded through the vise. Because the lower volume is stressed, its volume expands slightly and pushes the upper unstressed volume upward, which simulates the moment of the arrival of the S-wave frontline, pushing the room floor upward. A conductive tape pasted on the rock top surface is connected to a copper plate immediately below an electric field mill (EF-308T; Tierra Tecnica Corp.). When the top surface is charged, this charge is induced in the plate, thereby creating an electric field toward the sensor of the electric field mill. Consequently, the electric field mill detects the electrification on the sample as a function of the electric field, $q_p = \varepsilon E_d$, where q_p is the surface charge density on the copper plate, ε is the permittivity of air, and E_d is the detected electric field. This electric field mill outputs the 0.8-s averaged E_d for 1 s after signal analysis in 0.2 s. The two series of data are synchronized at the data recorder (8855; Hioki E.E. Corp.). The surface charge density on the rock top surface q_0 and the electric field on the surface E_0 are calculated using the surface area ratio between the copper plate S_p and the rock top surface S_0.

To minimize the effect of the sample block moving slightly at the application of the first load L from 0 kN, we first kept $L = 2$ kN for 30–60 min and then started further loading to levels up to 12 kN. Figure 5 portrays some results. The E_0 and p_0 values signify the differences from the values at the initial load $L = 2$ kN. Positive values denote an electric field with upward electric lines of force, which means that the rock top surface is positively charged. The charge is dissipated with fluctuation after unloading. Similar results were obtained when loading/unloading was repeated at 1–2 hour intervals. Figure 6a shows the plots of E_0 (and p_0) against the maximum load L_{max} applied to the lower block volume. Although the values are scattered, clear correlation exists between the load and electrification. If this trend is extrapolated linearly to 0–2 kN, the respective true values of E_0 and p_0 would be the values shown in the figure plus ca. 0.2 V/m and ca. 2×10^{-12} C/m^2.

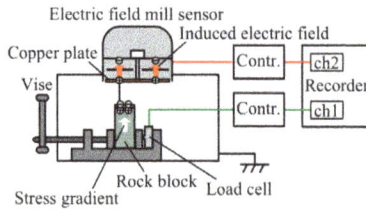

Fig. 4. Setup of non-uniform loading tests using rock blocks. Modified from Takeuchi et al. (2010)

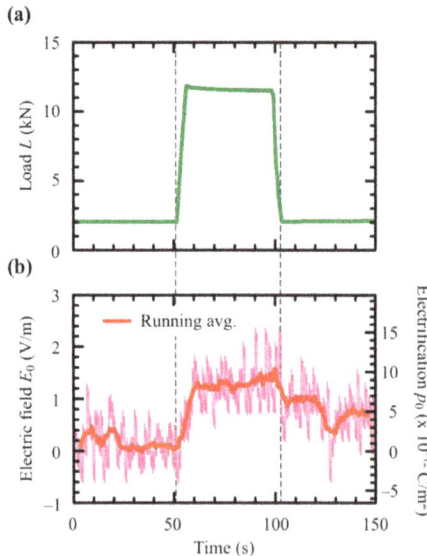

Fig. 5. Typical example of experimental measurements. (a) Loading profile L. (b) Electric field E_0 at the rock top surface and surface charge density calculated as p_0. E_0 and p_0 denote differences from the values at the initial load $L = 2$ kN. Modified from Takeuchi et al. (2010)

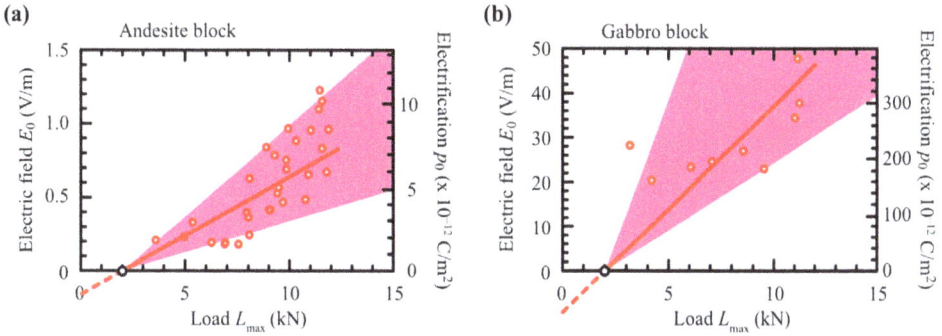

Fig. 6. Electric field E_0 and electrification p_0 measured on the top surface of the upper unstressed volume of the rock block shown against the maximum load L_{max} applied to the lower volume of the rock. (a) Quartz-bearing andesite. (b) Quartz-free gabbro. E_0 and p_0 denote differences from the values at the initial load $L = 2$ kN. Modified from Takeuchi et al. (2010)

The andesite under study involves some quartz. Therefore, we should consider first the piezoelectric effect as the cause of the signals detected in the laboratory. Here, we conducted similar experiments with a quartz-less gabbro block ($9.4 \times 2.0 \times 9.7$ cm^3) as a control. Results show that the gabbro showed the same electric field and electrification, even more intensely than the andesite (Figure 6b), which implies that the piezoelectric effect of quartz is not the prime cause of the signals detected in the laboratory. Randomly oriented piezo-dipoles in the andesite block would be canceled by one another. The interfacial electrokinetic effect of pore water is secondarily considered as the cause of the signals detected in the laboratory. However, we can also discount this effect because the samples were well air-dried. Now, we expect the stress-activation of positive holes (Takeuchi et al., 2010).

Minerals that form igneous rocks, including andesite, generally involve peroxy bonds that are ubiquitous lattice defects ($O_3X–OO–YO_3$ with X, Y = Si^{4+}, Al^{3+}, etc.). Figure 7 shows an example of an energy level of the peroxy bond in the case of quartz. The unoccupied $3\sigma_u^*$ level usually lies in the forbidden band (Figure 7a). When the lattice structure around this bond is mechanically deformed, the energy level shifts downward and becomes an accepter (Figure 7b). As an electron jumps in from a neighboring oxygen anion (O^{2-}), a hole appears at this site (Figure 7c), which is a positive hole whose state (O^-) can move through the valence band like charge carriers moving in any p-type semiconductor material (Figure 7d). This mechanism can construct a model below to explain the signals detected in the laboratory. When we stressed the lower volume of the andesite/gabbro block, positive hole charge carriers were activated in the volume. Positive holes attempted to diffuse into the upper unstressed volume and were simultaneously attracted electrically by electrons trapped at peroxy bonds. Consequently, an electric unevenness, i.e. an electric polarization, was formed in the lower volume. This polarization and a small part of the holes reaching the top surface charged the surface positively. As the load was released, positive holes slowly dissipated and recombined with electrons. As a result, the polarization and electrification dissipated with fluctuations.

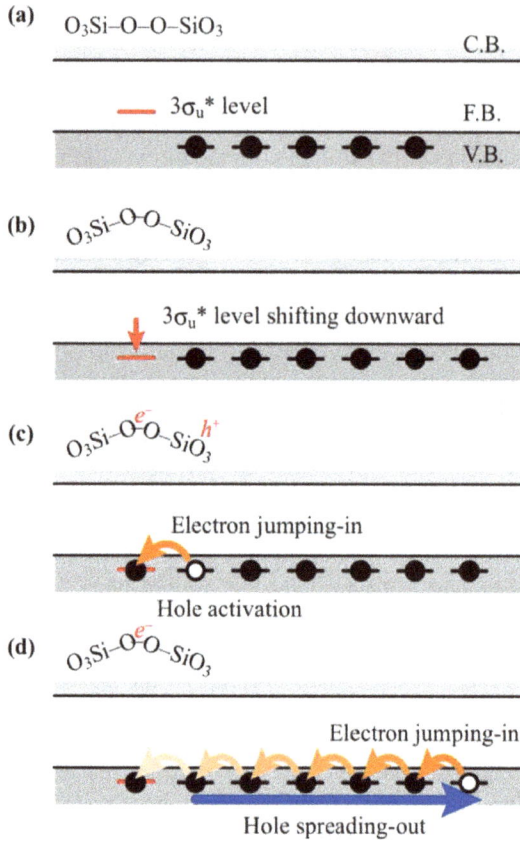

Fig. 7. Band model of stress-activation of a positive hole in the case of quartz. (a) Normal state. (b) Downward shift of the unoccupied $3\sigma_u^*$ level accompanying structure deformation around a peroxy bond under subjection of a load. (c) Jump-in of an electron and activation of a positive hole. (d) Movement of a positive hole through the valence band by successive electron-hopping steps. Modified from Takeuchi et al. (2010)

2.4 Short discussion

The stress activation of positive holes will occur also in the andesite bedrock surrounding the underground observation room at the arrival of seismic waves because the basic mechanism of activation and movement of positive holes is expected to be similar in the observation sites and laboratory. Considering the electrode system simplified as portrayed in Figure 2c, the transient current I_{ts} passing through the internal resistance of the recorder R_{in} during arriving of S-waves is given as $I_{ts} = (V_{rec} - V_{bkg})/R_{in}$, where V_{rec} is the recorded potential and V_{bkg} is the background level. Integration with time gives the charge induced on the electrode as ca. 10^{-9} C and its surface density as ca. 10^{-10} C/m^2. These values are larger than those obtained from laboratory experiments, but we should consider differences in conditions such as the stressed volume, the applied stress, the stress rate, etc.

3. On the ground surface

3.1 Observation sites and systems

Our observation sites on the ground surface in the early stage were located in Miyagi and Fukushima Prefectures, Japan (Figure 1).

(A) **Aobayama** (N38°15′, E140°50′): This site was located in a growth of miscellaneous trees in the middle of a slight slope toward a mountain stream on the Aobayama Campus of Tohoku University. Figure 8a shows the four pairs of reference electrodes (RE-5; M.C. Miller Co. Inc.) horizontally and vertically buried in the ground (Takeuchi et al., 1995, 1997a). Electric signals were transported to a pen-recorder with a 10-Hz low-pass filter through coaxial cables. Thereafter, the A/D converted data were stored on the HD of a PC at 1 Hz sampling. Figure 8b presents another system built later (Takeuchi et al., 2000). A pair of reference electrodes (RE-5; M.C. Miller Co. Inc.) was buried vertically with a 2-m distance separating them. A dodecagonal plate electrode (3 m diameter) was placed on 3 1.5-m-long ceramic insulators about 50 m from the electrodes. The plate electrode was connected to the ground through a data recorder. A seismometer was fixed on the floor of an observation room nearby. All electric signals were transported to the data recorder through coaxial cables. A PC stored the A/D converted data from the recorder at 100 Hz only from 5-s before the seismic trigger to 3-min after the trigger. This site was operational during October 1993 – March 1999.

(B) **Onagawa** (N38°26′, E141°29′): This site was located at the site of the Onagawa geomagnetic field observatory of Tohoku University, about 80 m from a cliff facing the Pacific Ocean (Takeuchi et al., 1997b). Two pairs of reference electrodes (RE-5; M.C. Miller Co. Inc.) were used. One pair was buried horizontally with a 40-m separation distance in the N-S direction at 0.5 m depth. Another was buried vertically at 0.5 m and 1.5 m depths. All electric signals were transported to a pen-recorder with a 10-Hz low-pass filter through coaxial cables. Thereafter, the A/D converted data were stored on the HD of a PC at 1 Hz sampling. This site was operational during October 1994 – March 1996. Electric shocks from lightning frequently halted data recording.

(C) **Tsukidate** (N38°43′, E141°02′): This site was located in the campus of Tohoku-Polytechnic College, ca. 10 m from nearby a college building (Takeuchi et al., 1997b). The ground level of the college buildings had been raised. There was one pair of reference electrodes (RE-5; M.C. Miller Co. Inc.) buried vertically at 7 m and 10 m depths. All electric signals were transported to a pen-recorder with a 10-Hz low-pass filter through coaxial cables. Thereafter, the A/D converted data were stored on the HD of a PC at 1 Hz sampling. This site was operational during December 1994 – December 1998. However, data have included strong pulses since March 1995 because of construction work on a new building ca. 10 m from the electrodes.

(D) **Iitate** (N37°42′, E140°40′): This site was located at the site of the Iitate telescopic observatory of Tohoku University, far from densely populated areas. Figure 9 portrays locations of five reference electrodes (RE-5; M.C. Miller Co. Inc.) buried vertically and horizontally in the ground (Takeuchi et al., 1997b). All electric signals were transported to a pen-recorder with a 10-Hz low-pass filter through coaxial cables. Thereafter, the A/D converted data were stored on the HD of a PC at 1 Hz sampling. This site was operational during November 1995 – November 1996. However, the system was often unstable and stopped data recording.

(a)   (b)

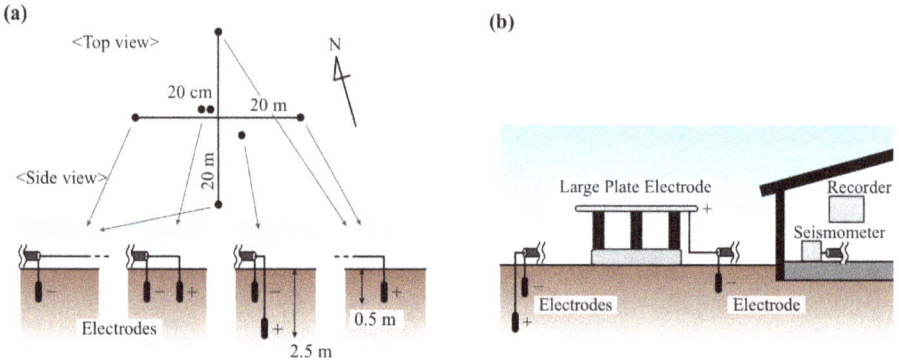

Fig. 8. (a) Observation system of (A) Aobayama site in the early stage. (b) System in the late stage. Modified from Takeuchi et al. (1995, 2000)

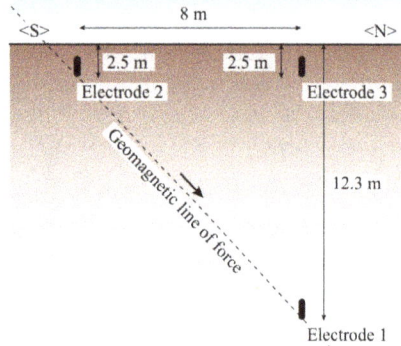

Fig. 9. Observation system of (D) Iitate site. Modified from Takeuchi et al. (1997b)

During these 5-year observations, it was confirmed that no electric signal was generated by minute displacements of the electrodes buried in the ground or by vibration of the coaxial cables. These sites detected electric signals with the waveform of damped oscillations at the arrival of seismic waves. Especially, (A) Aobayama site was so sensitive that it detected electric signals at the arrival of very weak seismic waves from a M7.6 earthquake that occurred in 1994, ca. 550 km below Vladivostok, Russia (Takeuchi et al., 1995, 1997a) and from the M7.2 Kobe earthquake that occurred in 1995, ca. 600 km from this site (Okubo et al., 2005). To minimize the effects of artificial noises overlapping the natural electrotelluric currents flowing horizontally, the vertically buried electrode pairs were more useful than those buried horizontally. Moreover, the vertical signals were 10 times stronger than the horizontal signals (Okubo et al., 2005). Based on these experiences, in the next stage, we selected burial of the electrode pairs vertical in the ground and renewed three observation sites in Akita Prefecture, Japan (Figure 1). The systems at each site were very similar. Hereinafter, we mainly treat observation data obtained at the three sites as described below.

(E) **Honjo** (N39°23', E140°04'): This site was located in a green belt on the Honjo Campus of Akita Prefectural University, which is surrounded by rice fields (Okubo et al., 2006b). No tall building existed around this site. Figure 10 depicts an outline of the system. One pair of reference electrodes (RE-5; M.C. Miller Co. Inc.) was buried vertically at 0.5 m and 2.5 m depths. A large aluminum plate electrode, with 4×4 m^2 area, was supported by five insulators at a height of about 4 m. The plate electrode was connected to the ground through a data recorder (DR-1021; TOA-DKK Corp.) with 1-MΩ input impedances on each channel. All electric signals were transported to the data recorder through coaxial cables. A PC controlled the recorder and stored data on its HD at 4 Hz. The data clock was synchronized to within 1 ms of the time accuracy using a GPS unit. Before starting ordinal observations, the large plate electrodes were tested under various weather conditions (e.g., heavy snow) from August 2000. Although the system functioned well in 2001–2006, the electrodes became too old to use in 2007. It became difficult to maintain and repair them. Therefore, we dismantled them in 2009.

(F) **Kyowa** (N39°40′, E140°23′): This site was located in a garden yard of a small recreation house of Akita Prefectural University (Okubo et al., 2006b) in a sparsely populated area. The system was identical to that of (E) Honjo site, as shown in Figure 10, except for the sampling rate: 10 Hz. The operation period was identical to that of (E) Honjo site.

(G) **Sennan** (N39°23′, E140°30′): This site was located on the sports ground of a former primary school surrounded by rice fields (Okubo et al., 2006b). No tall building existed around this site. The system resembled that at (F) Kyowa site. In addition to the two types of electrodes, this site had a seismometer (L-22D; Mark Products) fixed on the floor of the observation room nearby to detect the up–down component of the ground surface velocity, as shown in Figure 10. The operation period was identical to those of (E) Honjo and (F) Kyowa sites.

Hereinafter, we designate the electric signal detected between the pair of reference electrodes as "Earth potential difference (EPD)" and designate the electric signal detected between the large plate electrode and the ground as "atmospheric electricity (AE)".

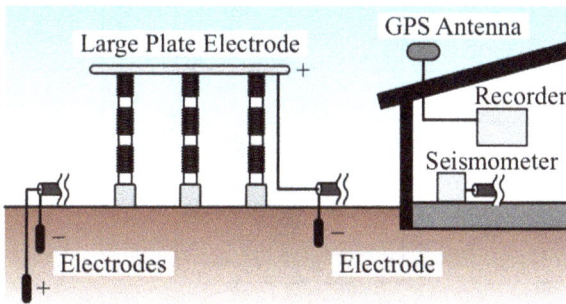

Fig. 10. Observation system of (E) Honjo, (F) Kyowa, and (G) Sennan sites. The seismometer was set up only at (G) Sennan site. Modified from Okubo et al. (2004)

3.2 Observed signals

A M6.3 earthquake occurred near the boundary between Iwate and Miyagi Prefectures at 13:02 on 2 December 2001, UT. The epicenter was (N39°23′, E141°16′). The focal depth was 130 km, which was extremely large relative to the distances among the three sites in Akita Prefecture, as were the hypocentral distances of the sites. Therefore, the local seismic intensities at the sites were similar.

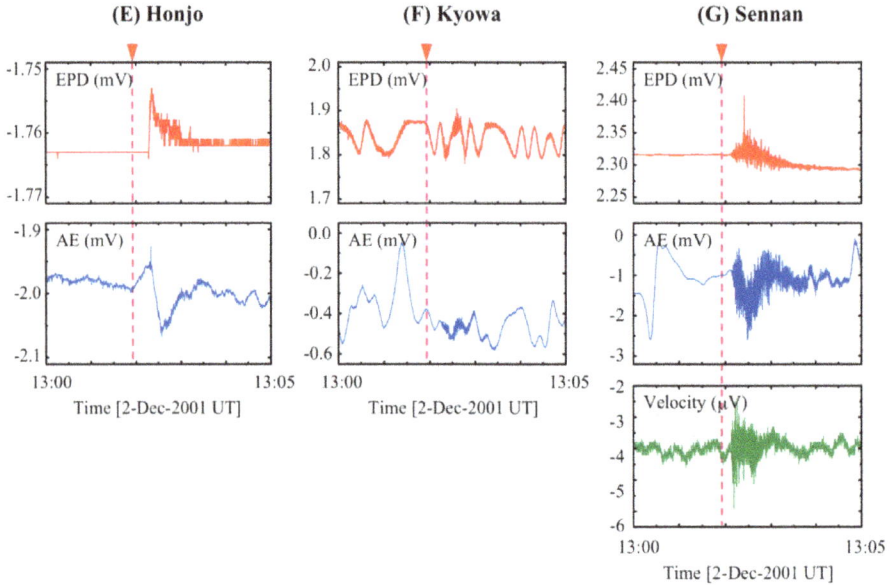

Fig. 11. Raw data recorded before and after the arrival of seismic waves at (E) Honjo, (F) Kyowa, and (G) Sennan sites. The seismometer was set up only at (G) Sennan site. Modified from Okubo et al. (2004)

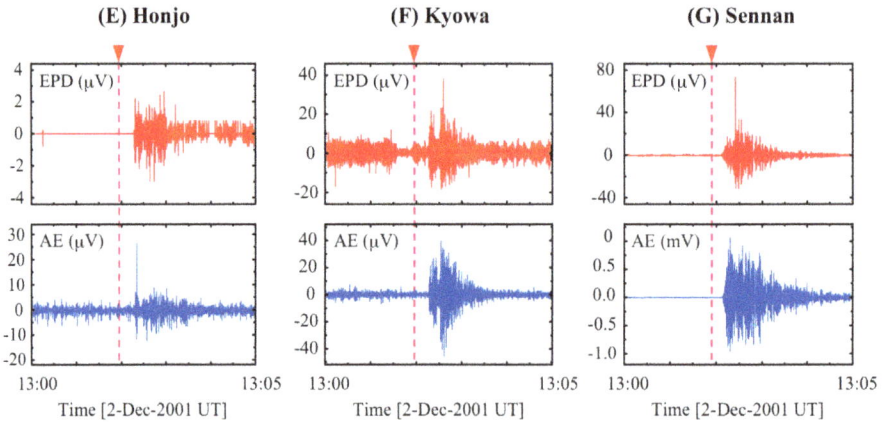

Fig. 12. Signals extracted from raw data presented in Figure 11 using the 1-s moving average method. Modified from Okubo et al. (2004)

Figure 11 portrays plots of raw data obtained before and after the arrival of seismic waves at respective sites. The upper rows show the EPD signal, the middle ones show the AE signal, and the lower one, only that for (G) Sennan site, shows the ground velocity. Dotted lines show the origin time of the earthquake. Signal oscillations are confirmed, although some are small against the background variations. To clarify the signals that we examined specifically, we adopted the moving average method. A time period of 1 s was used for calculation of the moving average. The original data were subtracted by the 1-s moving averages. Then the signals were extracted as shown in Figure 12. The EPD and AE signals with the waveform of damped oscillations are induced simultaneously at the arrival of seismic waves, although some still include high background noise. The waveforms of EPD and AE signals are similar to that of the up–down velocity in (G) Sennan site. The waveforms at (E) Honjo and (F) Kyowa sites will also probably resemble those of the local velocity.

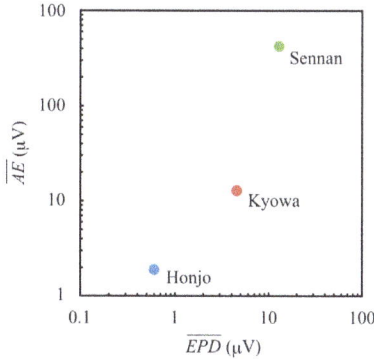

Fig. 13. Positive relation between the average of the earth potential difference ($\overline{EPD}$) and the average of the atmospheric electricity ($\overline{AE}$). Modified from Okubo et al. (2004)

Figure 13 shows the relation between the average amplitudes (*AA*) of the extracted EPD and AE signals, defined as

$$
AA = \sqrt{\frac{1}{30\,(\text{sec})} \sum_{\text{peak time}\,(t=0\,(\text{sec}))}^{t=30\,(\text{sec})} \left(signal + background\,(t)\right)^2} - \sqrt{\frac{1}{30\,(\text{sec})} \sum_{\text{no propagation}\,(t'=0\,(\text{sec}))}^{t'=30\,(\text{sec})} \left(background\,(t')\right)^2}. \tag{1}
$$

The summation in the first root is over a period of 30 s after the peak time of the extracted EPD and AE signals; another in the second root is over a period of 30 s when no seismic wave propagated. Hereinafter, we respectively designate the *AA* of the extracted EPD and AE signals as "$\overline{EPD}$" and "$\overline{AE}$". Figure 13 portrays a positive relation between $\overline{EPD}$ and $\overline{AE}$. Although the local seismic intensity is similar among the three sites described above, the $\overline{EPD}$ and $\overline{AE}$ at (E) Honjo site are smaller and those at (G) Sennan site are larger.

The (G) Sennan site was more sensitive than the other two sites in Akita Prefecture. It detected EPD and AE signals at the arrival of seismic waves from smaller earthquakes occurring in and around Akita Prefecture (Okubo et al., 2007). For example, the left columns in Figure 14 show plots of data obtained before and after the arrival of seismic waves from a M6.2 earthquake that occurred 12 km below the coast of Miyagi Prefecture at 22:14 on 25 July 2003, UT. To clarify the EPD and AE signals that we specifically investigated, we adopted the digital natural observation (D-NOB) method this time. This method is an analytical method used as a novel technique for signal analysis. Details of the concept and method are provided in the Appendix. The EPD and AE signals were extracted as shown in the right columns in Figure 14. Now it is clear that EPD and AE signals with the waveform of damped oscillations are induced simultaneously at the arrival of seismic waves. The waveforms of the EPD and AE signals resemble that of the ground up-down acceleration. These extracted waveforms resemble those presented in Figure 12.

Figure 15 shows relations among $\overline{EPD}$, $\overline{AE}$, and the average of amplitude of the up-down acceleration "$\overline{ACC}$" also obtained from Eq. 1 for earthquakes occurring in and around Akita Prefecture in 2003. The $\overline{EPD}$ is linearly related with $\overline{ACC}$ and has a roughly linear relation with $\overline{AE}$.

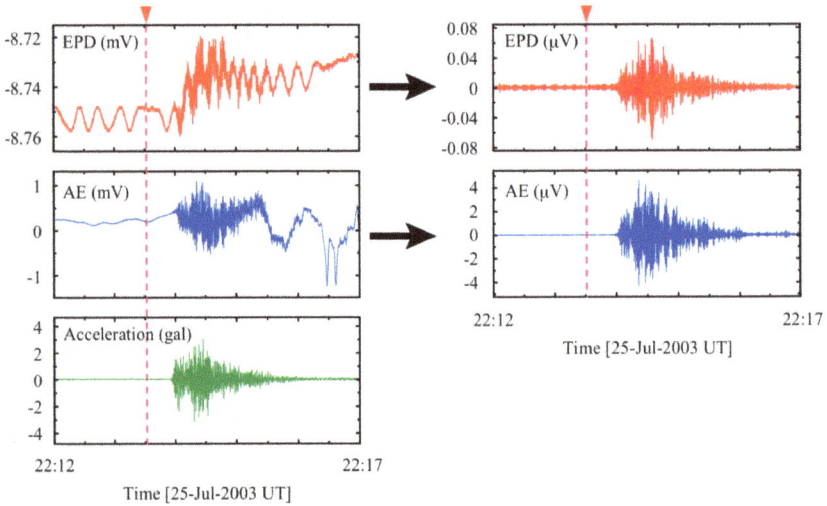

Fig. 14. (Left columns) Observation data recorded before and after the arrival of seismic waves at (G) Sennan site. (Right columns) Signals extracted from observation data in the left columns using the D-NOB method. Modified from Okubo et al. (2007)

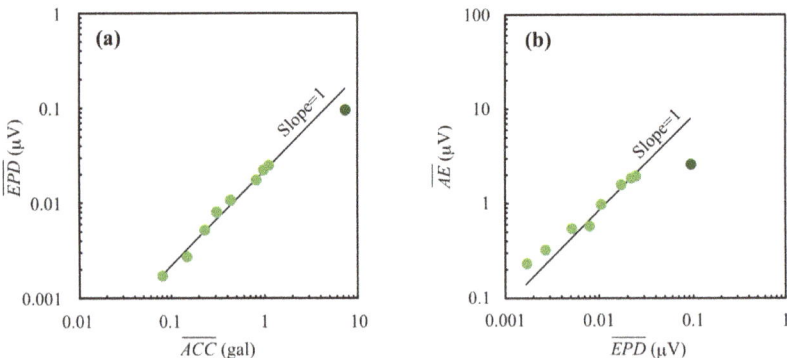

Fig. 15. (a) Positive relation between the average of the earth potential difference ($\overline{EPD}$) and the average of the ground up-down acceleration ($\overline{ACC}$). (b) Positive relation between the average of the atmospheric electricity ($\overline{AE}$) and $\overline{EPD}$. Linear lines with the slope of 1 are obtained from the plots, excluding the dark plots. Modified from Okubo et al. (2007)

3.3 Generation mechanisms

First, we discuss possible generation mechanisms of EPD signals. As possible mechanisms, we can list the four candidates again:

1. **Piezoelectric effect:** The soil at the observation sites includes quartz grains. Therefore, seismic waves cause electric polarization at each grain, which will generate an electric field in the soil if the summation of the polarizations is sufficiently large under a certain convenient condition. However, because piezoelectric polarizations at each quartz grain in the soil will generally be so small and random, its vector summation will be almost negligible. Therefore, this candidate is discounted as the prime factor.
2. **Interfacial electrokinetic effect:** The soil, at least deeper soil, generally includes pore water. When seismic waves push the wet soil, the pore water will flow and cause streaming potential as a result of the interfacial electrokinetic effect. This potential will generate an electric field in the soil.
3. **Radon gas:** Radon is widely distributed in rocks/soils of various types. When seismic waves arrive, radon gas will emanate from the ground. Radioactive decay of radon gas can increase the air ion concentration. This increment will increase the permittivity of the air and the capacitance of the large plate electrode, which will produce an apparent electromotive force between the plate electrode and the ground surface, both exposed to the vertical atmospheric electric field. Consequently, AE signals appear. However, radon gas cannot explain EPD signals.
4. **Positive hole:** In general, soils include mineral grains from igneous rocks. When seismic waves arrive, positive holes will be activated in the grains. An electric unevenness by mobile positive holes and trapped electrons will generate an electric field in the soil. However, if the soil is wet, then pore water may absorb the charges (Balk et al., 2009).

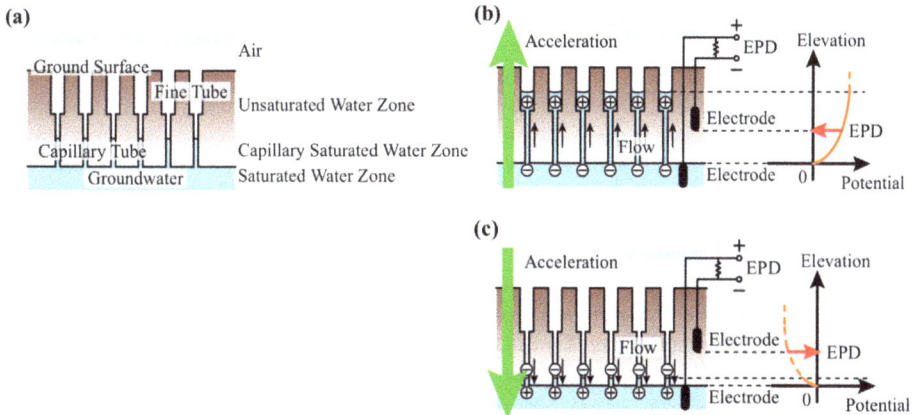

Fig. 16. Generation mechanism of the Earth potential difference (EPD) signals in a geohydraulic model. (a) Model of the near-surface soil layer from the viewpoint of water-saturation/unsaturation. (b) Negative EPD signal induced by streaming potential because of the upward flow of water along the tubes. (c) Positive EPD signal induced by streaming potential because of the downward flow of water along the tubes. Modified from Okubo et al. (2005)

All candidates may simultaneously cause phenomena. Their contribution ratios will be determined by complex conditions (e.g., the arrival direction of seismic waves, inhomogeneous bedrock, wet/dry conditions, etc.) and may be different each time. However, a linear relation exists between the $\overline{EPD}$ and $\overline{ACC}$ at (G) Sennan site, as depicted in Figure 15. Moreover, in those observations, the EPD waveform resembles that of the ground up-down acceleration. Therefore, the force applied to pore water varies similarly to the EPD waveform. These results demonstrate that the amplitude of EPD signals increases proportionally with the pressure difference applied to pore water. Consequently, in this section, we specifically examine the second candidate (interfacial electrokinetic effect) and propose a detailed model for EPD signals.

From the viewpoint of geohydraulics, the water content generally increases with increased depth in the near-surface soil layer with pores. The upper region is called the unsaturated water zone; the lower one is the saturated water zone (Smith, 1982; McCarthy, 2006). Here, we consider only the vertical component. Figure 16a portrays a network model of pores in the near surface soil layer as a bundle of tubes connecting the air and the saturated water zone. The upper part of the tubes is fine and the lower part has capillaries. The lower part of the capillary tubes maintains the pore water, named the capillary saturated water zone. The lower part of this system is the saturated water zone, where the network of pore water is completely connected. Negative ions adhere to the inner wall of the tubes, so that positive ions are predominant in the capillary water. This engenders the formation of electric potential difference between the center and wall of the tubes: the so-called zeta-potential.

Based on early observations and theoretical work, a close coupling dynamics is well known to exist between the seismic acceleration and the pore pressure or water level (Muire-Wood and King, 1993; Kano and Yanagidani, 2006; Yan et al., 2008). This coupling dynamics must be valid in our geohydraulic system presented in Figure 16a. As portrayed in Figure 16b, when acceleration is inflicted upward in this system, the water in the capillary tubes flows upward along the tubes. This water is positively charged. Therefore, flowing up of the water leads to positive electrification in the upper range of the tubes. However, the lower range of the tube charges is negative. Consequently, vertical electric polarizations are formed and an electric potential difference (called streaming potential) appears between the upper level of the water flowing up and the saturated water zone. No more charge exists above the upper level of the water. Therefore, the electric potential in this range is almost constant under an ideal condition. When the positive and negative terminals of electrodes are buried respectively in the saturated water zone and the unsaturated water zone, they will detect a negative EPD signal. As Figure 16c shows, when acceleration is inflicted downward in this system, the opposite phenomenon occurs and the electrode will detect a positive EPD signal. Consequently, when seismic waves induce vertical oscillation of this geohydraulic system, the electrodes detect EPD signal oscillation.

Next, we propose an AE signal generation mechanism. We first consider the apparent AE signals because of vibration of the large plate electrodes. However, if so, then the three plate electrodes with identical setup would show AE signals with equal (or equivalent) amplitude. This is contrary to the result portrayed in Figure 13. As Figures 12 and 14 show, the AE signal oscillation appears along with the EPD signal oscillation. Moreover, as shown in Figures 13 and 15, the $\overline{AE}$ exhibits a good positive relation with $\overline{EPD}$. Therefore, the generation mechanism of AE signals must couple with that of EPD. Figure 17 presents a schematic of a possible mechanism of AE signals. As shown in Figure 16, streaming potential oscillation appears in the unsaturated water zone. This oscillation is equivalent to the electric polarization oscillating vertical to the ground immediately under the ground surface. Their charges on the ground surface generate an atmospheric electric field vertical to the ground. Consequently, an electric potential appears between the large plate electrode and the ground surface. Charges move between the electrode and the ground via the input impedance of the recorder to cancel the potential difference oscillation, which is detected as an AE signal oscillation.

Actually, similarly oscillating AE signals are often detected at (H) Hosokura site in addition to the AE signal shown in Figure 3b in section 2. Although the floor of the underground observation room is dry, the andesite bedrock below the floor will involve pore water. When seismic waves induce oscillation of the capillary water, the induced streaming potential will generate the vertical electric field oscillating in the room.

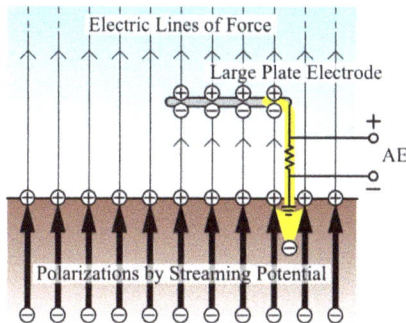

Fig. 17. Generation mechanism of the atmospheric electricity (AE) signals coupled with generation of streaming potential in the near-surface soil layer. Modified from Okubo et al. (2004)

3.4 Short discussion

According to the geohydraulic system shown in Figure 16, the EPD signal amplitude is expected to depend on the electrode positions. We assume six cases as shown in Figure 18.

a. When both electrodes are located horizontally parallel to each other in the unsaturated water zone (Figure 18a), EPD signals will be very small. Even if streaming potential appears at the electrodes, the potential difference between them will be small. For example, at (A) Aobayama site, the vertical EPD signal was 10 times larger than the horizontal one, in which each electrode was separated 2 m (Takeuchi et al., 1995, 1997a).
b. When one electrode is located in the unsaturated water zone and another is in the saturated water zone (Figure 18b), EPD signals will be large. The electrode pair is located at both ends of electric polarization because of streaming potential. Therefore, EPD takes the maximum.
c. When both electrodes are located in the saturated water zone (Figure 18c), EPD signals will be very small. The pore water network is completely connected. Therefore, this zone is sufficiently conductive and most streaming potential is canceled. In fact, at (A) Aobayama site, two electrodes at 10 m and 12 m depths, both probably in the saturated water zone, detected extremely small EPD signals (Takeuchi et al., 1995, 1997a).

d. When one electrode is located deeply in the unsaturated water zone and another is deeply in the saturated water zone (Figure 18d), EPD signals will be large. This case acts like case (b). For example, at (C) Tsukidate site, a pair of electrodes was buried at 7 m and 10 m depths. A borehole survey confirmed that the upper level of the saturated water zone was at 9 m depth. This site detected sufficiently large EPD signals against background variations (Takeuchi et al., 1995, 1997a, 1997b).

e. When both electrodes are located in the unsaturated water zone far from the deep saturated water zone (Figure 18e), EPD signals will be very small. No pore water flows up in the upper range of the thick unsaturated water zone. Therefore, little potential difference exists between the electrodes.

f. When one electrode is in the unsaturated water zone and another is near the saturated water zone (Figure 18f), EPD signals will not be so large. Until the lower electrode is submerged in the water flowing up, no significant potential difference exists between the electrodes.

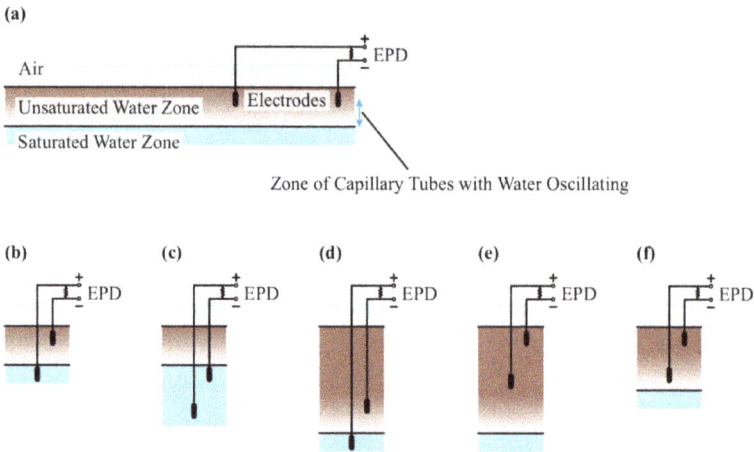

Fig. 18. Six cases of locations with the electrode pair buried and the saturated/unsaturated water zone levels. (a) Both electrodes are horizontally parallel to each other in the unsaturated water zone. (b) One electrode is in the unsaturated water zone; another is in the saturated water zone. (c) Both electrodes are in the saturated water zone. (d) One electrode is deeply situated in the unsaturated water zone; another is deeply situated in the saturated water zone. (e) Both electrodes are in the unsaturated water zone far from the deep saturated water zone. (f) One electrode is in the unsaturated water zone; another is near the saturated water zone. Modified from Okubo et al. (2005)

The significant difference of the $\overline{EPD}$ among the three sites in Akita Prefecture probably results from the difference in position of the buried electrodes and the saturated water zone.

(E) Honjo site was 6 m above the surrounding rice fields. The saturated water zone is expected to be a few meters deeper than the fields. Therefore, this site corresponds to case (e); the $\overline{EPD}$ was small.

(F) Kyowa site was on a flat land in the middle of mountains. Therefore, this site corresponds to case (f); the $\overline{EPD}$ was not so large.

(G) Sennan site was on the same level as the surrounding rice fields. The saturated water zone is expected to be a few meters deeper than the fields. Therefore, this site corresponds to case (b); the $\overline{EPD}$ was large.

4. Above the ground surface

According results of our laboratory experiments, as described in Section 2, when a large volume of dry igneous rock appears on the ground surface, a vertical atmospheric electric field will arise from positive holes activated on the ground surface at the arrival of seismic waves. Additionally, according to our geohydraulic system, as shown in Figures 16 and 17, greater ground acceleration can cause a larger charge density on the ground surface under a certain condition of the saturated/unsaturated water zones, as in case (b) of Figure 18, which engenders a larger atmospheric electric field vertical to the ground surface. This case will correspond to (G) Sennan site. However, the charge density on the ground surface cannot be large in case (e) of Figure 18 because part of the electric field vertical to the ground will be absorbed in the thick upper unsaturated water zone, which engenders a decrease of the charge density on the ground surface and a small AE signal. This case corresponds to (E) Honjo site.

If the surface charge induced on the ground surface is sufficiently large to cause corona discharge in the atmosphere, say ca. 5×10^{-5} C/m² (Lockner et al., 1983), then it may be detected as earthquake lightning. If the amplitude of the vertical atmospheric electric field induced by the surface charges is sufficiently large above a large area, say ca. 1 kV/m (Pulinets et al., 2000; Rapoport et al., 2004), then it may even disturb plasma in the ionosphere with detectable amplitudes. The amplitude of this field depends not only on the seismic wave amplitude but also on the underground water condition. Therefore, there is no guarantee that this field is strongest above the epicenters. This electric field will form from the epicenter. Therefore, no guarantee exists that the influence of this field on ionospheric disturbances is strongest at the epicenter.

However, such disturbances will be overwhelmed in further strong disturbances coming later. They are detected as abnormal changes in the total electron content obtained from analyses of GPS signals, so-called GPS-TEC (Liu et al., 2010; Rolland et al., 2010; Galvan et al., 2011). The most widely accepted causes of such GPS-TEC anomalies are Rayleigh waves and tsunamis. Their vertical motion triggers acoustic gravity waves in the neutral atmosphere. These waves reach the ionosphere in 6–7 min and interact with plasma in the ionosphere. Moreover, the flow of the tsunami, which is a conductive mass moving in a magnetic field, generates a surrounding weak secondary magnetic field because of magnetohydrodynamic interaction (Tyler, 2005). We defer to other reports in the literature for details because such postseismic phenomena are beyond the scope of this chapter.

5. Summary

To make steady progress in the study of seismo-electromagnetic precursors, our group has believed that it is important, first of all, to prove the existence of co-faulting and coseismic phenomena. In accordance with our beliefs, we built observation sites on and in the ground (intermittently 8 sites in total) and have observed electric signals with (1) pairs of reference electrodes in the ground for EPD signals, (2) condenser-like plate electrodes insulated from the ground for AE signals, and/or (3) seismometers at some sites. As a result, we gained the following knowledge related to the signals.

1. **In the early stage** (on the ground surface): Results showed that EPD and AE signals were generated at the arrival of seismic waves. Their amplitudes were positively related with the local seismic intensity. The vertical amplitude of EPD was 10 times larger than the horizontal amplitude.
2. **In the late stage** (on the ground surface): We again detected EPD and AE signals. We extracted weak signals from the background fluctuation using the moving average method and the D-NOB method. We evaluated their amplitudes, which showed that $\overline{EPD}$ and $\overline{ACC}$ were linearly related and $\overline{EPD}$ and $\overline{AE}$ were roughly in a linear relation. We proposed their generation models based on streaming potential caused by vertically oscillating pore water.
3. **In the current stage** (under the ground surface): We detected electric signals probably because of the generation of a vertical electric field in the underground room. We expected positive electrification of the floor as a possible cause. We conducted non-uniform loading tests in a laboratory using a rock block quarried from the site and verified the positive electrification.

Our studies described in this chapter represent great advancement for the steady progress of the study of seismo-electromagnetic precursors. The next step, we think, is to understand the dynamic electric and electromagnetic fields in the ground further. To do so, it is necessary to set up multiple observation sites under, on, and above the ground surface simultaneously, which can detect three-dimensional dynamic electric and electromagnetic phenomena from focal zones. Unfortunately, our observation sites on the ground surface are no longer in operation. However, we developed the underground observation site, (H) Hosokura. Now this site has various other sensors: an electric field mill, two air ion counters (for positive and negative ions), an air temperature-humidity probe, reference electrodes, and two fluxgate magnetometers (Takeuchi et al., 2009; Okubo et al., 2011).

Finally, we summarize electric and electromagnetic (and magnetic) phenomena described in this chapter as shown below.

a. **Preseismic signals:** Quasi-static electromagnetic changes appear from the focal zone before its failure (Figure 19a). We can predict earthquakes if we can detect them with conviction. However, detection of such direct signals with scientific evidence is extremely rare. One reason is probably that their amplitude is extremely small. Another is that these changes probably cannot penetrate the water-saturated near-surface sediment layer. Placing observation sites below the layer, like (H) Hosokura site, will support the detection of these signals. Although some reports in the literature describe ionospheric disturbances occurring before earthquakes (Hayakawa et al., 2010; Oyama et al., 2008), we defer to the literature for details because such preseismic phenomena are beyond the scope of this chapter.
b. **Co-faulting signals:** Quasi-static electromagnetic changes, probably with further strong amplitudes, appear from the focal zone during failure (Figure 19b). Although detection of such direct signals using scientific evidence is very rare, probably for the reasons described above, one report describes that fluxgate magnetometers at (H) Hosokura site detected changes in the geomagnetic field during failure (Okubo et al., 2011). On the other hand, seismic waves induce pore water oscillation in the near-surface sediment

layer and cause the streaming potential. Formation of the vertical atmospheric electric field caused by the streaming potential starts spreading out from the epicenter in association with seismic wave propagation. Moreover, a tsunami is generated and gravity waves spread upward.

c. **Coseismic signals:** Local electric and electromagnetic phenomena are generated under, on, and above the ground surface at the arrival of seismic waves (Figure 19c). This chapter specifically addresses these signals. The streaming potential is generated in the near-surface sediment layer; the vertical atmospheric electric field is generated above the ground. They are detected as EPD and AE signals. On the other hand, gravity waves from seismic waves propagate upward.

d. **Postseismic signals:** Gravity waves from the seismic waves and the tsunami motion disturb the ionosphere with detectable amplitudes from satellites after calm of the main shock. On the other hand, the tsunami flow generates a magnetic field around it because of magnetohydrodynamic interaction.

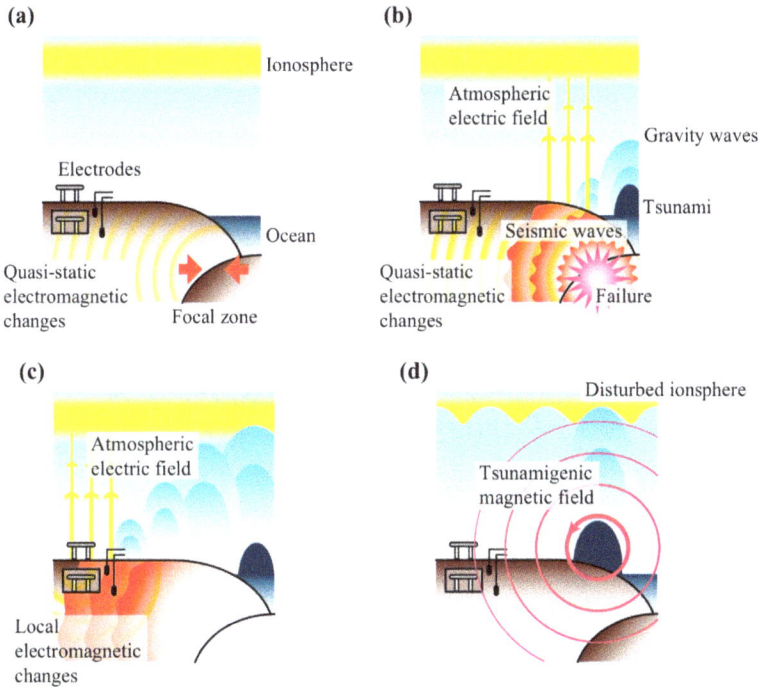

Fig. 19. Electric and electromagnetic phenomena classified in four stages before and after the occurrence of earthquakes: (a) Preseismic, (b) Co-faulting, (c) Coseismic, and (d) Postseismic

6. Appendix

The digital natural observation (D-NOB) method is a kind of linear transformation technique used to address the instantaneous nature of a signal waveform (Iijima, 2000, 2001). Linear transformation simplifies the quantitative evaluation of the output result. The outline of this method is the following:

First, an identical operator I and a delay operator D are defined respectively as

$$I f(n) = f(n) \text{ and} \tag{2}$$

$$D f(n) = f(n-1), \tag{3}$$

where $f(n)$ are arbitrary time series data. Using them, two operators are also defined as

$$\Gamma = (1-\lambda)I + \lambda D \text{ and} \tag{4}$$

$$\Lambda = \lambda I - \lambda D, \tag{5}$$

where λ $(0 < \lambda < 1)$ is a weighting factor, normally 0.5. When an operator is assumed as

$$X_m^{(M)} = \Gamma^{M-m}\Lambda^m, \tag{6}$$

with $m = 0, 1, 2, ..., M$, the following equation is obtained.

$$n_m^{(M)}(n) = X_m^{(M)}f(n). \tag{7}$$

The $n_m^{(M)}(n)$ are time series data of the Mth degree and the mth order, called a fundamental observation value (FOV) corresponding to the time series data $f(n)$. They represent the instantaneous variation of the observed waveform at time n. From Eq. 7, the FOV is

$$n_m^{(M)} = \sum_{l=0}^{M} \binom{M}{l}(1-\lambda)^{M-l}\lambda^l \left\{ \left(\frac{\lambda}{1-\lambda} \right)^m k^{(M)}(m,l) \right\} f(n-l), \tag{8}$$

with $m = 0, 1, 2, ..., M$, where

$$k_m^{(M)}(m,l) = \frac{1}{\binom{M}{l}} \sum_{r=0}^{l} (-1)^r \binom{m}{r}\binom{M-m}{l-r}\left(\frac{1-\lambda}{\lambda} \right)^r, \tag{9}$$

with $l = 0, 1, 2, ..., M$ and where $\binom{M}{m}$ represents the binomial coefficients.

Eq. (7) changes completely from $n_m^{(M)}(n)$ of the Mth order to the original data $f(n)$ using

$$f(n) = \sum_{m=0}^{M} \binom{M}{m}n_m^{(M)}(n). \tag{10}$$

Eq. (10) is called the NOB inverse transform. Eqs. (7) and (10) are collectively designated as the NOB-transform.

7. References

Balk, M., Bose, M., Ertem, G., Rogoff, D.A., Rothschild, L.J. & Freund, F.T. (2009). Oxidation of water to hydrogen peroxide at the rock-water interface due to stress-activated electric currents in rocks. *Earth and Planetary Science Letters*, Vol. 283, pp. 87–92.

Eftaxias, K., Frangos, P., Kapiris, P., Polygiannakis, J. Kopanas, J. & Peratzakis, A. (2004). Review and a model of pre-seismic electromagnetic emissions in terms of fractal electrodynamics. *Fractals*, Vol. 12, pp. 243–273.

Fraser-Smith, A.C., Bernardi, A. & McGill, P.R. (1990). Low-frequency magnetic field measurements near the epicenter of the Ms 7.1 Loma Prieta earthquake. *Geophysical Research Letters*, Vol. 17, pp. 1465–1668.

Galvan, D.A., Komjathy, A., Hickey, M.P. & Mannucci, A.J. (2011). The 2009 Samoa and 2010 Chile tsunamis as observed in the ionosphere using GPS total electron content. *Journal of Geophysical Research*, Vol. 116, pp. A06318.

Hayakawa, M., Kasahara, Y., Nakamura, T., Muto, F., Horie, T., Maekawa, S., Hobara, Y., Rozhnoi, A.A., Solovieva, M. & Molchanov, O.A. (2010). A statistical study on the correlation between lower ionospheric perturbation as seen by subionospheric VLF/LF propagation and earthquakes. *Journal of Geophysical Research*, Vol. 115, pp. A09305.

Iijima, T. (2000). *Theory of Natural Observation Method*, Morikita Publishing Company, Tokyo, pp. 194 (in Japanese).

Iijima, T. (2001). *The Digital Natural Observation Method*, Morikita Publishing Company, Tokyo, pp. 142 (in Japanese).

Kano, Y. & Yanagidani, T. (2006). Broadband hydroseismograms observed by closed borehole wells in the Kamioka mine, central Japan: Response of pore pressure to seismic waves from 0.05 to 2 Hz. *Journal of Geophysical Research*, Vol. 111, pp. B03410.

Liu, J.Y., Tsai, H.F., Lin, C.H., Kamogawa, M., Chen, Y.I., Lin, C.H., Huang, B.S., Yu, S.B. & Yeh, Y.H. (2010). Coseismic ionospheric disturbances triggered by the Chi-Chi earthquake. *Journal of Geophysics*, Vol. 115, pp. A08303.

Lockner, D.A., Johnston, M.D.S. & Byerlee, J.D. (1983). A mechanism to explain the generation of earthquake lights. *Nature*, Vol. 302, pp. 28–33.

McCarthy, D.F. (2006). *Essentials of Soil Mechanics and Foundations: Basic Geotechnics, 7th Edition*, Prentice Hall, Upper Saddle River, pp. 864.

Muire-Wood R. & King, G.C.P., 1993. Hydrological signatures of earthquake strain. *Journal of Geophysical Research*, Vol. 98, pp. 22,035–22,068.

Okubo, K., Yamamoto, K., Takayama, M. & Takeuchi, N. (2005). Generation mechanism of Earth potential difference signal during seismic wave propagation and its observation condition. *IEEJ Transactions on Fundamental and Materials*, Vol. 125, pp. 614–618 (in Japanese with English abstract).

Okubo, K., Sato, S., Ishii, T. & Takeuchi, N. (2006a). Observation of atmospheric electricity variation signals during underground seismic wave propagation. *IEEJ Transactions on Electrical and Electronic Engineering*, Vol. 1, pp. 182–187.

Okubo, K., Yamamoto, K., Takayama, M. & Takeuchi, N. (2006b). Conditions of atmospheric electricity variation during seismic wave propagation. *Electrical Engineering in Japan*, Vol. 157, pp. 1–9.

Okubo, K., Takayama, M. & Takeuchi, N. (2007). Electrostatic field variation in the atmosphere induced by earthpotential difference variation during seismic wave propagation. *IEEE Transactions on Electromagnetic Compatibility*, Vol. 49, pp. 163–169.

Okubo, K., Takeuchi, N., Utsugi, M., Yumoto, K. & Sasai, Y. (2011). Direct magnetic signals from earthquake rupturing: Iwate-Miyagi earthquake of M7.2, Japan. *Earth and Planetary Science Letters*, Vol. 305, pp. 65–72.

Oyama, K., Kakinami, Y., Liu, J.Y., Kamogawa, M. & Kodama, T. (2008). Reduction of electron temperature in low-latitude ionosphere at 600 km before and after large earthquakes. *Journal of Geophysical Research*, Vol. 113, pp. A11317.

Pulinets, S.A., Boyarchuk, K.A., Hegai, V.V., Kim, V.P. & Lomonosov, A.M. (2000). Quasielectrostatic model of atmosphere-thermosphere-ionosphere coupling. *Advances in Space Research*, Vol. 26, pp. 1209–1218.

Rapoport, Y., Grimasky, V., Hayamawa, M., Ivchenko, Juarez-R, V.D., Koshevaya, S. & Gotynyan, O. (2004). Change of ionospheric plasma parameters under the influence of electric field which has lithospheric origin and due to radon emanation. *Physics and Chemistry of the Earth*, Vol. 29, pp. 579–587.

Rolland, L.M., Occhipinti, G., Lognonne, P. & Loevenbruk, A. (2010). Ionospheric gravity waves detected offshore Hawaii after tsunamis. *Geophysical Research Letters*, Vol. 37, pp. L17101.

Saradjian, M.R. & Akhoondzadeh, M. (2011). Thermal anomalies detection before strong earthquakes ($M > 6.0$) using interquartile, wavelet and Kalman filter methods. *Natural Hazards and Earth System Sciences*, Vol. 11, pp. 1099–1108.

Smith, G. N. (1982). *Elements of Soil Mechanics for Civil and Mining Engineers, 5th Revised Edition*, Collins, London, pp. 512.

Takeuchi, A., Okubo, K., Watanabe, S., Nakamura, Y. & Takeuchi, N. (2009). Electric and Ionic Environmental Circumstances Interacting at Hosokura Underground Mine in Northeast Japan. *IEEJ Transactions on Fundamental and Materials*, Vol. 129, pp. 870–874.

Takeuchi, A., Futada, Y., Okubo, K. & Takeuchi, N. (2010). Positive electrification on the floor of an underground mine gallery at the arrival of seismic waves and similar electrification on the surface of partially stressed rocks in laboratory. *Terra Nova*, Vol. 22, pp. 203–207.

Takeuchi, N., Narita, K., Ono, I., Goto, Y. & Chubachi, N. (1995). Measurement of seismic wave with vertically induced Earth potential difference. *Transactions of IEE of Japan*, Vol. 115-C, pp. 1548–1553 (in Japanese with English abstract).

Takeuchi, N., Chubachi, N. & Narita, K. (1997a). Observation of earthquake waves by the vertical earth potential difference method. *Physics of the Earth and Planetary Interiors*, Vol. 101, pp. 157–161.

Takeuchi, N., Chubachi, N., Narita, N. Honma, N. & Takahashi, T. (1997b). Characteristics of vertical earth potential difference signals. *Transactions of IEE of Japan*, Vol. 117-C, pp. 554–560, (in Japanese with English abstract).

Takeuchi, N., Okubo, K. & Honma, N. (2000). Potential variations of metal electrode in the air during seismic wave propagation. *Transactions of IEE of Japan*, Vol. 120-C, pp. 1409–1415 (in Japanese with English abstract).

Tyler, R.H. (2005). A simple formula for estimating the magnetic field generated by tsunami flow. *Geophysical Research Letters*, Vol. 32, pp. L09608.

Uyeda, S., Nagao, T. & Kamogawa, M. (2009). Short-term earthquake prediction: Current status of seismo-electromagnetics. *Tectonophysics*, Vol. 470, pp. 205–213.

Varianatos, P. (2005). *The Physics of Seismo Electric Signals*, Terra Scientific Publishing Company, Tokyo, pp. 338.

Yan, R., Chen, Y., Gao, F. W. & Huang, F.Q. (2008). Calculating skempton constant of aquifer from volume strain and water level response to seismic waves at Changping seismic station. *Acta Seismologica Sinica*, Vol. 21, pp. 148–155.

3

Electric Displacement by Earthquakes

Antonio Lira and Jorge A. Heraud
Pontificia Universidad Católica del Perú
Peru

1. Introduction

Electromagnetic emission in the atmosphere usually occurs in relation with charge acceleration between clouds and the Earth's surface. Lightning is the best known electromagnetic emission in nature and takes place in thunderstorms. We can also ask the question, if it is possible to have electromagnetic emission -for example, a flash of light- in the atmosphere originating in the Earth´s interior. The answer is affirmative and in this chapter we will try to describe some theories about it.

2. Electromagnetic observations

Electromagnetic emission is a secondary effect which can take place in the atmosphere caused by earthquakes (Richter 1958). Among the secondary effects of earthquakes electromagnetic emission is the brightest area of seismology. In the report on the great Chilean earthquake of 1960, Warwick associated 18 MHz radio emission (Fig. 1.) with rock fracture (Warwick et al. 1982). The first evidence with hard data of co-seismic electromagnetic radiation was found during the Matsushiro earthquake swarm between 1965 and 1967 in Japan. Color and black and white photographs were taken by many observers and reproduced by Derr (Derr 1973). One of these photographs is shown here (Fig. 2.). An excellent film of earthquake lights was made during the Peruvian earthquake of 15 August 2007 (Fig. 3.) by a television cameraman, http://www.youtube.com/watch?v=SHmHsP1gd8I. We found a time difference correlation between seismic waves and light flashes in Lima, 150 km from the epicenter (Lira 2008; Heraud and Lira 2011).

3. Theories of earthquakes lights

One of the older theories to explain the relation between earthquakes and light emissions is the piezoelectric theory, due to Finkelstein (Finkelstein et al. 1973). This theory involves the idea that earthquake lightning could be caused by piezoelectric fields produced in rocks by seismic waves. This piezoelectric theory has several disadvantages: first, the electrical resistivity of the rock would need to be of the order of 10^9 Ωm. Second, high-frequency pressure waves would be necessary. Third, as an alternative it would be necessary to have localized high-conductivity channels in high-resistivity surroundings from rock layers at depths of the order of 10 km to Earth's surface.

Mitzutani proposed an electrokinetic theory to provide a possible means of earthquake prediction (Mitzutani et al. 1976). Diffusion of fluid through rocks into a dilatant focal

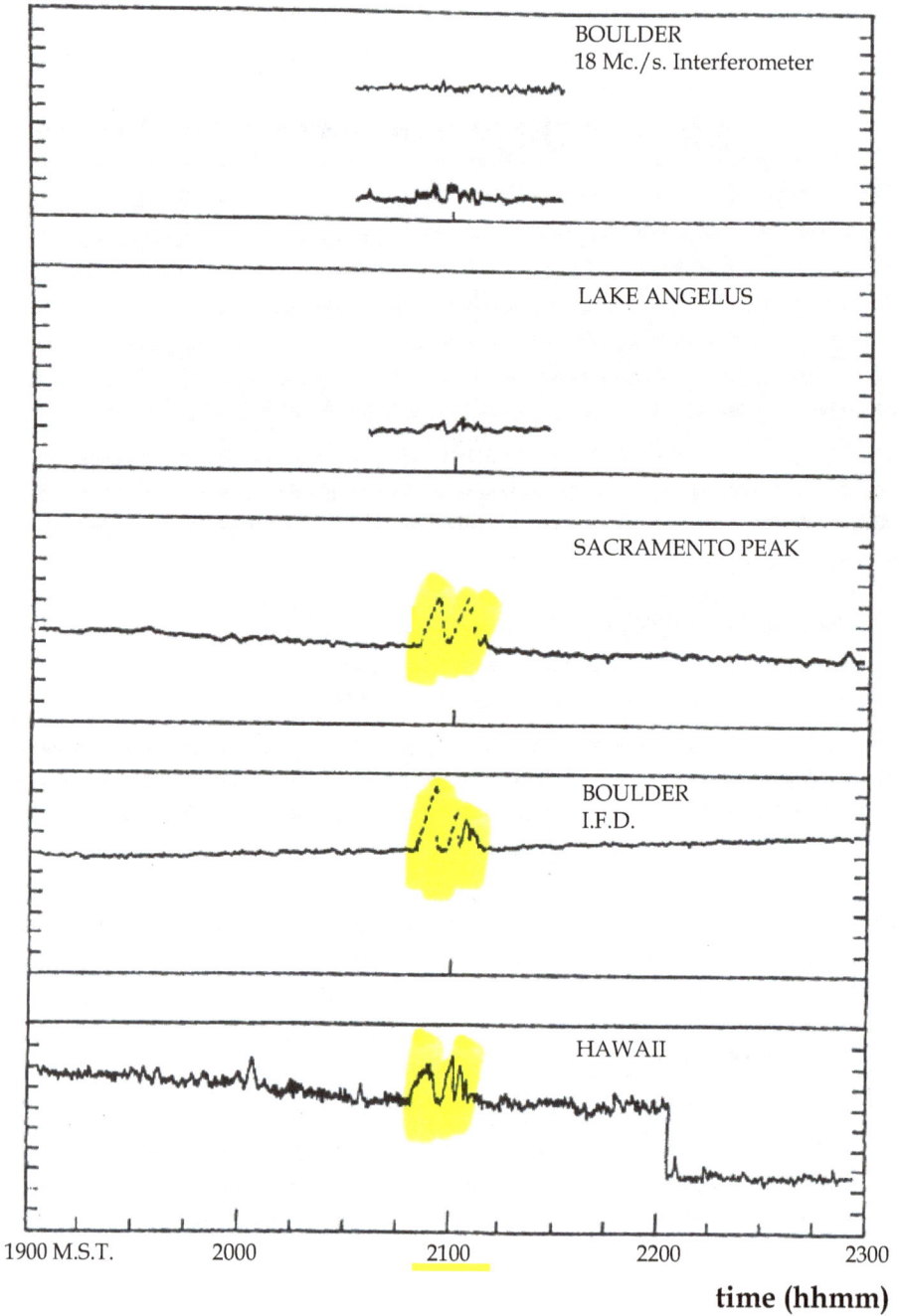

Fig. 1. Radio emission seen 6 days prior to the great Chilean earthquake of 1960 (Warwick et al. 1982)

region preceding or following an earthquake would cause significant variation of electric field near the focal region and may be related to earthquake lightning. Mitzutani obtained

$$\text{grad V} / \text{grad P} = 10^2 - 10^3 \text{ volt/kbar}, \tag{1}$$

where V is the streaming potential and P the pressure of the fluid (Fig. 4.). For a M6.4 earthquake he estimated

$$\text{grad P} = 1 - 10^2 \text{ bar/km} \tag{2}$$

So we have an electric potential gradient of 1 - 1000 volts for 10 km, the linear dimension characteristic to the earthquake. This voltage is not big enough for earthquake lightning.

Fig. 2. Photograph of a light in the sky during the Matsushiro earthquake swarm between 1965 and 1967 (Derr 1973)

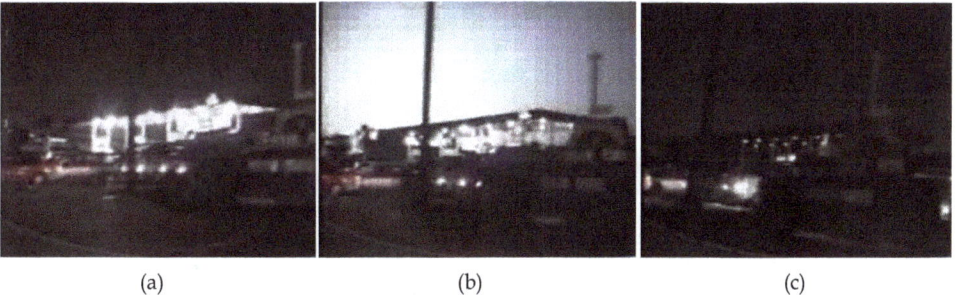

| (a) | (b) | (c) |

Fig. 3. Lights in the sky of Lima filmed during the Peruvian earthquake of 2007. (a) Before a light. (b) During a light. (c) After a light. (Lira 2008)

Lockner proposed a friction-vaporization theory to explain the generation of earthquake lights (Lockner et al. 1983). In this theory a central conductor, a few centimeters wide on the fault axis surrounded by a low conductivity sheath of rock containing vaporized pore water arises through the earthquake. This central conductor would collect charge in the shear zone, and because it is hundreds of meters deep and only some centimeters wide, it will concentrate the charge along its edges, where the curvature is at its highest. If the conductor is shallow enough, the charge concentrated along its top edge would produce an intense electric field at the Earth's surface, and would induce coronal discharge. This theory has a problem: if the high heated vaporized water reaches the Earth's surface it will come out like a steam jet, even though no reported observation has ever confirmed this. However, water vapour expelled at Enceladus, a satellite of Saturn, was photographed 2009 by the Cassini spacecraft, as shown in Fig. 5. and reported from Hartogh (Hartogh et al. 2011).

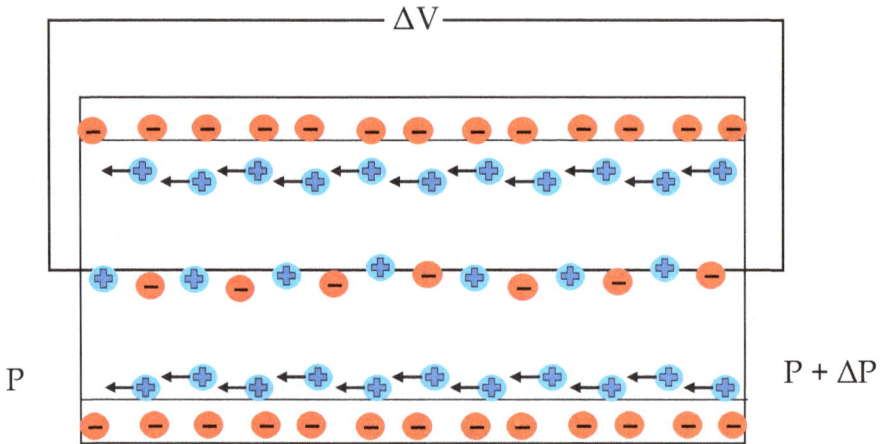

Fig. 4. Potential difference ΔV and pressure difference ΔP in a capillary

Fig. 5. Photograph of vapour water jets at Saturn's satellite Enceladus, taken by the Cassini spacecraft (Photo: NASA/JPL/Space Science Institute)

One of the more sophisticated theories to account for the electrification of the Earth's surface by earthquakes, is that of F. T. Freund (Freund 2007). In this theory the Earth's crust works like a semiconductor diode and the physical "battery" is driven by pressure. Positive holes and electrons are the charge carriers, and they are activated when rocks are subjected to stress, as shown in Fig. 6. The interesting part is that the flow of positive holes, thus generated, occurs also through the uncompressed rock beyond the stressed volume and flow by diffusion even through gravel and sand. According to Freund, his model still does not address the case of faults filled with water.

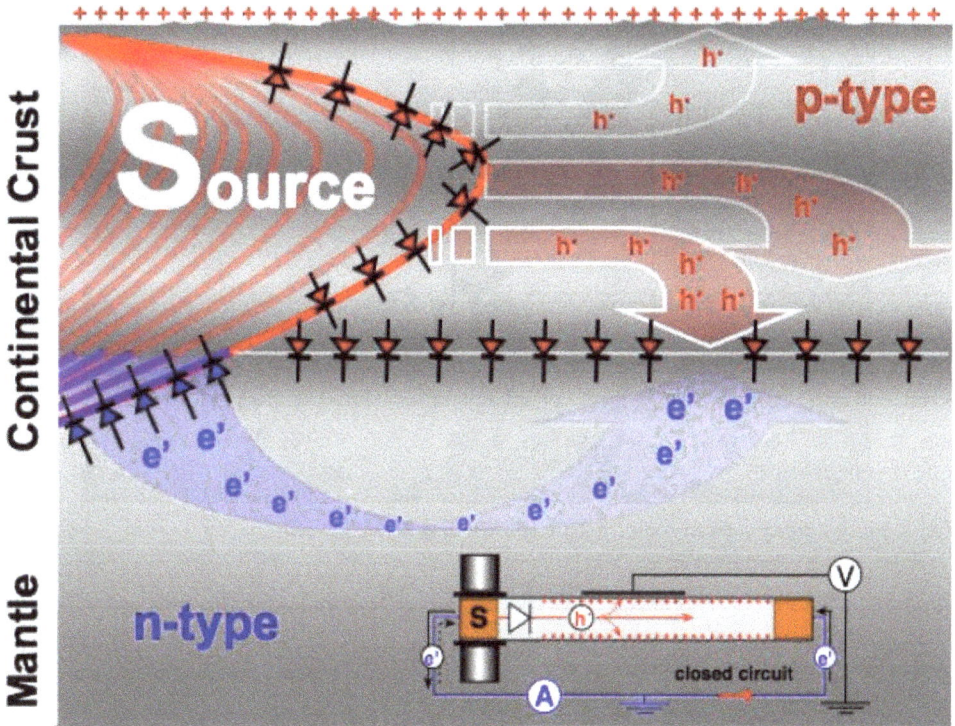

Fig. 6. Freund's theory for the generation of positive holes (Freund 2007)

4. Laboratory investigation of electromagnetic emission by earthquakes

Laboratory experiments show that pressure on rocks produces an electrical potential (Freund 2003) and electromagnetic radiation (Brady and Rowell 1986), as shown in Fig. 7.

As seen in the last section, different theories have been proposed to explain the electrification of the Earth's surface by earthquakes, but the actual process in the crust of the Earth remained unknown until Akihiro Takeuchi, at an underground mine in Japan, observed electrification on the floor of a gallery at the arrival of seismic waves (Takeuchi et al. 2010), as schematically shown in Fig. 8. He discovered, that the arrival of S waves is synchronized with an electric pulse, that flows across the internal resistance of a voltmeter, as shown in Fig. 9.

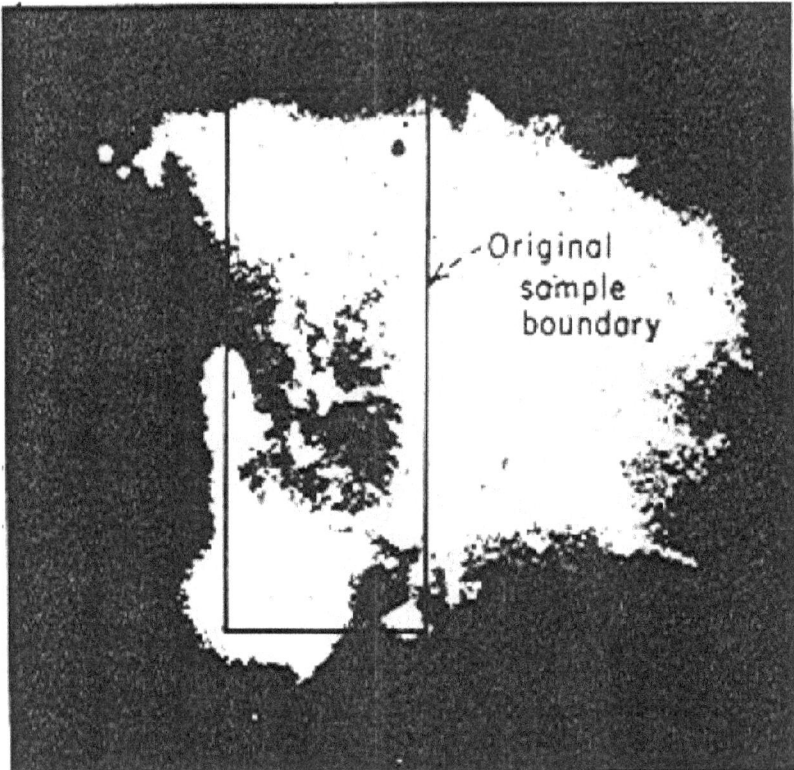

Fig. 7. Light emission from rock fracture (Brady and Rowell 1986)

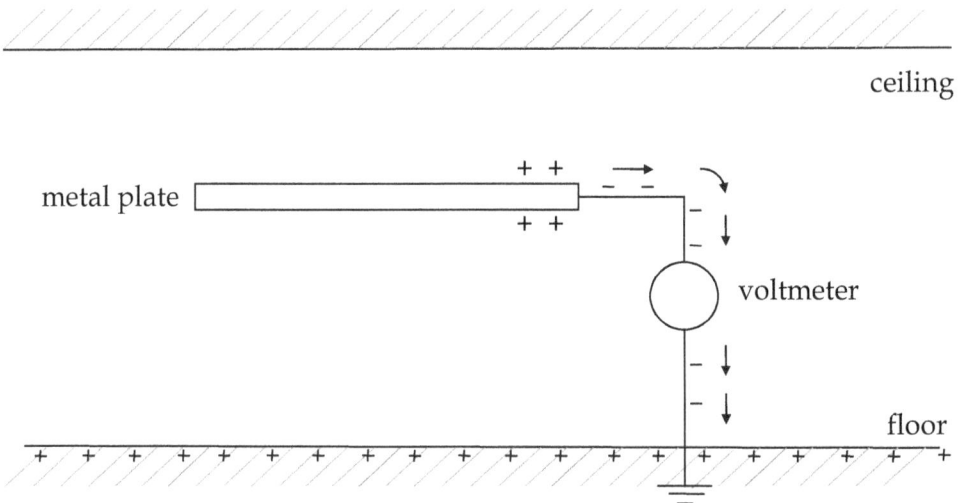

Fig. 8. Measuring the electrification on the floor of an underground mine gallery through a voltmeter at the arrival of seismic waves

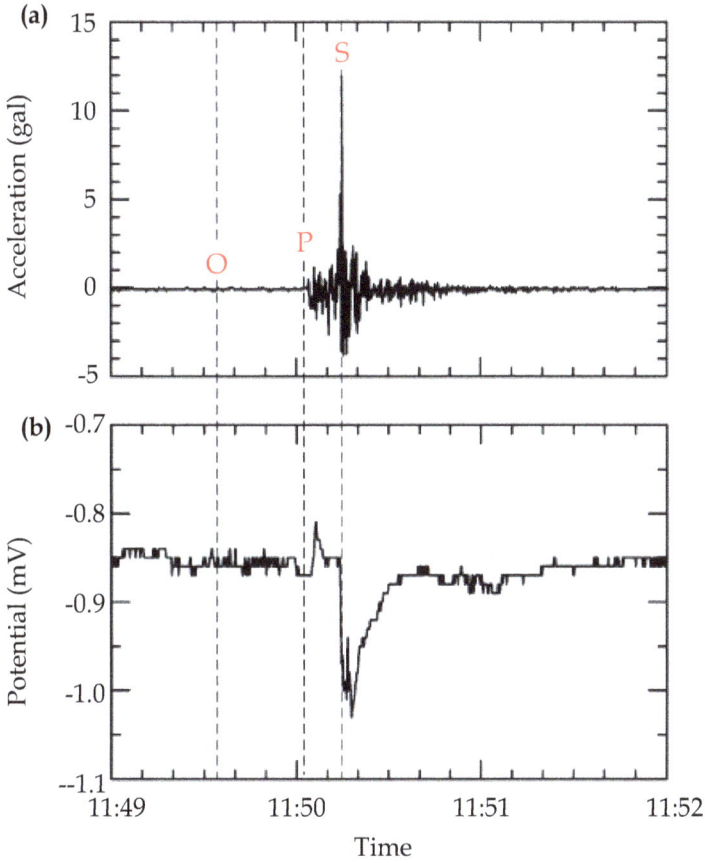

Fig. 9. Electric pulse (b) detected at the arrival of a S wave (a) (Takeuchi et al. 2010)

5. Electric currents in the crust of the Earth

The crust of the Earth is composed of a great variety of igneous, metamorphic and sedimentary rocks. The electrical resistivity of crustal rocks may vary over several orders of magnitude ($10^{-1} - 10^5$ Ωm), as shown in Fig. 10., depending on a wide range of petrological and physical parameters.

From an electrical point of view we will suppose that the Earth's crust consists of a number of plane strata of different materials. In this case, insulators and imperfect conductors form a series electrical circuit, and the total electric resistance becomes

$$R_T = \sum R(\text{insulators}) + \sum R(\text{imperfect conductors}) \tag{3}$$

Under the action of external electric forces no conduction current is produced, except when disruptive discharge occurs. We shall now suppose, for the sake of simplicity, that the Earth's crust consists of an isotropic homogeneous insulator with permittivity ε, and that in the interior of the Earth, at a distance d away from the Earth's surface, an electric charge q is generated due to an earthquake, which we would consider as a point charge, as shown in Fig. 11.

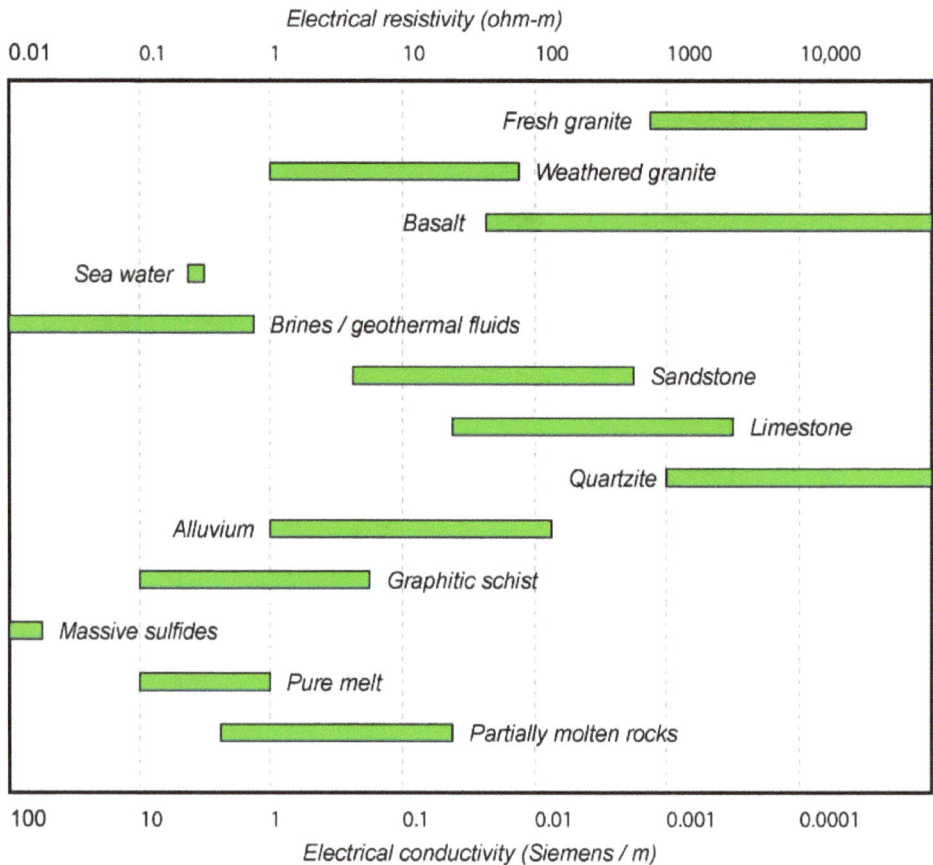

Fig. 10. Electrical resistivity of rocks (Geophysics 2009)

The electric field of the charge q can not produce conduction current through the Earth, but the electric charge is displaced within the Earth generating a polarization charge on the Earth's surface. The polarization charge density at a point P on the Earth's surface is:

$$\sigma_{pol} = q \, \varepsilon_0 \, (\varepsilon - \varepsilon_0) \, d \; / \; 2\pi \, \varepsilon \, (\varepsilon + \varepsilon_0) \, (d^2 + r^2)^{3/2} \tag{4}$$

where ε_0 is the permittivity of free space and approximately that of the atmosphere, and is given by

$$\varepsilon_0 = 8.85 \times 10^{-12} \; C^2/Nm^2 \tag{5}$$

The displacement of the electric charge evidently constitutes an electric current, which is produced as the orbits of the electrons are displaced to some extent under the influence of the electric field of the charge q. This current, however, can only exist until the electric charges reach the Earth's surface. The electric polarization at the surface of the Earth disappears when the electric charge in the interior of the Earth is removed.

atmosphere

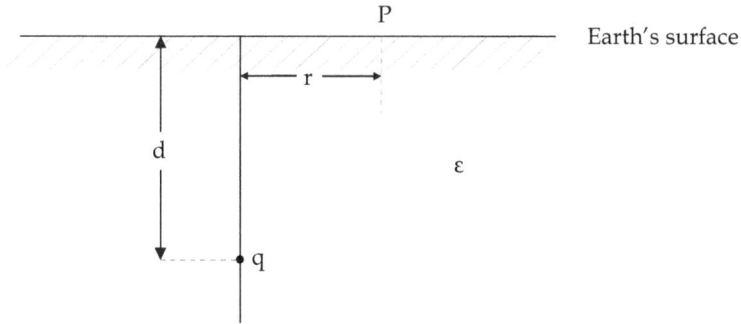

Fig. 11. Point charge q embedded in the interior of the Earth

A negative point charge –q embedded in the interior of the Earth would induce a negative charge distribution on the Earth's surface, as shown in Fig. 12. The lines of force are shown qualitatively in Fig. 13. The electric field strength on the Earth's surface is:

$$E = q \ / \ 4\pi \ \varepsilon_0 \ (d^2 + r^2) \tag{6}$$

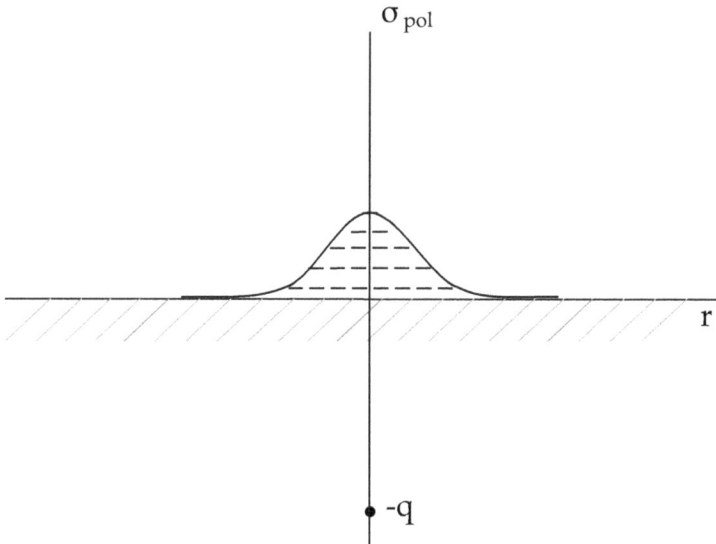

Fig. 12. The distribution of polarization charge on the Earth´s surface due to a point charge - q in its interior

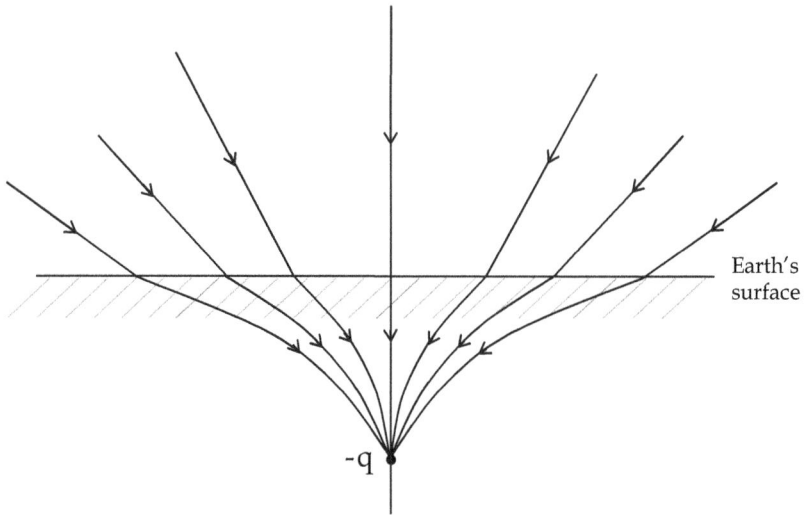

Fig. 13. Lines of force for a point charge -q in the interior of the Earth

Then σ_{pol} and E have their maximum values on the Earth's surface directly above the charge –q for r=0. The electromagnetic emission in the atmosphere is produced if at any point of the Earth's surface a limit of the electrical potential is reached at which a sudden electrical discharge through the atmosphere occurs.

6. Experiments on the electric displacement of soil, water and granite

Here, we report experimental research to explain how electric charge can be generated on the Earth's surface by earthquakes.

In the first experiment a dry soil block, 4 centimeters high, 10 centimeters wide and 8 meters long, is perfectly discharged. Then, one end of the block is charged with an electrical induction machine, as shown in Fig. 14. We could detect with an electroscope a polarization charge at the surface of the block until a distance of 5 meter.

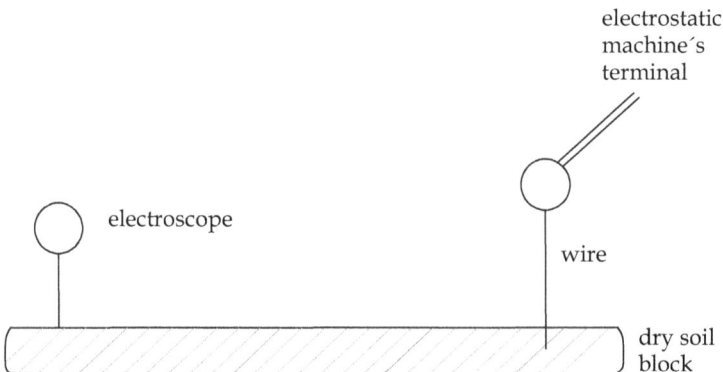

Fig. 14. A polarization charge is induced at the surface of a dry soil block

In the second experiment a vessel, 14 centimeters high and with a diameter of 25 centimeters, filled with distilled water is perfectly discharged. Then, the bottom of the vessel is connected by a wire to the machine and charged, as shown in Fig. 15. With the electroscope we found a polarization charge at the surface of the water.

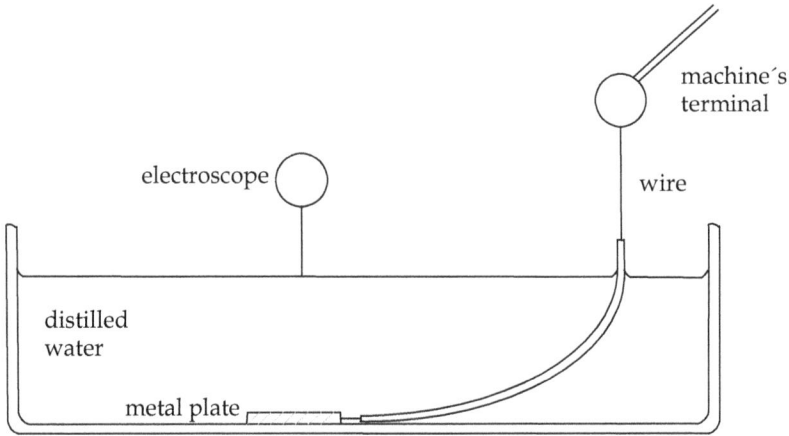

Fig. 15. A polarization charge is induced at the surface of distilled water

In the third experiment a red granite block, 2 centimeters high, 10 centimeters wide and 16 centimeters long, is perfectly discharged. Then, the upper surface of the block is connected by a wire to the machine and charged, as shown in Fig. 16. Again, we could detect a polarization charge at the lower surface of the block with the electroscope.

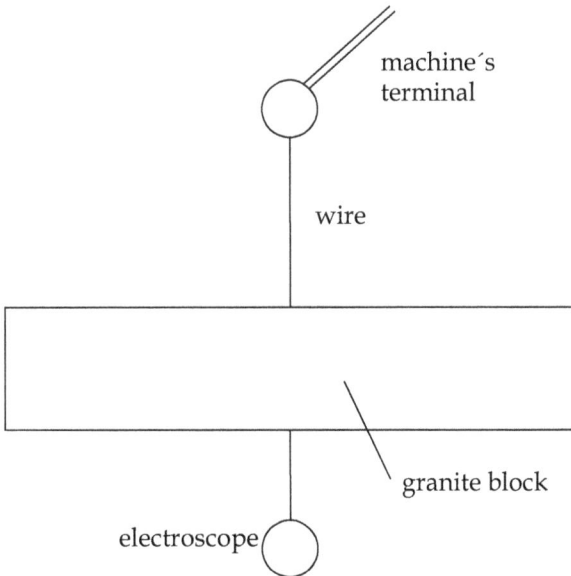

Fig. 16. A polarization charge is induced at the lower surface of the granite block

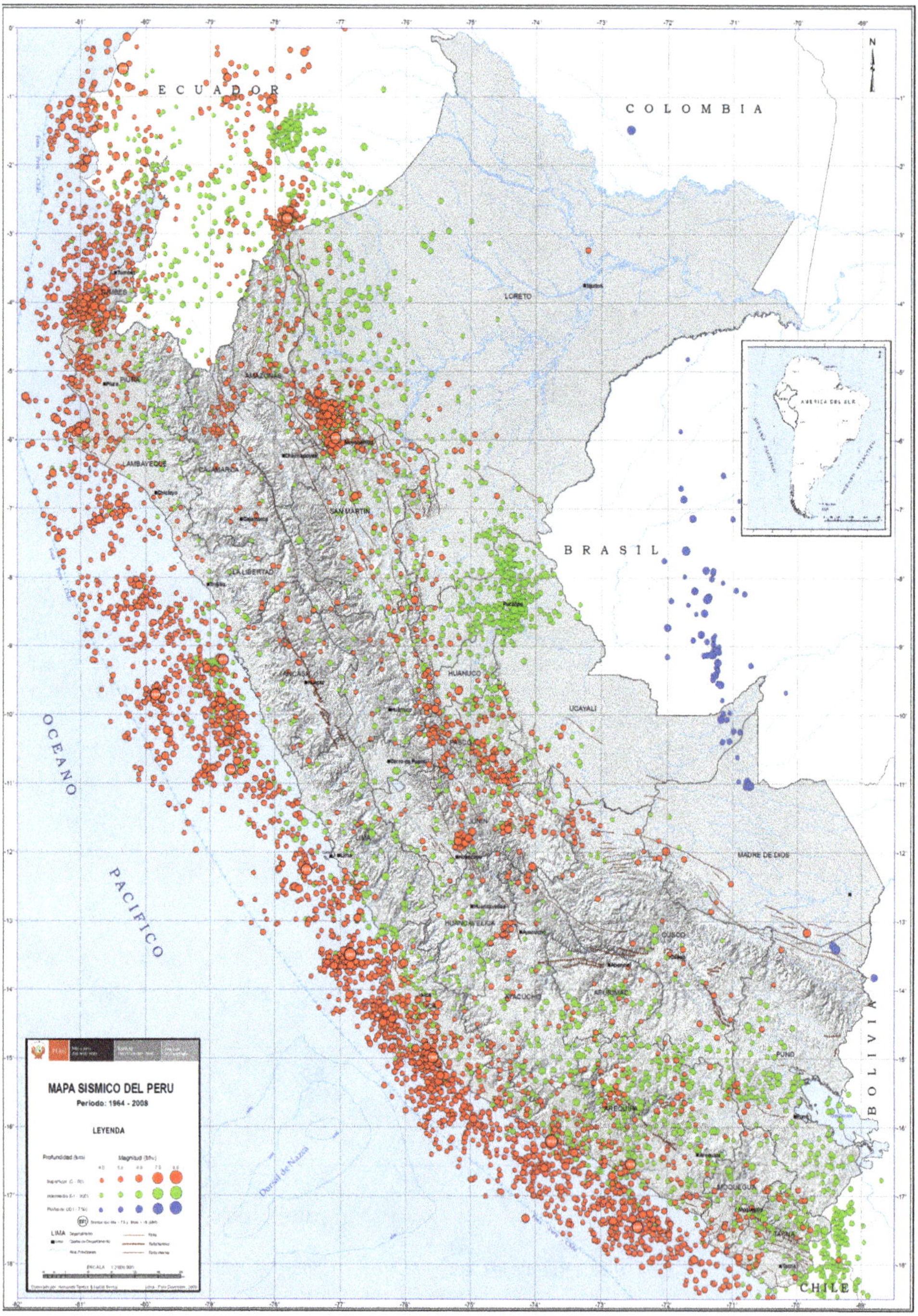

Fig. 17. Seismicity of Peru (Tavera and Bernal 2009)

7. Conclusions

Table 1. gives the dielectric constants of soil, water and granite. Using these, we would expect a surface polarization charge under the action of an electric field in these materials. In all studied cases our experimental researches confirmed this.

	Dielectric Constant, $\kappa = \varepsilon / \varepsilon_0$
Dry soil	14
Distilled water	81
Red granite	5

Table 1. Material characteristics

Using the relation (6), we could expect electrical discharge through the atmosphere if the interior of the Earth is charged with electricity according to

$$q \, / \, 4\pi\varepsilon_0 d^2 \geq 3 \ (10^6) \ \text{V/m} \tag{7}$$

Using the relation (4), we would also expect electrical discharge through the atmosphere if the polarization charge at the Earth's surface is

$$\sigma_{pol} \geq [0.53 \ (10^{-4}) \ \varepsilon_0 \ (\varepsilon - \varepsilon_0) \, / \, \varepsilon \ (\varepsilon + \varepsilon_0)] \ \text{C/m}^2 \tag{8}$$

Although the upper crust of the Earth can contain an insulator stratum, it is possible to have electric currents due to induction phenomena. A polarization charge can therefore be generated at the Earth's surface, it is only necessary that somewhere in the interior of the Earth is charged with electricity, and this can occur with an earthquake.

8. Future research

Considering the seismicity of Peru (Fig. 17.), we observe that the shallow earthquakes (depth < 60 km) take place near the coast (red dots). Light emissions during the Peruvian earthquake of 2007 were also observed in this area. These facts may be correlated in so far as the nearer the electric charge q is to the Earth's surface, the bigger is the polarization charge density as well as the electric field strength on the Earth's surface.

One next step is to conduct an experiment, that connects a grounded rod to an electroscope in order to detect the surface polarization charge during an earthquake at the San Lorenzo Island (situated 4 km in front of Lima's coast), where pre-seismic and co-seismic luminescence have been reported.

9. Acknowledgment

Figures have been drafted by J. Velásquez.

10. References

Brady, B. T. and Rowell, G. A., 1986, Laboratory investigation of the electrodynamics of rock fracture, *Nature*, Vol. 321, 488-492.

Derr, J. S., 1973, Earthquake lights: A review of observations and present theories, *Bulletin of the Seismological Society of America*, Vol. 63, No. 6, 2177-2187.

Finkelstein, D., Hill, R. D., and Powell J. R., 1973, The Piezoelectric Theory of Earthquake Lightning, *Journal of Geophysical Research*, Vol. 78, No. 6, 992-993.

Freund, F. T., 2003, Rocks That Crackle and Sparkle and Glow: Strange Pre-Earthquake Phenomena, *Journal of Scientific Exploration*, Vol. 17, No. 1, 37-71.

Freund, F. T., 2007, Pre-earthquake signals – Part II: Flow of battery currents in the crust. *Natural Hazards and Earth System Sciences*, Vol. 7, No. 5, 543-548.

Geophysics 223, B1.6, http://www.ualberta.ca/~unsworth/UA-classes/223/notes223/223B1-2009.pdf, 2009.

Hartogh, P., Lellouch, E., Moreno, R., Bockelée-Morvan, D., Biver, N., Cassidy, T., Rengel, M., Jarchow, C., Crovisier, J., Helmich, F. P. and Kidger, M., 2011, Direct detection of the Enceladus water torus with Herschel, *Astronomy and Astrophysics*, Vol. 532, No. L2, 1-6.

Heraud, J. A. and Lira, A., 2011, Co-seismic luminescence in Lima, 150 km from the epicenter of the Pisco, Peru earthquake of 15 August 2007, *Natural Hazards and Earth System Sciences*, Vol. 11, No. 4, 1025-1036.

Lira, A., 2008, Time Difference Correlation between Seismic Waves and Earthquake Lights, *Seismological Research Letters*, Vol. 79, No. 4, 500-503.

Lockner, D. A., Johnston, M. J. S., and Byerlee J. D., 1976, A mechanism to explain the generation of earthquake lights, *Nature*, Vol. 302, 28-32, 1983.

Mitzutani, H., Ishido, T., Yokokura T. and Ohnishi S., 1976, Electrokinetic phenomena associated with earthquakes, *Geophysical Research Letters*, Vol. 3, Nr. 7, 365-368.

Richter, C. F., 1958, Other Secondary Effects of Earthquakes, *Elementary Seismology*, W. H. Freeman and Company, San Francisco, 132-133.

Takeuchi, A., Futada, Y., Okubo, K. and Takeuchi N., 2010, Positive electrification on the floor of an underground mine gallery at the arrival of seismic waves and similar electrification on the surface of partially stressed rocks in laboratory, *Terra Nova*, Vol. 22, No. 3, 203-207.

Tavera H. and Bernal I., 2009, Mapa Sísmico del Perú, Periodo: 1964-2008, *Instituto Geofísico del Perú*.

Warwick, J. W., Stoker, C., Meyer, T. R., 1982, Radio Emission Associated With Rock Fracture: Possible Application to the Great Chilean Earthquake of May 22,1960, *Journal of Geophysical Research*, Vol.87, No. B4, 2851-2859.

4

Wavelet Spectrogram Analysis of Surface Wave Technique for Dynamic Soil Properties Measurement on Soft Marine Clay Site

Sri Atmaja P. Rosyidi[1] and Mohd. Raihan Taha[2]
[1]Universitas Muhammadiyah Yogyakarta
[2]Universiti Kebangsaan Malaysia
[1]Indonesia
[2]Malaysia

1. Introduction

The surface wave measurement is one of in-situ seismic methods based on the dispersion of Rayleigh waves (R-waves) which is used to determine dynamic soil properties, i.e., the shear wave velocity (V_S), shear modulus (G), damping ratio (D) and depth of each layer of the soil profile. Much of the basis of the theoretical and analytical work of this method for soil investigation has also been developed (Stokoe et al., 1994). Seismic data used in surface wave analysis are non-stationary in nature, i.e., varying frequency content in time. Especially in the low frequency range measurement, i.e., in soft soil deposit, the interested frequency of surface wave can be relatively low, i.e., less than 20 Hz. In these frequency values, the noisy signals from the natural or man-made sources may disturb the identical frequency level of the surface wave signals generated from the source. Therefore, a time-frequency decomposition of a seismic signal is needed to obtain the correct information of phase spectrum generated from signal transformation. In most of surface wave methods, the data analysis from time to frequency domain has been carried out by using Fourier transformation. However, some information of non-stationary seismic data in analysis maybe lost due to any arbitrary periodic function of time with period which is expressed as sum a set of sinusoidal in Fourier transform. The Fourier analysis is unable to preserve the time dependence. In addition, it also can not describe the evolutionary spectral characteristics of non-stationary processes. Thus, a new tool, i.e., wavelet analysis is required which allows time and frequency localization of the signals in the surface wave measurement beyond customary Fourier analysis.

Based on processing signal data at different scales or resolutions, wavelet analysis is becoming an important tool for identifying and analyzing localized variations of signal power, particularly it is well-suited for approximating data with sharp discontinuities within a time series. By decomposing a time series into time-frequency spectrum (TFW), one is able to determine both the dominant modes of variability and how those modes vary in time. The wavelet transform has been used in numerous studies, i.e., Meyers et al. (1993), Liu (1994), Weng & Lau (1994), Wang & Wang (1996) for climate and meteorological studies, Foufoula-Georgiou and Kumar (1995), Capilla (2006), Rosyidi et al. (2009) for geophysics, and Chik et al. (2009) and Rosyidi (2011) for civil engineering applications, respectively. Theoretical aspect of wavelets is given in many essential literatures such as Daubechies (1992), Mallat (1999), Soman and Ramachandran (2005). From their studies, the wavelet analysis has been successfully proven as an interactive technique for analyzing the waveforms and non-stationary characteristic of generated seismic signals from the surface wave measurements. Capilla (2006) used Haar wavelet transform application on the detection of micoseismic signal arrival. The results showed that the seismic series is able to be derived from an ensemble of subprocesses operating at characteristic scales and with time dependent variability from the wavelet approach where the discontinuous Haar wavelet decomposition shows the capability for efficiently extracting and locating emphasis the higher frequency in time sudden transitions associated with the transient events. Rosyidi et al. (2009) also conducted the study on the identification and reconstruction of the wave response spectrum from seismic surface wave propagation on a Malaysian residual soil using time-frequency analysis of the continuous wavelet transforms. Their results showed that the wavelet analysis is useful in spectral analysis, time-frequency decomposition for the identification of transient events in non-stationary signal and filtering of noisy signals in seismic surface waves records. In surface wave measurements, the application of wavelet analysis has been started by Kim and Park (2002) who used a harmonic wavelet transform for determining dispersion curve in the spectral analysis of surface wave (SASW) method. Their results showed that a new procedure based on wavelet transform was proposed for calculating the phase and group velocities at each frequency independent of remaining frequency components using the information around the time at which the signal energy. The method was also less affected by noise and near field effect than the phase unwrapping method that used as a common procedure in the surface wave measurement. Kritski et al. (2007) proposed a mathematical model to establish a relationship between the continuous wavelet transform of a signal and its propagated counterpart in a dispersive and attenuating medium. Their results showed that the wavelet model is able to estimate both phase and group velocities, as well as the attenuation coefficient. In addition, Shokouhi et al. (2003) explained the advantages of the wavelet approach in the SASW measurement, i.e., detection and characterization of cavities and objects buried in the ground and characterization of layer interfaces, with respect to layer dipping and abrupt interface changes.

The aim of this research is to improve the capability of in-situ surface wave measurement by developing the wavelet spectrogram analysis of surface waves (WSASW) technique for measurement of the soil dynamic properties, i.e., the shear wave velocity, shear modulus, and damping ratio at soft marine clay soils sites. This technique has capability to reconstructed spectrograms of noisy seismic waves and produces the enhanced phase data to develop the phase velocity dispersion curve. In soft soil site, the environmental noises are dominant in the recorded seismic signals due to the wave frequency of interest are identical to the frequency level of noisy signals. Therefore, the time-frequency wavelet spectrum is employed to localise the interested response spectrum of surface waves. A filtration procedure is also proposed in order to remove the noisy signals from the seismic records which were captured during field measurement.

In this research, a test site at Radio Televsyen Malaysia (RTM), Kelang, Malaysia was selected as location of measurement (Fig. 1). The site is a fairly flat open paddy field and an on going construction was seen about 500 m away from the site. The site is generally an original ground and the soil mass is mainly of greyish clay. The regional geology of the site has been classified as recent quaternary of dominantly alluvial deposits of soft marine clay with traces of organics. The soil descriptions from the two boreholes at the location have also shown that the soil type found were quite similar with the geology classification, i.e., greyish clay with decayed wood at most of the soil layers of the subsoil stratum (Fig. 2).

Location of Study
Kelang, Malasia

Map source: The World Factbook (2009)
Central Intelligence Agency, Washington, D.C.
http://www.cia.gov/library/publications/the-world-fact book/index.html

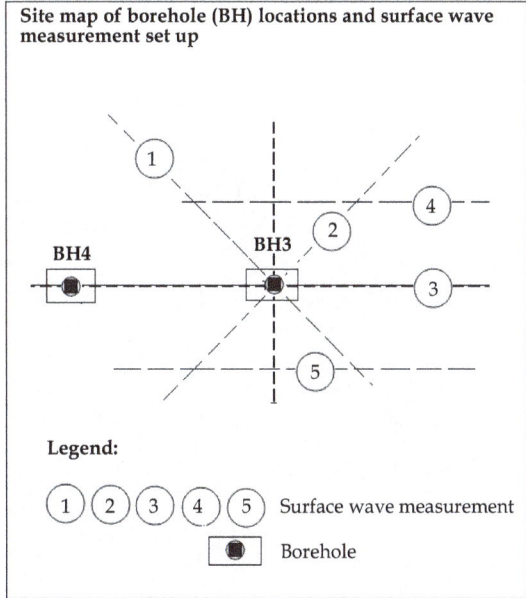

Site map of borehole (BH) locations and surface wave measurement set up

Legend:

① ② ③ ④ ⑤ Surface wave measurement

▣ Borehole

Fig. 1. Location of study

2. Research method

In the proposed WSASW technique, there are four main stages as described in the following sections.

2.1 Field measurement

In this study, the seismic signal data were collected by using the SASW field measurement set up. There are three important set ups in the SASW measurement for soft soil location, i.e., adequate wave frequencies produced from the various impact sources, capability of receivers or geophones to receive the interested frequency and the appropriate acquisition unit or spectrum analyser used in the measurement. A set of impact hammer sources of various frequencies was used to generate R waves on the soil surface. The propagation of the waves were detected using two receiving geophones and the analog signals were then transmitted to a spectrum analyser which consisted of acquisition box and transferred digitally to a notebook computer (Fig. 3).

Sledge hammers of 8 and 12 kg were used as transient impact sources in the SASW measurement. One of advantages using an impact (transient) source is to generate and measure a broad range of frequencies simultaneously. However, the frequency content is often limited and it is also important to realise that different transient sources generate energy over different frequency ranges. Prior to the experiments, a pilot study on frequency range test on transient hammers used in this study was carried out. The hammer generated surface waves over different frequency range with adequate amplitude and they were able to be detected by the receiver. For a typical soil deposit, the highest frequency necessary is in the order of 200 to 800 Hz (Nazarian, 1984). Therefore, the selected sledge hammers in this research are appropriate to be used for sampling the soft soil layer up to approximately 5 meter of depth.

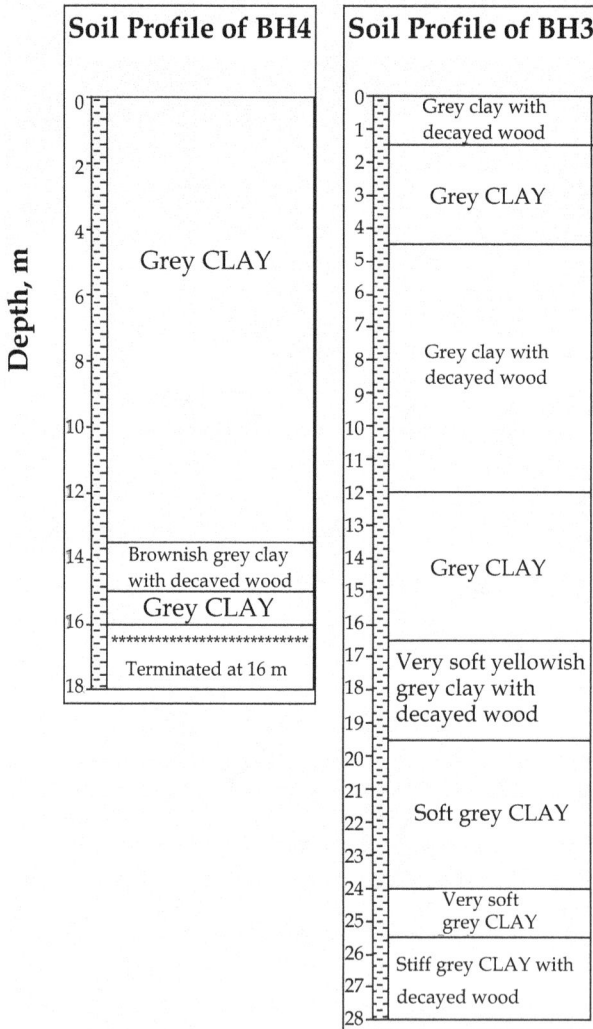

Fig. 2. Borehole data of BH3 and BH4 from location of study (Fig. 1)

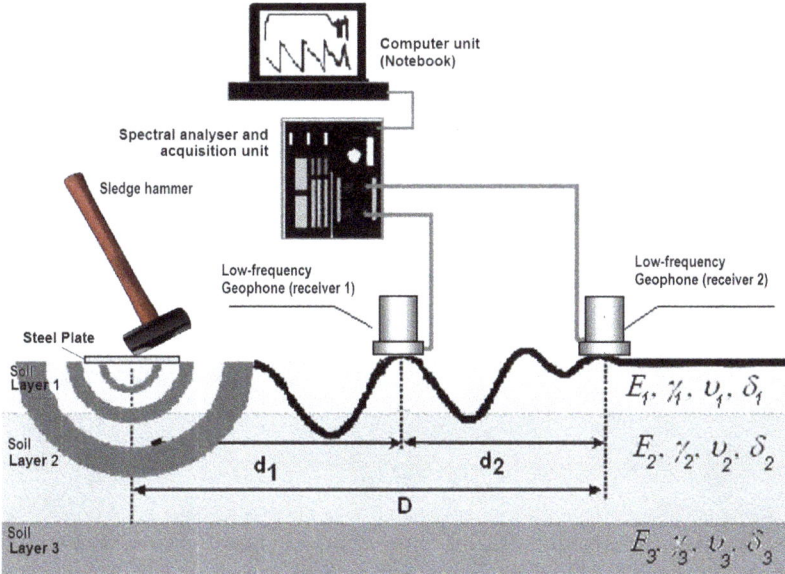

Fig. 3. SASW measurement setup applied on the soil sites

Vertical geophones of 1 Hz used in this study only receive the vertical displacement of the generated signal from the impact sources as the interested component in the measurement. Several configurations of the receiver and the source spacings were required in order to sample different depths. The measurement configuration of the SASW test used in this study is the midpoint receiver spacings. In addition, the short receiver spacings with a high frequency source were used to sample the shallow layers of the soil profile. Larger receiver spacings with a set of low frequency sources were employed to sample the deeper layers. The distance between the source and the near receiver was set up equal to the distance between the receivers (Fig. 3). This configuration is adequate for reducing the near-field effect (Heisey et al., 1982; Ganji et al., 1998).

2.2 Development of experimental phase velocity dispersion curve

An experimental phase velocity dispersion curve from all receiver spacings in one configuration measurement was generated based on the phase angle data from the both signals received by geophones. However, in fact, the phase angle is difficult to be interpreted from the noisy signals because it should be analyzed from huge amounts of non-stationary seismic data in nature i.e. varying frequency content in time. Therefore, the time-frequency localization is needed to provide accurate information of wave spectrum. In this study, the time-frequency analysis of continuous wave transform (CWT) was employed for localizing and filtering the interested wave spectrum.

2.2.1 Continuous wavelet transform (CWT)

The continuous wavelet transform (CWT) technique is becoming a common tool to analyse localised variation of power within a time series for non-stationary signal, i.e., seismic signals. In the technique, wavelets dilate in such a way that the time component also changes for different frequency. When the time window component of the wavelet decreases or increases, the frequency component of the wavelet is shifted towards high or low frequencies, respectively. Therefore, as the frequency resolution increases, the time resolution decrease and vice versa, which is called as multi resolution analysis, analysing the signal at different frequencies with different resolutions (Mallat, 1989). CWT offers good spectral and poor temporal resolution at low frequency, which is useful for low frequency analysis with long duration signals, and good temporal and poor spectral resolution at high frequency which is valuable for high frequency signals with short duration. This optimal time-frequency resolution property makes the CWT technique useful for non-stationary seismic analysis.

A wavelet is defined as a function of $\psi(t) \in L^2(\Re)$ (L^2 is the set of square integrable function) with a zero mean, which is finite energy signals in both time and frequency. By dilating and translating the wavelet $\psi(t)$, a family of wavelets can be produced as:

$$\psi_{\sigma,\tau}(\tau) = \frac{1}{\sqrt{\sigma}} \psi\left(\frac{t-\tau}{\sigma}\right) \tag{1}$$

where σ is the dilation parameter or scale and τ is the translation parameter ($\sigma, \tau \in \Re$ and $\sigma \neq 0$). The CWT is defined as the inner product of the family wavelets $\Psi_{\sigma,\tau}(t)$ with the signal of f(t) which is given as:

$$F_W(\sigma,\tau) = \int_{-\infty}^{\infty} f(t) \frac{1}{\sqrt{\sigma}} \bar{\psi}\left(\frac{t-\tau}{\sigma}\right) dt \tag{2}$$

where $\bar{\psi}$ is the complex conjugate of ψ and $F_W(\sigma,\tau)$ is the time-scale map. The convolution integral from equation 2 can be computed in the Fourier domain. To reconstruct the function f(t) from the wavelet transform, Calderon's identify (Daubechies, 1992) can be used and is obtained as:

$$f(t) = \frac{1}{C_\psi} \int_{-\infty}^{\infty} \int_{-\infty}^{\infty} F_W(\sigma,\tau) \psi\left(\frac{t-\tau}{\sigma}\right) \frac{d\sigma \ d\tau}{\sigma^2 \sqrt{\sigma}} \tag{3}$$

$$C_\psi = 2\pi \int \frac{|\hat{\psi}(\omega)|^2}{\omega} d\omega < \infty \tag{4}$$

where $\hat{\psi}(\omega)$ is the Fourier transform of $\psi(t)$. The integrand in equation 4 has an integrable discontinuity at $\omega = 0$ and implies that $\int \psi(t) dt = 0$. In this study, the mother wavelet of the Morlet wavelet was used. The shape of the Morlet wavelet is a Gaussian-windowed complex sinusoid. It is defined in the time and frequency domains as follows:

$$\Psi_0(t) = \pi^{-\frac{1}{4}} e^{imt} e^{-t^2/2} \tag{5}$$

$$\hat{\psi}_0(s\omega) = \pi^{-\frac{1}{4}} H(\omega) e^{-(s\omega-m)^2/2} \tag{6}$$

where m is the wavenumber, and H is the Heaviside function. The time and frequency domain plot of Morlet wavelet is shown in Fig. 4. In Figure 4a, the Morlet wavelet is shown within an adjustable parameter m of 7 which is used in this study. This parameter can be used for an accurate signal reconstruction of seismic surface waves in low frequency. The Gaussian's second order exponential decay used in time resolution plot results in the best time localisation.

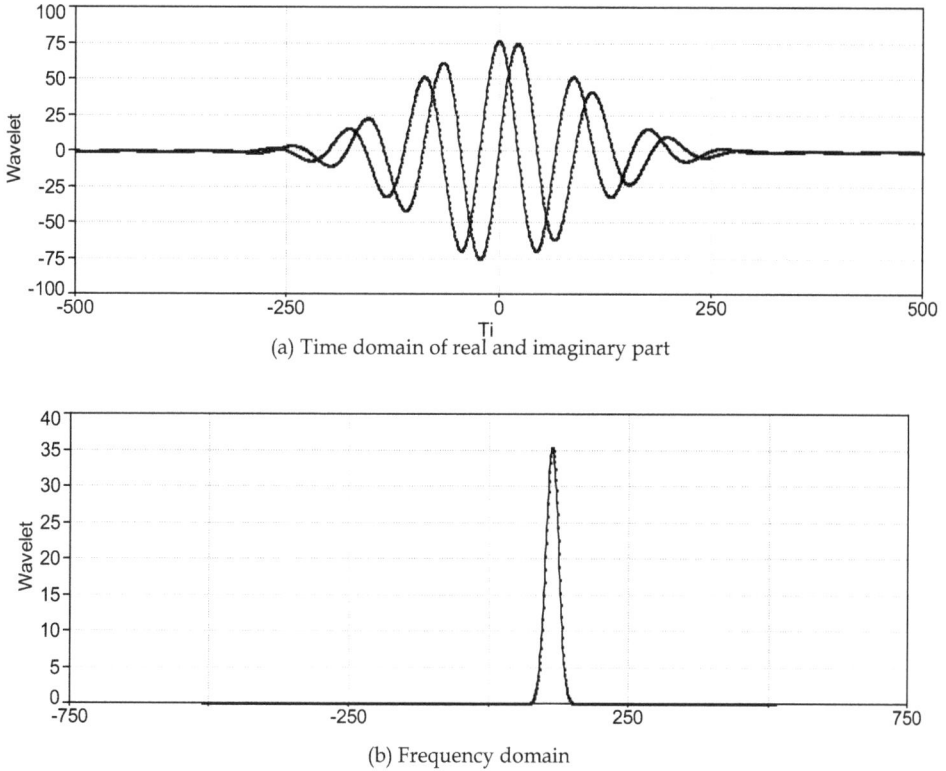

(a) Time domain of real and imaginary part

(b) Frequency domain

Fig. 4. Time and frequency domain plot of Morlet wavelet

2.2.2 Proposed procedure of the CWT technique in surface waves analysis

A flow chart of procedure of the CWT technique in the WSASW method is described in Figure 5. The detail procedure is discussed in the following section.

1. Select the wavelet function and a set of scale, s, to be used in the wavelet transform. The different wavelet function may influence the time and frequency resolution. In this study, a Morlet wavelet function was selected as a mother wavelet in the CWT filtering.
2. Develop the wavelet scalogram by implementing the wavelet transform (equation 2) using computed convolution of the seismic trace with a scaled wavelet dictionary. Scalogram is a local time-frequency energy density which measures the energy of the signals in the Heisenberg box of each wavelet. The detail discussion of scalogram can

refer to Mallat (1989). Wavelet scale is calculated as fractional power of 2 using the formulation (Torrence & Compo, 1998):

$$s_j = s_0 2^{j\delta_j}, j = 0,1,...,J \tag{7}$$

$$J = \delta j^{-1} \log_2\left(\frac{N\delta_t}{s_0}\right) \tag{8}$$

where, s_0 is smallest resolvable scale = $2\delta t$, δt is time spacing, and J is largest scale.

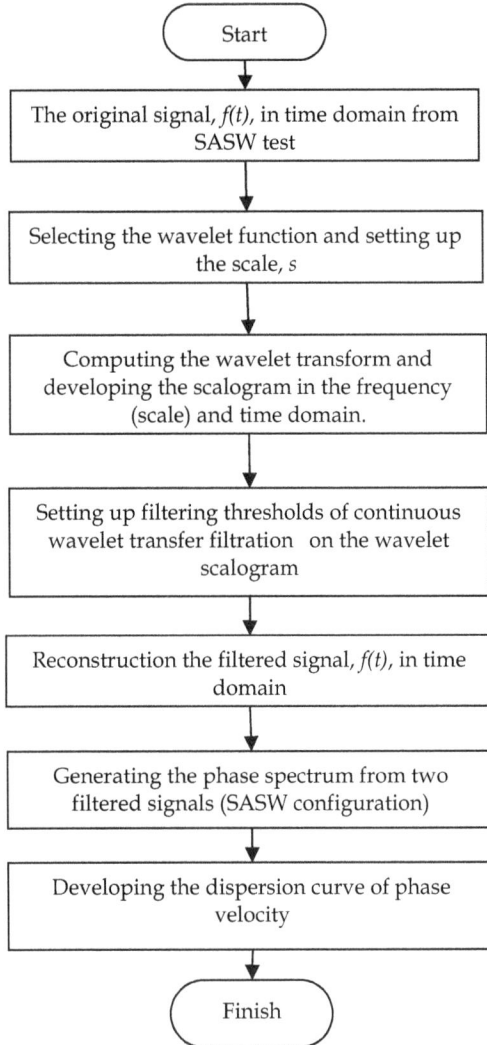

Fig. 5. Flow chart of CWT Filtering procedure

3. Convert the scale dependent wavelet energy spectrum (scalogram) of the signal to a frequency dependent wavelet energy spectrogram in order to compare directly with Fourier energy spectrum.

4. Perform the CWT filtration on the wavelet spectrogram by obtaining the time and frequency localization thresholds. Wavelet spectrogram is developed from the scalogram which allows the filtration technique implemented directly to the spectrum. In this study, the CWT filtration was developed by a simple truncation filter concept which only considers the passband and stopband. Threshold values in time and frequency domain are then set as the filter values between passband and stopband. It allows a straight filtering in each of the dimensions of times, frequencies and spectral energy. The noisy or unnecessary signals can be eliminated by zeroing the spectrum energy and consequently, they are fully removed when reconstructing the time domain signal. Thus, the interested spectrum of signals are to be passed when the spectrum energy is maintained in original value. A design of the CWT filtration is proposed by Rosyidi (2009) and can be written as:

$$f(s) = \begin{cases} 0, & 1 \le s \le F_l \\ 1, & F_l \le s \le F_h \\ 0, & F_h \le s \le N \end{cases} \tag{9}$$

$$f(u) = \begin{cases} 0, & 1 \le u \le T_l \\ 1, & T_l \le u \le T_h \\ 0, & T_h \le u \le N \end{cases} \tag{10}$$

5. The value of 1 means the spectrum energy is passed and the value of 0 represents the filtration criteria when the spectrum energy is set as 0.

6. Reconstruct the time series of seismic trace using equation 3.

7. Calculate the phase different from reconstructed signals at each frequency to develop the phase spectrum for the experimental dispersion curve. The phase data can be calculated from:

$$\phi_n(s) = \arctan\left(\frac{\Im\{s^{-1}W_n^{XY}(s)\}}{\Re\{s^{-1}W_n^{XY}(s)\}} \right) \tag{11}$$

where,

$$W_n^{XY}(s) = W_n^X(s)W_n^{Y*}(s) = \text{wavelet cross spectrum} \tag{12}$$

8. Finally, by extracting the data of the phase angle from the phase spectrum, a composite experimental dispersion curve can be calculated by the phase difference method. The time of travel between the receivers for each frequency can be calculated by:

$$t(f) = \frac{\phi(f)}{(360f)} \tag{13}$$

where f is the frequency, $t(f)$ and $\phi(f)$ are, respectively, the travel time and the phase difference in degrees at a given frequency. The distance of the receiver (d) is a known parameter. Therefore, the Rayleigh wave velocity, V_R or the phase velocity at a given frequency is simply obtained by:

$$V_R = \frac{d}{t(f)} \tag{14}$$

and the corresponding wavelength of the Rayleigh wave, L_R may be written as:

$$L_R(f) = \frac{V_R(f)}{f} \tag{15}$$

By repeating the procedure outlined above and using equation 13 through 15 for each frequency value, the R wave velocity corresponding to each wavelength is evaluated and the experimental dispersion curve is subsequently generated.

2.3 Shear wave velocity profile

An inversion analysis was used to generate the shear wave velocity profile. In the inversion process, a profile of a set of a homogeneous soil layer extending to infinity in the horizontal direction was assumed. The last layer is usually taken as a homogeneous half-space. Based on the initial profile, a theoretical dispersion curve was constructed using an automated forward modeling analysis involving 3-D dynamic stiffness matrix method (Kausel & Röesset, 1981). In the model, displacements and stresses (or traction) of the propagation of the waves on a horizontal surface can be expanded using the Fourier series in the circumferential direction and in terms of cylindrical function (Bessel, Neuman or Hankel functions) in the radial direction.

For axisymmetric loading only one Fourier series term is needed (the 0 term), and the radial and vertical displacements (U and W) can be expressed by:

$$U(r) = qR \int_{k=0}^{\infty} \overline{u} J_1(kR) J_1(kr) dk \tag{16}$$

$$W(z) = qR \int_{k=0}^{\infty} \overline{w} J_1(kR_0) J_0(kr) dk \tag{17}$$

where J_0 and J_1 = the zero and the first order Bessel function, k = the wave number, r = the radial distance from the source, R = the radius of the disk, q = the magnitude of the uniformly distributed load; $\overline{u}$ and $\overline{w}$ = functions of k for a harmonic load at the surface with wavelength $2\pi/k$. Kausel & Röesset (1981) showed that the displacement $\overline{u}$ and $\overline{w}$, in Equation 16, can be written as:

$$\overline{u} = \sum_{i=1}^{2n+2} u_{i1} w_{i1} \frac{k}{k_i (k^2 - k_i^2)} \tag{18}$$

$$\overline{w} = \sum_{i=1}^{2n+2} w_{i1}^2 \frac{k}{k_i (k^2 - k_i^2)} \tag{19}$$

For a system of n layers over a half-space, u_{i1} and w_{i1} denote the horizontal and vertical displacements at the surface in the ith mode and can be found from the corresponding mode shape. By substituting equation 18 and 19 to 16 and 17, respectively the integral can be evaluated analytically in closed form. This solution is particularly convenient when dealing with a large number of layers as in the case when it is desired to obtain a detailed variation of the soil properties. Subsequently, the theoretical dispersion curve generated using the 3-D model was ultimately matched to the experimental dispersion curve based on lowest root mean square (RMS) error with an optimisation technique from Joh (1996).

2.4 Development of soil shear modulus and damping ratio profile

The soil shear modulus profile can be obtained by linear elastic model involving the parameter of the shear wave velocity obtained from inversion process as mentioned in previous section. The soil shear modulus is calculated from the following equation (Kramer, 1996):

$$G = \frac{\gamma V_S^2}{g} \tag{20}$$

where G = the dynamic shear modulus, V_S = the shear wave velocity, g = the gravitational acceleration; and γ = the total unit weight of the material. Nazarian & Stokoe (1986) explained that the modulus parameter of material is maximum value at a strain below about 0.001 %. In this strain range, modulus of the materials is also taken as constant.

In order to measure the soil attenuation from signals recorded from field measurement, the spectrogram attenuation model developed by Rosyidi (2009) was employed in the analysis. The decrease in amplitude (energy density) of the vertical component of the R-wave with distance due only to geometric configuration is also called the radiation damping or geometric spreading. An effective soil damping ratio of R-wave in layered medium can be defined from the attenuation analysis and the value is frequency dependent. Its value may become very high for the first few modes of vibration. The attenuation (α) of R-wave can be performed by the spectrogram attenuation model proposed by Rosyidi (2009) as follows:

$$\ln\left[\frac{W_f^{R_2}(u,s)}{W_f^{R_1}(u,s)}\right] = \ln\left[\left\{\frac{R_1}{R_2}\right\}^n \{G(R) \cdot G(I) \cdot K(R)\} e^{\{-\alpha(f)(R_1-R_2)\}}\right] \tag{21}$$

where, R_1 dan R_2 = geophones distance from the sources (if using two geophones), $W_f^{R_1}(u,s)$ dan $W_f^{R_2}(u,s)$ = spectrogram magnitude response for geophone 1 and 2 respectively, $G(R)$ = geometric spreading factor, $G(I)$ = instrumentation correction factor and $K(R)$ = correction for refracted and transmitted waves.

Finally, the experimental attenuation curves can be used in the inversion process aimed in estimating the variation of soil shear damping ratio with depth. The inversion process is carried out using the SURF code (Herrmann, 1994), based on a weighted, damped, least-squares algorithm. Experimental attenuation curve consists of surface-wave attenuation data at different frequencies obtained from equation 21. The amplitude variation with distance can be used to obtain the experimental attenuation curve.

The solution to the inversion problem of estimating the dissipation factors is based on (Aki & Richards, 1980):

$$\alpha(f) = \frac{2\pi f}{V_R^2} \left\{ \sum_i^N V_{P,i} \left(\frac{\partial V_R}{\partial V_P} \right)_i D_{P,i} + \sum_i^N V_{S,i} \left(\frac{\partial V_R}{\partial V_S} \right)_i D_{S,i} \right\} \tag{22}$$

where V_P, V_S and V_R are the P-, S- and Rayleigh-wave velocities, respectively. The suffix i refers to the ith layer and the summation is carried out over N layers of the stratified soil model. D_P and D_S are the damping ratio values for P- and S-waves, respectively.

In homogeneous media with high values of Poisson's ratio (v),the influence of the P-wave damping ratio on Rayleigh-wave attenuation is very small (Viktorov, 1967). For layered media, the influence of D_P on Rayleigh-wave attenuation is negligible for values of V_P/V_S greater than 2 (i.e. v > 1/3) as described in Xia et al. (2002). These values are typical for saturated soils, and in many temperate regions the water table is usually shallow, thus it is reasonable to perform the inversion of Rayleigh-wave attenuation assuming a constant value of the ratio D_P/D_S or assuming that no bulk loss is present (Herrmann, 1994, Foti, 2004). It is important to point out that the relationship between the attenuation of Rayleigh waves and the dissipative properties of each layer is influenced by the shear-wave velocity profile of the medium. The detail procedure of the inversion process for obtaining the shear damping ratio is discussed by Herrmann (1994), Lai and Rix (1998) and Foti (2004).

3. Results and discussion

3.1 Seismic data and spectrum analysis

Figure 5 shows two examples of the recorded signals from averaging multiple impacts from the field seismic measurement at soil test site. The signals were received by two geophones in 8 m receiver spacing. From the recorded signals, it can be recognised that higher amplitude is measured for the fundamental mode of R- wave amplitude. It is also noted that the decreasing signal magnitude is identified as the R-wave attenuation in the soil layer which is an important characteristic for energy decrement. The waveform of seismic signal recorded (Fig. 5) is transient and-non stationary. Weak recorded signal of seismic wave particularly in channel 2 is also identified as an effect of environmental noise which maybe produced from ground noise and man-made vibrations. This means that either the input signals or behaviour of system at different moments in time was not identical.

When the signals were transformed into frequency domain using FFT (fast Fourier transform), time-dependent behaviour of the seismic waves and noisy events were lost (Fig. 6). The energy content of these events which are present at different times and frequency would not be picked up by conventional Fourier analysis. In the other words, the conventional spectral analysis of non-stationary signal of seismic waves cannot describe the local transient event due to averaging duration of signals. It also cannot instantly separate the event of true seismic waves from noisy signals. Consequently, it is difficult to capture the correct phase information in the transfer function of both signals (Fig. 7).

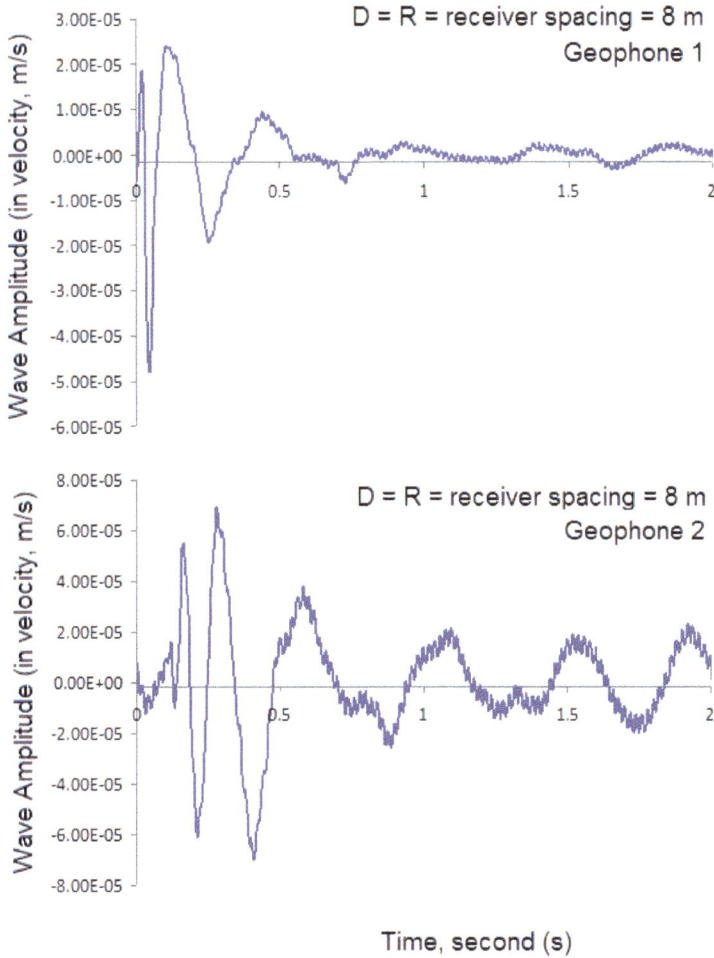

Fig. 5. The time signals from 8 m receiver spacing from the measurement

3.2 CWT filtering and analysis in the WSASW

In order to enhance the pattern of non-stationary seismic wave signals from noisy signal, both signals were then transformed in time-frequency resolution by the CWT. This time and frequency analysis of CWT was employed to overcome the identification problem of spectral characteristics of signals. Fig. 8 and 9 present the CWT spectrogram of the time signal from geophone 1 and geophone 2, respectively (Fig. 5), which was constructed by using a mother wavelet of Morlet.

Three main energy events at different frequency bands were clearly detected which may result in both low and high mode of seismic and noisy signals (Figure 8 and 9). It can be seen that coherent low frequency energy was found in the range of up to 2 - 10 Hz in both CWT spectrograms (event B, C, D, E). This spectrum range is clearly captured and identified

Frequency, Hz

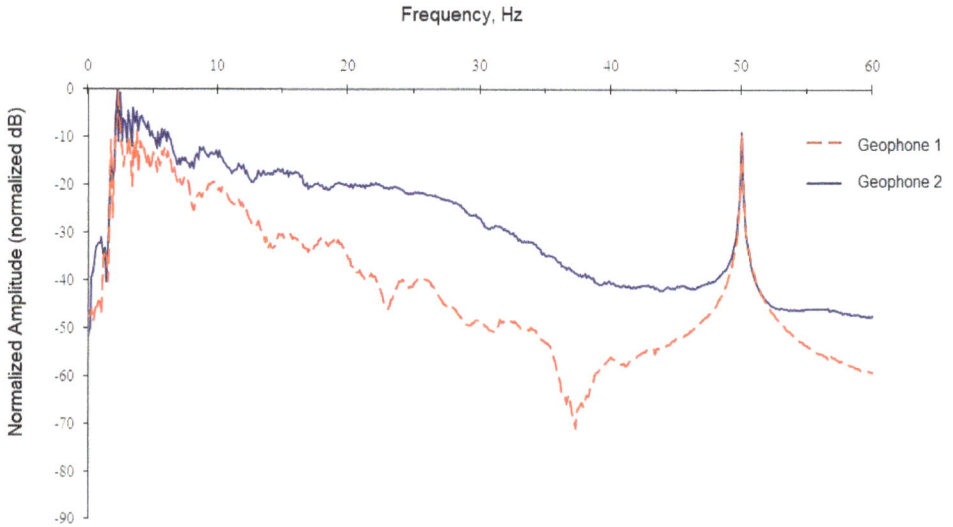

Fig. 6. FFT spectrum of the time signals from 8 m receiver spacing of the measurement

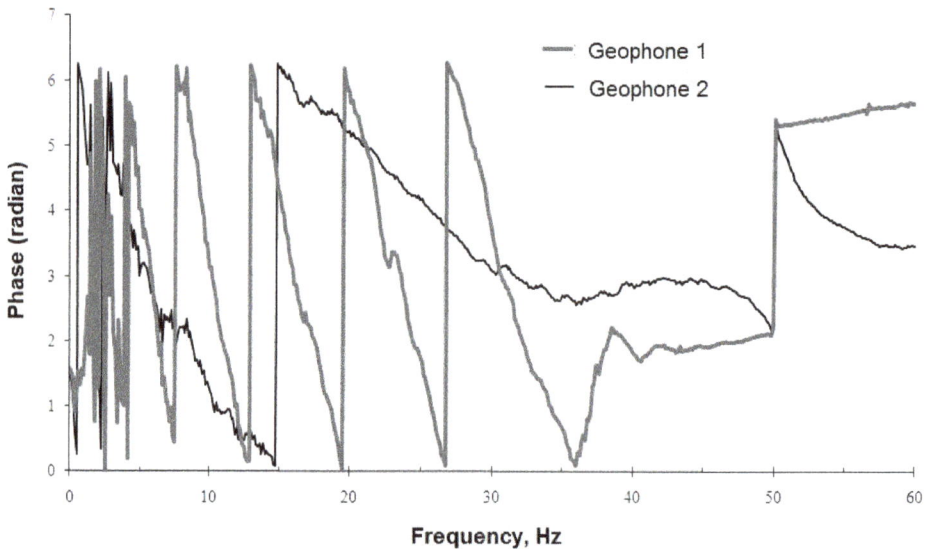

Fig. 7. Phase spectrum of each time signals from 8 m receiver spacing

as dominant noisy signals or ground rolls. Another noisy signal received during measurement was generated from the electrical devices and generator which has constantly the frequency content of 50 Hz (event F). The spectrum events of surface wave signals are recognized at event A with the frequency level of 4 to 35 Hz with arrival time of 0.012 to 0.50 s which consist of high magnitude of energy.

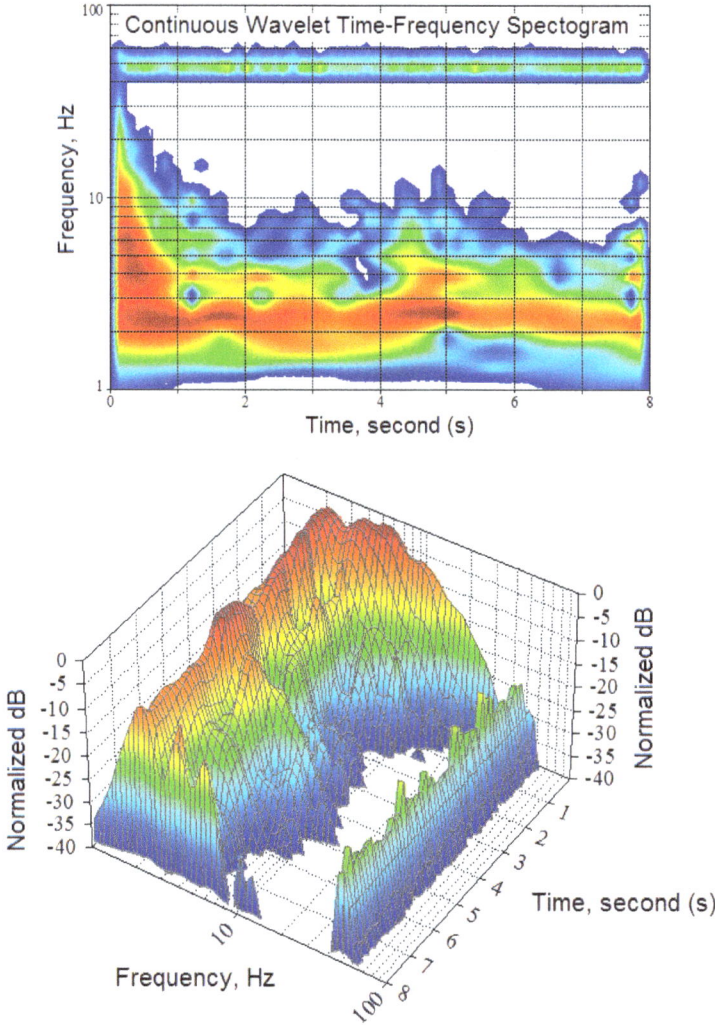

Fig. 8. The CWT spectrogram for signals received by geophone 1

In order to separate the original seismic wave, the wavelet spectrogram filtration was then implemented. There are two primary ways to set the thresholds for wavelet filtering. The first is to define a region of time-frequency space. This is primarily used to isolate and reconstruct signal components. The time and frequency fields define limits in spectrogram filtering. In this study, the time and frequency range of noise signal was set as threshold of wavelet (equation 9 & 10). It means that the noisy signals are removed from the spectrogram and only the interested seismic wave signals remain. Table 1 shows an example of the threshold parameters of time-frequency used in filtering criteria for the signals from 8 m receiver spacing of SASW measurement. Consequently the inverse wavelet transform returns a denoised seismic signal from the filtered spectrogram of interest. Demonstration of the wavelet analysis in denoising and reconstructing the recorded seismic signals is shown in Fig. 10. Particularly for seismic signal recorded on channel 2, the reconstructed waveform of denoised signal improves the signal pattern of the seismic surface waves. The highest amplitude shown at first phase of signals are recognised as low frequency energy from the noisy signals or ground rolls based on the spectrogram analysis, therefore, it should be filtered.

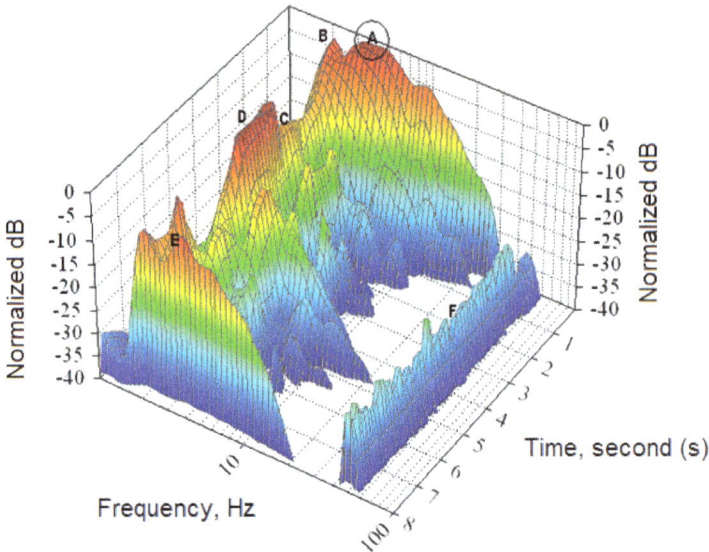

Fig. 9. The CWT spectrogram for signals received by geophone 2

Threshold	Time (s)		Frequency (Hz)	
	T_1	T_2	F_1	F_2
Geophone 1	0.030	0.650	2.97	7.08
Geophone 2	0.002	0.440	2.75	6.05

Table 1. Time and frequency threshold in CWTF

The phase spectrum from denoised signals from the surface wave measurement was then constructed by equation 11. Compared to the phase spectrum from original signals, the enhanced phase spectrum from the CWT filtration provides the better phase information versus frequency range without noisy interference needed in the surface wave analysis (Fig. 11). It shows that the CWT and wavelet filtering is an effective tool for identifying, denoising and reconstructing the noisy seismic surface waves measured on the soil profile. Finally, based on the phase different method (Equation 13 – 15), a phase velocity dispersion curve from enhanced phase spectrum can be obtained. Fig. 12 presents the dispersion curves obtained from CWTF compared to the original dispersive velocity data (only produced from masking process without any filtration), and dispersion curves from the continuous surface wave (CSW) measurement and the impulse response filtration (Joh, 1996) analysed from WinSASW.

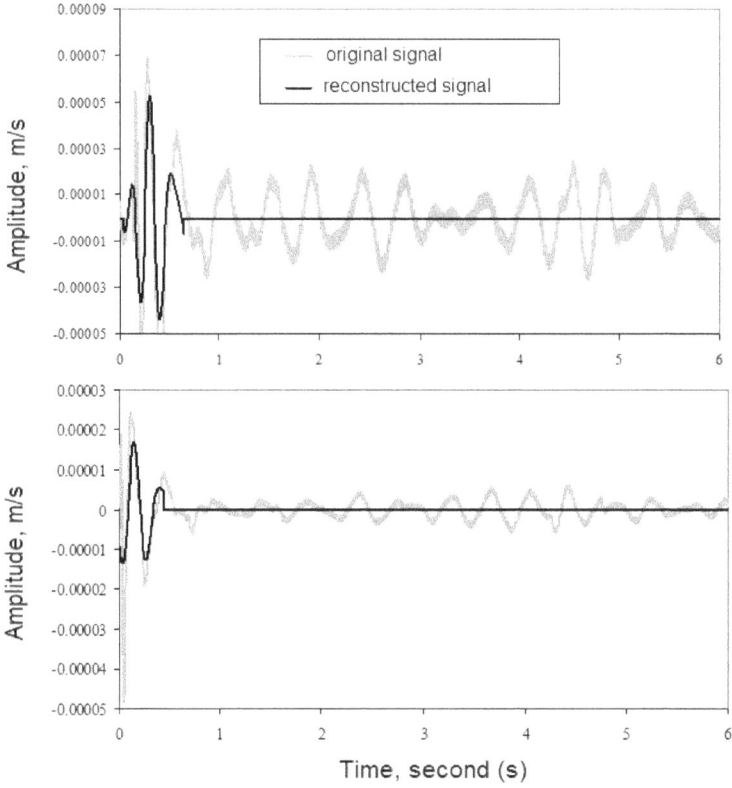

Fig. 10. Reconstructed signals from the CWTF

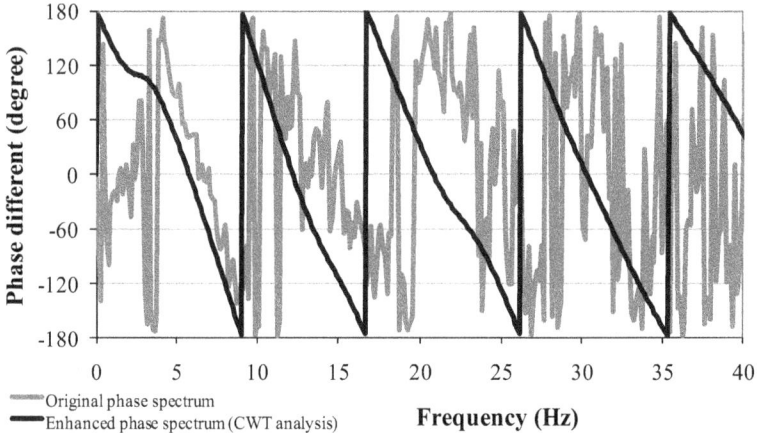

Fig. 11. Phase spectrum from two signals obtained by CWTF and spectrogram analysis
compared to the original phase spectrum

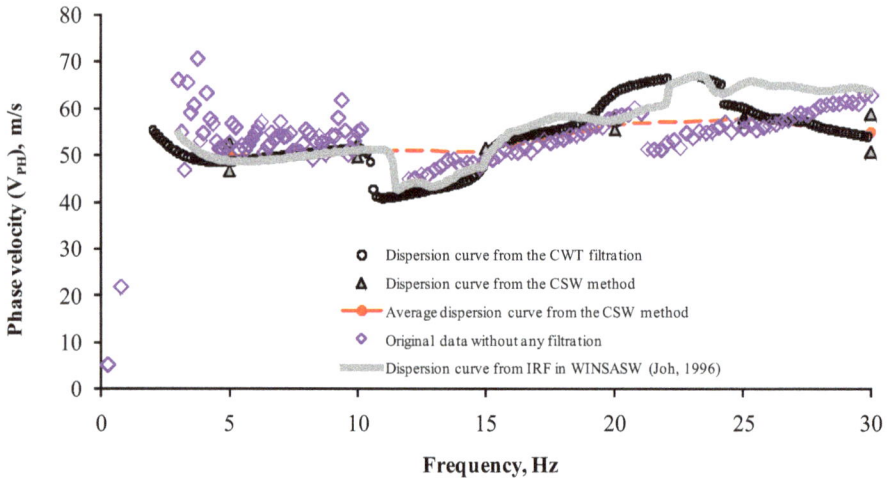

Fig. 12. Comparison between dispersion curves from the CWTF in the WSASW method and the dispersion curves from the original phase data (without any filtrations), developed by the CSW method and by impulse response filtration (IRF) built in WINSASW (Joh, 1996)

3.3 Shear wave velocity evaluation

The actual shear wave velocity of the soil profile is produced from the inversion of the experimental dispersion curve. In the inversion process, a profile of a homogeneous layer extending to infinity in the horizontal direction is assumed. An example shear wave velocity profile of soil site from this study is shown in Fig. 13. The average inverted shear wave velocity of soil layer for RTM Kelang test sites was found to be 54. 90 m/s with a range of 38.52 to 103.53 m/s. Using the shear wave velocity parameter, the soil material in this study could be evaluated and classified as soft clay (marine clay). The result shows that the soil classification based on the shear wave velocities is also reasonably in agreement with the laboratory tests.

As part of the validation on the results of the shear wave velocity profile obtained from this study, a steady state method or also well known as the continuous surface wave (CSW) measurement was carried out at same locations. In CSW measurement, a set of low frequency content generated from harmonic vibration source was set up in the range of 5 until 30 Hz with the fixed receiver spacing of 1 m. In this frequency level, the observed soil profile can be investigated until 6 m of depth. The comparison between a shear wave profile determined by WSASW analysis and CSW method is also shown in Figure 13. This comparison shows that the shear wave velocity profile by a WSASW technique is in a good agreement with value of the shear wave velocity determined by CSW method.

3.4 Shear modulus evaluation

Based on the shear wave velocity profile, the shear modulus (G) profile of RTM Kelang site can be calculated and the result is given in Fig. 14. The result of G is also compared with the shear modulus calculated using Hardin & Drnevich (1972) model and the shear modulus obtained from the CSW measurement.

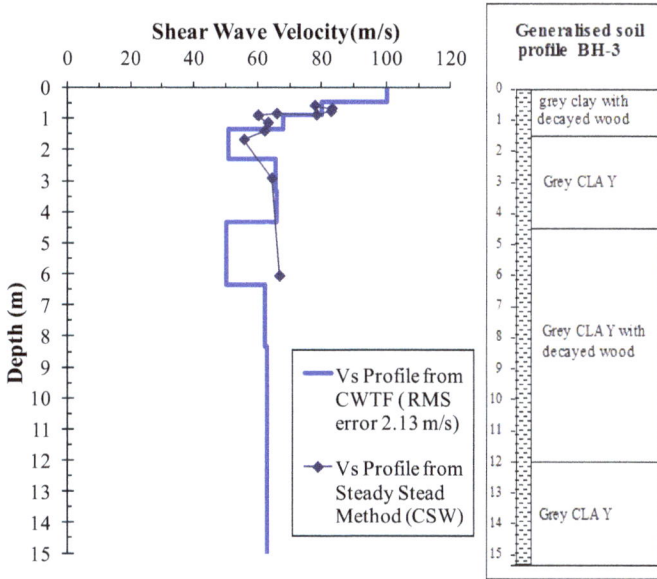

Fig. 13. A shear wave velocity profile of investigated soil at RTM Kelang site and
comparison with the borehole log

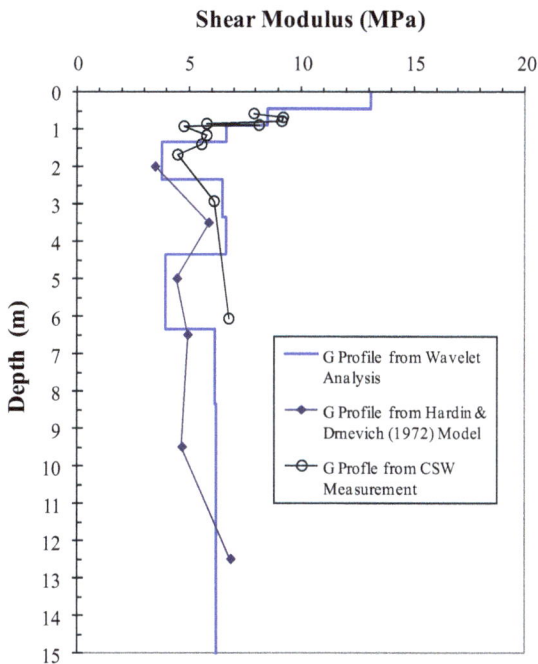

Fig. 14. A shear modulus profile of investigated soil at RTM Kelang site

From Fig. 15 and Table 2, the soil parameters of physical properties and effective soil stresses used for the shear modulus calculation using Hardin & Drnevich (1972) is presented. The soil parameters were obtained from the laboratory tests on soil samples collected from the drilling at the observed depth. Mathematical equation developed by Hardin & Drnevich (1972) can be written as:

$$G_{maks} = \frac{A(OCR)^k P_A^{1-n} \left(\sigma_m'\right)^n}{F(e)}$$
(23)

where,

A	= the dimensionless elastic stiffness coeficient,
σ_m'	= mean effective soil stress (obtained from Fig. 15)
P_A	= atmosphereic pressure,
n	= exponent soil constant equal to 0.5,
$F(e)$	= $0.3 + 0.7e^2$, and e = void ratio,
k	= exponent soil constant depending on the plasticity index of soil,
OCR	= over consolidation ratio.

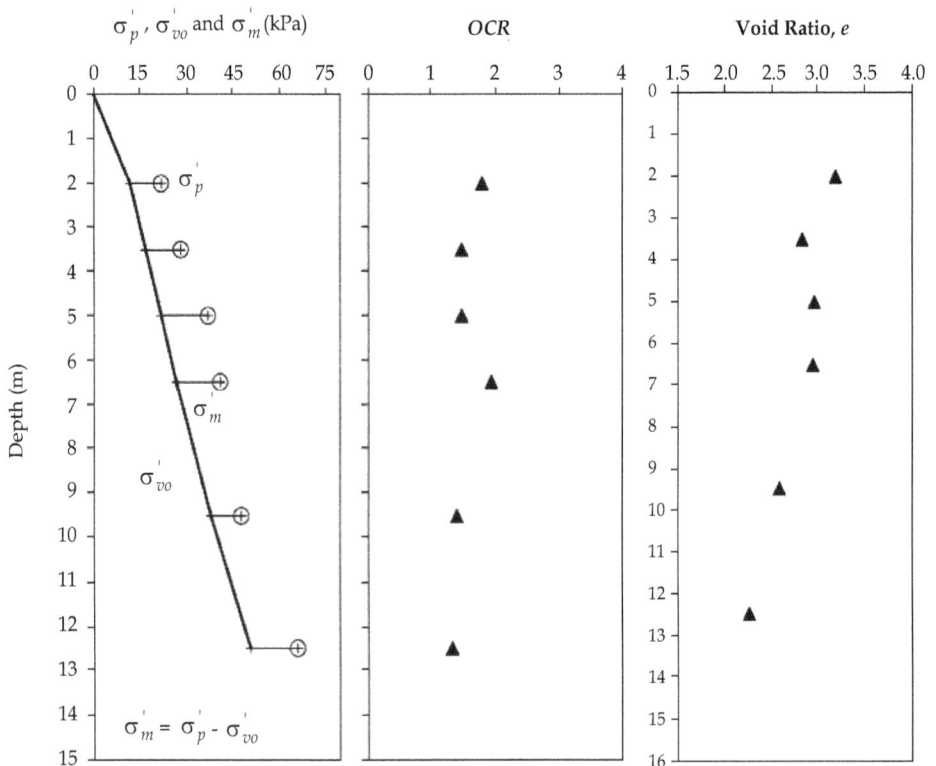

Fig. 15. Profile of soil stress, over consolidation ratio (OCR) and void ratio parameter at observed depth from the site

Shear modulus of soil profile calculated from Hardin & Drnevich (1972) based on laboratory soil parameter is shown in Fig. 15. From the CSW measurement, shear modulus can also be obtained based on Eq. 20. Comparing between these methods, it can be observed that a good agreement of the soil shear modulus at observed location was obtained by the wavelet spectrogram analysis of surface waves (WSASW), Hardin and Drnevich (1972) model and the continuous surface wave measurement (Fig. 14).

Depth (m)	σ'_m (kPa)	n	k	$F(e)$	OCR	$G = G_{maks}$ (MPa)
2	10	0.5	0.4765	7.468	1.8	3.50
3.5	20	0.5	0.3165	5.403	1.5	5.88
5	12	0.5	0.4415	5.788	1.5	4.47
6.5	14	0.5	0.3165	5.788	1.9	4.95
9.5	10	0.5	0.311	4.675	1.4	4.69
12.5	16	0.5	0.366	4.003	1.3	6.88

Table 2. Soil parameters used for shear modulus calculation using Hardin & Drnevich (1972) model at the site

3.5 Shear damping ratio evaluation

Fig. 16 shows the wavelet spectrum of 80 cm receiver spacing for signals received by geophone 1 and 2, respectively. The decay factor curve of the R-wave for the experimental data is then obtained from the plot of the ratio of the second signal magnitude from spectrogram (w_2) over the first signal magnitude (w_1) versus frequency (Fig. 17) where the curve shows a general trend of frequency dependency. A regression analysis is then performed on the experimental data to obtain the decay factor. The theoretical regression analysis of attenuation derived from equation 21 can then be written as:

$$\ln\left[\frac{W_f^{R_2}(u,s)}{W_f^{R_1}(u,s)}\right] = -\alpha(f)(\Delta R) + k = -2\alpha_0(f) + k \tag{24}$$

The best-fit curve is then established between the decay factor of the experimental data and the regression analysis equation by trial and error for different values of α_0 from visual best-fit evaluation of the two curves. From Fig. 17, the best-fit value of frequency-independent attenuation coefficient of the soil is calculated as 5×10^{-3} s/m at frequency of 3 to 20 Hz. The frequency range of the attenuation coefficient of the R-waves in the soil layers was chosen using the bandwidth criteria. The root mean square error for this fitting curve is found to be 0.27.

By repeating the procedure for attenuation analysis in each frequency value for all seismic data, the experimental attenuation curve is subsequently generated. An example of attenuation versus frequency curve at the soil site is presented in Fig. 18a. By knowing the experimental attenuation profile, the shear damping ratio can be obtained by inversion process as mentioned in Section 2.4. In the inversion analysis, the soil model is typically assumed as the homogeneous linear elastic layers over a halfspace with model parameter of shear wave velocity, shear damping ratio and thickness for each layer. In this study, due to shear damping ratio is unknown data and there is no prior information from the previous

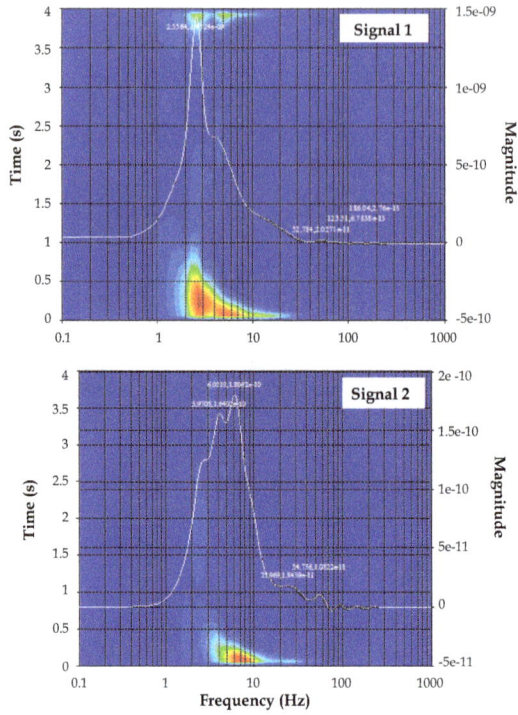

Fig. 16. The wavelet spectrum of 80 cm receiver spacing for signals received by geophone 1 and 2

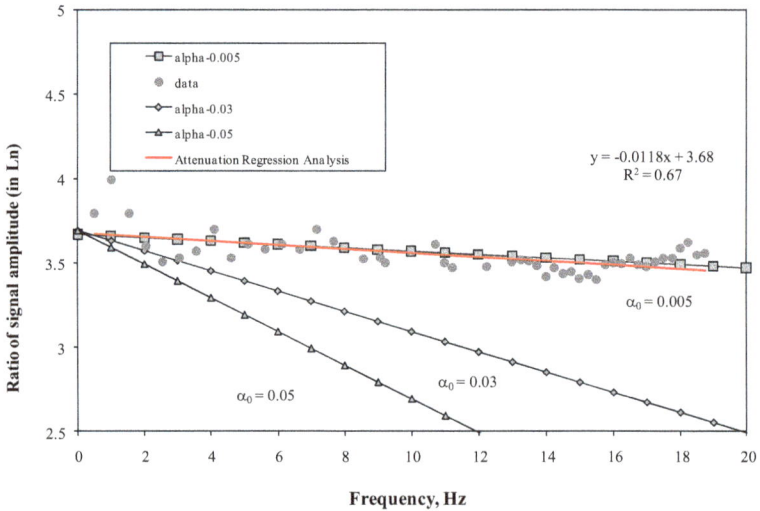

Fig. 17. The best-fit frequency-independent attenuation coefficient of the soil based on attenuation regression analysis

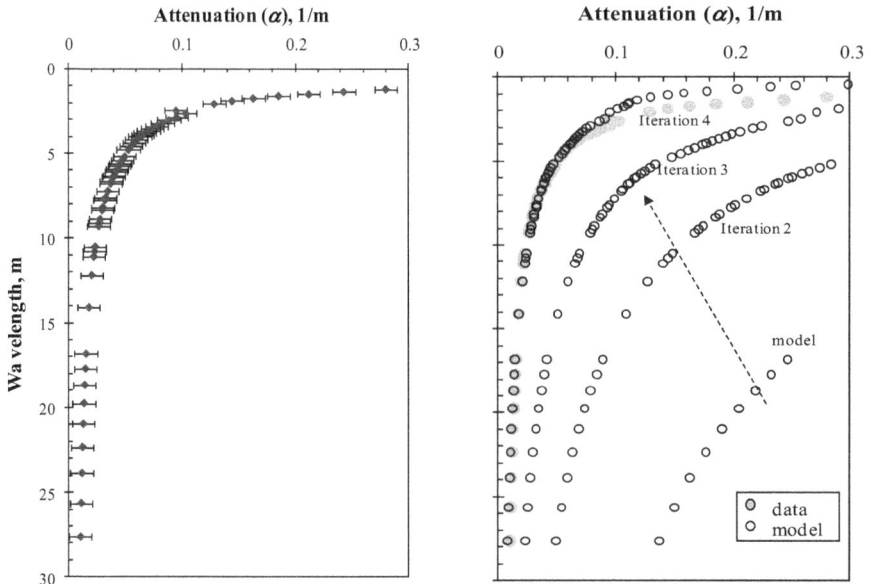

Fig. 18. (a) An example of experimental attenuation versus frequency curve at the soil site and (b) iteration process in the inversion analysis for fitting of theoretical attenuation curve to the experimental curve

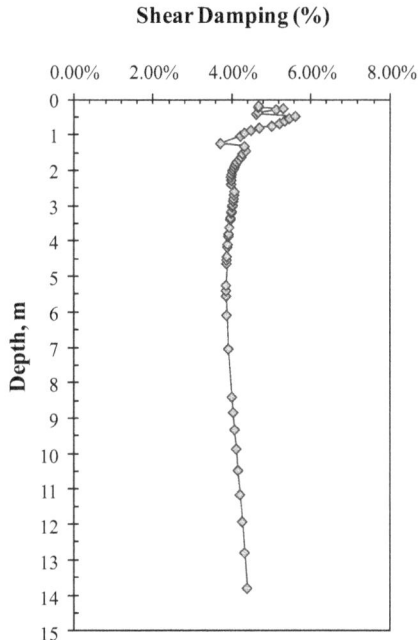

Fig. 19. Final shear damping ratio for the soil site

field or laboratory soil data, therefore, it should be assumed with rational values for soil model parameter. The inversion analysis is proceed by using Herrmann (1994) code based on a weighted, damped, least-squares algorithm. Fig. 18b shows comparisons between the experimental attenuation curves and the theoretical attenuation data calculated from equation 22. Iteration processes were conducted to match between both experimental and theoretical curves. Fig. 19 presents the final profiles of shear damping ratio for the last iteration of the inversion process with lowest RMS error.

4. Conclusion

In this paper, an improved seismic method of the wavelet spectrogram analysis of surface waves (WSASW) technique for measurement of the soil dynamic property at soft soil site is presented. The identification, denoising and reconstruction technique of the wave response spectrum from seismic surface wave propagation on a residual soil using time-frequency analysis of continuous wavelet transforms is also proposed. The mother wavelet of Morlet was used for providing good resolution of spectrogram at low frequency and is also effective in the detection of low frequency noises. The spectrogram could be used to clearly identify the various events of interest of the seismic surface waves and noisy signals. Based on the generated spectrogram, the thresholds for CWT filtration could be easily obtained. Consequently, the denoised signals of the seismic surface waves were able to be reconstructed by inverse wavelet transform considering the thresholds of the interested spectrum.

A good agreement was obtained between the dispersion curve obtained from the phase spectrum based on the CWT filtration used in WSASW method compared to the dispersion curve analysed from the IRF technique and the experimental dispersion curve from the CSW measurement. The technique is also able to evaluate the soil dynamic properties, i.e., shear wave velocity, shear modulus and damping ratio properties at soft clay soil site as performed in this study. Comparison between the shear wave velocity and shear modulus obtained using WSASW method compared to that of the WSASW and Hardin & Drnevich (1972) model were found to be good match. Finally, the WSASW technique based on wavelet analysis is a potential tool and useful for identification and evaluation of the transient events in non-stationary signals produced from the seismic measurement at the soft soil sites.

5. Acknowledgment

The support for this work was provided in part by Hibah Bersaing Grant 2009 SP2H No. 058/SP2H/PP/DP2M/IV/2009 from the Ministry of National Education, Indonesia and research project with Jabatan Kerja Raya (JKR) Malaysia. These supports are gratefully acknowledged. Authors would also like to grateful appreciate to Mr. Khairul Anuar Mohd. Nayan and Mr. Mecit Kurt for their assistances in the field measurement.

6. References

Aki, K. & Richards, P.G. (1980). *Quantitative Seismology*, Vol.1, W.H. Freeman & Co., ISBN: 978-0716710585, San Fransisco.

Capilla, C. (2006). Application of the Haar Wavelet Transform to Detect Microseismic Signal Arrivals, *Journal of Applied Geophysics*, Vol. 59, No.1, (May 2006), pp. 36–46, ISSN: 0926-9851.

Chik, Z., Islam, T., Mustafa, M.M., Sanusi, H., Rosyidi, S.A. & Taha, M.R. (2009). Surface
Wave Analysis Using Morlet Wavelet in Geotechnical Investigations, *Journal of
Applied Sciences*, Vol.9, No.19, pp. 3491-3501, ISSN: 18125654.

Daubechies, I. (1992). *Ten Lecturers on Wavelets*, ISBN: 978-0898712742, Society of Industrial
and Applied Mathematics, Pennsylvania.

Foti, S. (2004). Using Transfer Function for Estimating Dissipative Properties of Soils from
Surface-Wave Data, *Near Surface Geophysics*, Vol.2, (November 2004), pp.231-240,
ISSN: 1569-4445.

Foufoula-Georgiou, E. & Kumar P. (1995). Wavelets in Geophysics: An Introduction, In:
Wavelet in Geophysics, E. Foufoula-Georgiou & P. Kumar (Eds.), Academic Press,
ISBN: 9780122628504, California.

Ganji, V., Gucunski, N., & Nazarian, S. (1998). Automated Inversion Procedure for Spectral
Analysis of Surface Waves, *Journal of Geotechnical & Geoenvironmental Engineering*,
Vol.124, No.8, (August 1998), pp. 757-770, ISSN: 1090-0241.

Hardin, B.O. & Drnevich, V.P. (1972). Shear Modulus and Damping in Soils: Measurement
and parameter effects. *Journal of Soil Mechanics and Foundations Division*, Vol.98,
No.6, (June 1972), pp. 603-624, ISSN: 0044-7994.

Heisey, J.S., Stokoe II, K.H. & Meyer, A.H. (1982). Moduli of Pavement System from Spectral
Analysis of Surface Wave, *Transportation Research Record*, No.852, pp. 22-31, ISSN:
0738-6826.

Herrmann, R.B. (1994). *Computer Programs in Seismology*, User's Manual, St.Louis University,
Missouri, USA.

Joh, S.H. (1996). *Advance in Interpretation and Analysis Technique for the Spectral Analysis of Surface
Wave (SASW) Measurements*, Ph.D. Dissertation, the University of Texas at Austin.

Kausel, E. & Röesset, J.M. (1981). Stiffness Matrices for Layered Soils, *Bulletin of the Seismological
Society of America*, Vol.71, No.6, (December 1981), pp. 1743-1761, ISSN: 0037-1106.

Kim, D.-S. & Park, H.-C. (2002). Determination of Dispersion Phase Velocities for SASW
method Using Harmonic Wavelet Transform, *Soil Dynamics and Earthquake
Engineering*, Vol.22, No.8, (September 2002), pp.675-684, ISSN: 0267-7261.

Kramer, S.L. (1996). *Geotechnical Earthquake Engineering*, ISBN: 978-8129701930, Prentice-Hall,
New Jersey.

Kritski, A., Vincent, A. P., Yuen, D. A. & Carlsen, T. (2007). Adaptive Wavelets for
Analyzing Dispersive Seismic Waves, *Geophysics*, Vol.72, No.1, (February 2007), pp.
V1-V11, ISSN: 0016-8033.

Lai, C.G. & Rix, G.J. (1998). *Simultaneous Inversion of Rayleigh Phase Velocity and Attenuation
for Near-Surface Site Characterization*, Report No.GIT-CEE/GEO-98-2, Georgia
Institute of Technology.

Liu, P.C. (1994). Wavelet Spectrum Analysis and Ocean Wind Waves. In *Wavelets in
Geophysics*, E. Foufoula-Georgiou & P. Kumar (Eds.), 151-166, Academic Press,
ISBN: 9780122628504, California.

Mallat, S. (1989). A Theory for Multiresolution Signal Decomposition: The wavelet
Representation, *IEEE Transactions on Pattern Analysis and Machine Intelligence*,
Vol.11, pp.674-693, ISSN: 0162-8828.

Meyers, S.D., Kelly, B.G. & O'Brien, J.J. (1993). An Introduction to Wavelet Analysis in
Oceanography and Meteorology: With Application to the Dispersion of Yanai
Waves, *Monthly Weather Review*, No.121, pp.2858-2866, ISSN: 0027-0644.

Nazarian, S. & Stokoe II, K.H. (1986). *In Situ Determination of Elastic Moduli of Pavement Systems by Spectral-Analysis-of-Surface-Wave Method (Theoretical Aspects)*, Research Report 437-2, Center of Transportation Research, Bureau of Engineering Research, the University of Texas at Austin.

Nazarian, S. (1984). *In Situ Determination of Elastic Moduli of Soil Deposits and Pavement Systems by Spectral-Analysis-of-Surface-Wave method*, Ph.D. Dissertation, the University of Texas at Austin.

Rosyidi, S.A., Taha, M.R., Ismail, A. & Chik, Z. (2009). Enhanced Signal Reconstruction of Surface Waves on SASW Measurement Using Gaussian Derivative Wavelet Transform, *Acta Geophysica*, Vol.57, No.3, (September 1999), pp. 616-635, ISSN: 1895-6572.

Rosyidi, S.A.P (2009). Wavelet Analysis of Surface Wave for Evaluation of Soil Dynamic Properties, Ph.D. Thesis, the National Universiti of Malaysia, Bangi.

Rosyidi, S.A.P. (2011). Use of Wavelet Analysis and Filtration on Impulse Response for SASW Measurement In PCC Slab of Pavement Structure, *Contemporary Topics on Testing, Modeling, and Case Studies of Geomaterials, Pavements, and Tunnels, Geotechnical Special Pulblication 215*, American Society of Civil Engineers, pp. 74-82, ISBN: 978-0-7844-7626-0.

Shokouhi, P., Gucunski, N. & Maher, A. (2003). Application of Waveles in Detection of Cavities under Pavements by Surface waves, *Transportation Research Record* No.1860, pp. 57-65, ISSN: 0738-6826.

Soman, K.P. & K.L. Ramachandran (2005). *Insight into Wavelets from Theory to Practice*, Prentice-Hall of India, ISBN: 9788120340534, New Delhi.

Stokoe II, K.H., Wright, S.G., Bay, J.A. & Röesset, J.M. (1994). Characterization of Geotechnical Sites by SASW Method. In *Geophysical Characterization of Sites*, R.D. Woods (Ed.), 15-25,Oxford Publishers, ISBN: 97818811570363, New Delhi.

Torrence, C. & Compo, G.P. (1998). A Practical Guide to Wavelet Analysis, *Bulletin of the American Meteorological Society*, Vol.79, No.1, (January 1998), pp. 61-78, ISSN: 0003-0007.

Viktorov I.A. (1967). *Rayleigh and Lamb Waves: Physical Theory and Applications*, Plenum Press, ISBN: 0306302861, New York.

Wang, B. & Wang, Y. 1996. Temporal Structure of the Southern Oscillation as Revealed by Waveform and Wavelet Analysis, *Journal of Climate*, No.9, No.7, (July 1996), pp. 1586-1598, ISSN: 0894-8755.

Weng, H. & Lau, K.M. (1994). Wavelets, Period Doubling and Time-Frequency Localization With Application to Organization of Convection Over the Tropical Western Pasific, *Journal of the Atmospheric Sciences*, Vol.51, No.17, (Spetember 1994), pp. 2523-2541, ISSN: 0022-4928.

Xia J., Miller, R.D., Park, C.B. & Tian, G. (2002). Determining Q of Nearsurface Materials from Rayleigh waves, *Journal of Applied Geophysics*, Vol.51, No.2-4, (December 2002), pp.121–129.

Quasi-Axisymmetric Finite-Difference Method for Realistic Modeling of Regional and Global Seismic Wavefield — Review and Application

Genti Toyokuni[1], Hiroshi Takenaka[2] and Masaki Kanao[1]
[1]*National Institute of Polar Research*
[2]*Department of Earth and Planetary Sciences, Faculty of Sciences, Kyushu University*
Japan

1. Introduction

In this chapter, we describe recent developments of forward-modeling techniques for accurate and efficient computation of the realistic seismic wavefield. Our knowledge on the Earth's interior has been enhanced by mutual progress in observation and numerical methods. Since the first time-recording seismograph was built in Italy in 1875 (Shearer, 1999), the recorded seismic dataset has been growing at an almost exponential rate. Such a massive amount of seismic waveform data should be interpreted with consideration of the seismic source mechanism and Earth's inner structure, which explain each crest or trough in observed waveform traces. This interpretation can be achieved by forward modeling of seismic waveforms. In addition, recent progress in computation capacity has enabled investigation of the Earth's inner structure via waveform inversion, an inverse problem minimizing the difference between observed and synthetic seismograms. This method requires iterative computations of synthetic seismograms for each structural model renewal in the minimization process, so we need a forward modeling technique that produces accurate waveforms with small computation time and memory.

Writing mathematically, forward modeling (forward problem, modelization problem, or simulation problem) predicts error-free values of observable parameters $\mathbf{d}$ corresponding to a given model $\mathbf{m}$, i.e., this theoretical prediction can be denoted

$$\mathbf{m} \mapsto \mathbf{d} = \mathbf{g}(\mathbf{m}), \tag{1}$$

where $\mathbf{d} = \mathbf{g}(\mathbf{m})$ is a short notation for a set of equations $d_i = g_i(m_1, m_2, \cdots)$ $(i = 1, 2, \cdots)$ using the model parameters $\mathbf{m} = \{m_1, m_2, \cdots\}$. The operator $\mathbf{g}(\cdot)$ is called the forward operator, which expresses our mathematical model of the physical system under study (Tarantola, 2005). The forward modeling of seismic waveforms is therefore a theoretical method that applies a set of theoretical equations to determine what given seismographs would measure with respect to a preset combination of source and structure. Basically, the forward modeling of seismic waves solves the elastodynamic equation for a given source mechanism and structural model, including a set of density and elastic parameters.

2. Various numerical methods

The Earth's interior is strongly laterally heterogeneous. Since purely analytical methods do not provide solutions to the governing equations of seismic wavefields for such complex media, we are forced to use numerical modeling methods to predict realistic wavefields. Numerical methods transform an original differential problem into a system of algebraic equations, so that a continuous function in a differential equation must be represented by a finite system of numbers that must be stored in computer memory. Each numerical method is specific in how it represents a solution using a finite set of numbers, and how it approximates derivatives.

In recent years, there have been remarkable developments in numerical simulation techniques, associated with progress in computer architecture. Simulation of elastic wave propagation requires solution of the elastodynamic equation, consisting of the equations of motion and the constitutive laws under prescribed boundary and radiation conditions. If there is a need to satisfy these three strictly, we are obliged to solve them by analytical means, although it is nearly impossible except in special cases. In most cases, we have to rely on numerical modeling methods that approximate these three relations numerically. From a practical viewpoint, the numerical simulation methods for seismic wave motion can be classified into three groups: (1) Domain methods; (2) Boundary methods; and (3) Hybrid methods (Takenaka et al., 1998). Domain methods numerically approximate the elastodynamic equation, as well as boundary and radiation conditions. Solutions are reached by solving linear equations resulting from complete discretization of a medium, solution, and differential operators in time (or in frequency), throughout the spatial domain. Therefore, domain methods are applicable to modeling wavefields in arbitrary heterogeneous structures, since the medium parameters are distributed on numerical grid points. On the other hand, boundary methods can satisfy the elastodynamic equation and radiation conditions analytically, and they only discretize the boundary conditions. In these methods, the differential equations and boundary conditions are transformed into boundary integral equations involving unknown functions, which are then discretized and solved by various numerical techniques. Hybrid methods are combinations of several different methods among domain or boundary methods. This chapter treats computation by domain methods.

The domain methods contain various numerical methods, such as finite-difference method (FDM), pseudospectral method (PSM), finite-element method (FEM), and spectral-element method (SEM). These methods can be classified based on the kind of formulation they solve, e.g., strong formulation, weak formulation, etc. It is impossible to choose the best method among them, since a single method can hardly satisfy all demands. Therefore, the choice of suitable numerical method should be problem-dependent. Here, we explain numerical schemes based on the FDM, the most orthodox and user-friendly method of seismic wave computation. The FDM solves the elastodynamic equation in strong form, by replacing partial derivatives in space and time with finite-difference approximations, only at grid points in the computational domain.

The FDM grid distribution can be classified based on whether all wavefield variables are approximated at the same grid position. On a conventional grid, all variables are approximated at the same grid position. On an alternative staggered grid, each displacement and/or particle-velocity component, as well as stress component, has its own grid position with several exceptions, such as overlapping of three normal-stress components. The advantage of the staggered grid is its robustness for structures with high contrast of Poisson's ratio.

The elastodynamic equation (equations of motion and constitutive laws), together with the initial and boundary conditions, completely describe a problem of seismic wave propagation. If we keep the equations of motion separate from the constitutive laws, we can speak of the displacement-stress formulation. If we use particle velocity in the equations of motion, keep the constitutive laws, and add the definition of particle velocity, we obtain the displacement-velocity-stress formulation. If we apply a time derivative to the constitutive laws instead of adding the definition of particle velocity, then we have the velocity-stress formulation. If we eliminate the stress-tensor components by substituting the constitutive laws into the equations of motion, we get the displacement formulation (e.g., Moczo et al., 2007). In this chapter, we mainly deal with the velocity-stress formulation. The formulation for 3-D computations for general elastic media are described with mathematical expressions in Cartesian coordinates (x_1, x_2, x_3) as follows $(i, j, k, l \in \{1, 2, 3\})$:

$$\rho \frac{\partial v_i}{\partial t} = \frac{\partial \sigma_{ij}}{\partial x_j} + f_i, \tag{2}$$

$$\frac{\partial \sigma_{ij}}{\partial t} = c_{ijkl} \frac{\partial \epsilon_{kl}}{\partial t} - \dot{M}_{ij}, \tag{3}$$

where t is time, $\rho(\mathbf{x}); \mathbf{x} = (x_1, x_2, x_3)$ is the density; $c_{ijkl}(\mathbf{x})$ is the component of the elastic tensor. In addition, $v_i(\mathbf{x}, t)$, $f_i(\mathbf{x}, t)$, $\sigma_{ij}(\mathbf{x}, t)$, $\epsilon_{ij}(\mathbf{x}, t)$, and $\dot{M}_{ij}(\mathbf{x}, t)$ are components of the particle velocity vector, body force vector, stress tensor, strain tensor, and first-order time derivative of the moment tensor, respectively. We have used the summation convention over repeated suffixes.

In general orthogonal curvilinear coordinates (c_1, c_2, c_3), the elastodynamic equation corresponds to Eqs. (2), and (3) can be given as follows, together with the strain-velocity relation without using the summation convention (Aki & Richards, 2002):

$$\rho \frac{\partial v_i}{\partial t} = \frac{1}{h_1 h_2 h_3} \sum_{j,k} \left[\left(\frac{\sigma_{jk} h_1 h_2 h_3}{h_k} \right) \left(\frac{\delta_{ik}}{h_j} \frac{\partial h_k}{\partial c_j} - \delta_{jk} \sum_p \frac{\delta_{ip}}{h_p} \frac{\partial h_j}{\partial c_p} \right) + \delta_{ij} \frac{\partial}{\partial c_k} \left(\frac{\sigma_{jk} h_1 h_2 h_3}{h_k} \right) \right] + f_i, \tag{4}$$

$$\frac{\partial \sigma_{ij}}{\partial t} = \lambda \delta_{ij} \sum_p \frac{\partial \epsilon_{pp}}{\partial t} + 2\mu \frac{\partial \epsilon_{ij}}{\partial t} - \dot{M}_{ij}, \tag{5}$$

$$\frac{\partial \epsilon_{ij}}{\partial t} = \frac{1}{2} \left[\frac{h_i}{h_j} \frac{\partial}{\partial c_j} \left(\frac{v_i}{h_i} \right) + \frac{h_j}{h_i} \frac{\partial}{\partial c_i} \left(\frac{v_j}{h_j} \right) \right] + \frac{\delta_{ij}}{h_j} \sum_s \frac{v_s}{h_s} \frac{\partial h_i}{\partial c_s}, \tag{6}$$

where $h_i; i \in \{1, 2, 3\}$ are scaling functions peculiar to the coordinate system, and all variables related to the medium and wavefields are the same as in the Cartesian case, except that the $\mathbf{x}$ dependence should be replaced with $\mathbf{c} = (c_1, c_2, c_3)$ dependence. We consider special cases of these equations in cylindrical and spherical coordinates in the following sections.

With simulation of seismic wave motion by the domain methods, two "dimensions", i.e., the spatial dimension (heterogeneity) of a medium and the dimensionality of wavefields, become important. The heterogeneity of a medium is defined as the number of independent variables considering material parameters as spatial functions, whereas the dimensionality of wavefields is defined as the number of spatial variables in all independent variables of wavefields. We should choose these two dimensions for a compromise between accuracy and efficiency of waveform computation. Takenaka (1993; 1995) developed a notation describing the combination as (m, n) dimension, where m is the dimensionality of wavefields and n is the dimension of a medium. The $(3, 3)$ dimensional modeling, called 3-D modeling, calculates 3-D seismic wavefields in a 3-D structural model. The 3-D modeling provides accurate results, since it can treat the most realistic situation. However, full 3-D modeling up to a realistic high frequency is still computationally intensive and costly, even on parallel hardware. On the other hand, $(2, 2)$ dimensional modeling (2-D modeling) calculates 2-D wavefields in a 2-D structural model, which requires relatively small computational resources compared to 3-D modeling. Because of the severe computational requirement, waveform computation by domain methods had long been restricted to 2-D modeling, although the wave behavior of out-of-plane motion is underestimated. Moreover, 2-D modeling cannot correctly model geometrical spreading effects and the pulse shape in 3-D.

In the 1990s, an alternative $(3, 2)$ dimensional modeling called 2.5-D modeling, which calculates 3-D wavefields in a medium varying only in two dimensions, was introduced in seismology. It first assumes the structure to be invariant in one direction, and then applies a spatial Fourier transform to the 3-D wave equation in this direction. The resulting equations in a mixed coordinate-wavenumber domain consist of independent sets of 2-D equations for each wavenumber, such that numerical computations of these equations followed by an inverse Fourier transform over wavenumber generate 3-D synthetic seismograms. One can thus correctly model the 3-D geometrical spreading effects and pulse shape for all phases, and it makes possible a direct comparison between real and synthetic waveform data. Nevertheless, associated with computations for all discrete wavenumbers, 2.5-D modeling requires a long computation time, comparable to 3-D modeling, although it has a memory requirement only slightly greater than 2-D modeling.

3. Axisymmetric modeling

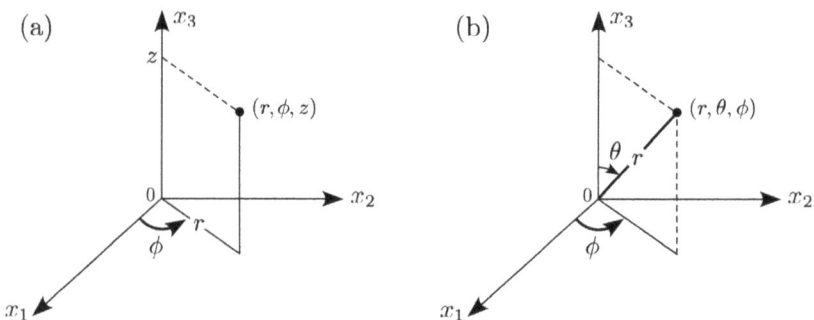

Fig. 1. Configuration of the coordinate systems. (a) Cylindrical coordinates (r, ϕ, z), and (b) spherical coordinates (r, θ, ϕ).

A more economical technique for modeling 3-D seismic wavefields is to approximate the structural model as rotationally symmetric along the vertical axis, include a seismic source, and then solve the wave equation in cylindrical or spherical coordinates. This method, called axisymmetric modeling, can correctly model 3-D geometrical spreading effects and pulse shape, with computation time and memory comparable to 2-D modeling. In the following subsections, we write the elastodynamic equations for axisymmetric modeling in both cylindrical coordinates (r, ϕ, z) and spherical coordinates (r, θ, ϕ), as shown in Figure 1, comparing them with the 3-D equations.

3.1 Cylindrical coordinates

Cylindrical coordinates (r, ϕ, z) are defined as $0 \leq r < \infty, 0 \leq \phi \leq 2\pi, -\infty < z < \infty$, which replace notations in Eqs. (4)–(6) as follows.

$$c_1 = r, \quad c_2 = \phi, \quad c_3 = z, \quad h_1 = 1, \quad h_2 = r, \quad h_3 = 1. \tag{7}$$

3.1.1 For 3-D modeling

Substituting Eq. (7) into Eqs. (4)–(6), we can get the 3-D elastodynamic equation in cylindrical coordinates:

$$\rho \frac{\partial v_r}{\partial t} = f_r + \frac{1}{r}\frac{\partial}{\partial r}(r\sigma_{rr}) + \frac{1}{r}\frac{\partial \sigma_{r\phi}}{\partial \phi} + \frac{\partial \sigma_{rz}}{\partial z} - \frac{\sigma_{\phi\phi}}{r}, \tag{8}$$

$$\rho \frac{\partial v_\phi}{\partial t} = f_\phi + \frac{1}{r}\frac{\partial}{\partial r}(r\sigma_{r\phi}) + \frac{1}{r}\frac{\partial \sigma_{\phi\phi}}{\partial \phi} + \frac{\partial \sigma_{\phi z}}{\partial z} + \frac{\sigma_{r\phi}}{r}, \tag{9}$$

$$\rho \frac{\partial v_z}{\partial t} = f_z + \frac{1}{r}\frac{\partial}{\partial r}(r\sigma_{rz}) + \frac{1}{r}\frac{\partial \sigma_{\phi z}}{\partial \phi} + \frac{\partial \sigma_{zz}}{\partial z}, \tag{10}$$

$$\frac{\partial \sigma_{rr}}{\partial t} = (\lambda + 2\mu)\frac{\partial v_r}{\partial r} + \frac{\lambda}{r}\frac{\partial v_\phi}{\partial \phi} + \lambda\frac{\partial v_z}{\partial z} + \frac{\lambda}{r}v_r - \dot{M}_{rr}, \tag{11}$$

$$\frac{\partial \sigma_{\phi\phi}}{\partial t} = \lambda\frac{\partial v_r}{\partial r} + \frac{\lambda + 2\mu}{r}\frac{\partial v_\phi}{\partial \phi} + \lambda\frac{\partial v_z}{\partial z} + \frac{\lambda + 2\mu}{r}v_r - \dot{M}_{\phi\phi}, \tag{12}$$

$$\frac{\partial \sigma_{zz}}{\partial t} = \lambda\frac{\partial v_r}{\partial r} + \frac{\lambda}{r}\frac{\partial v_\phi}{\partial \phi} + (\lambda + 2\mu)\frac{\partial v_z}{\partial z} + \frac{\lambda}{r}v_r - \dot{M}_{zz}, \tag{13}$$

$$\frac{\partial \sigma_{r\phi}}{\partial t} = \mu\left\{ r\frac{\partial}{\partial r}\left(\frac{v_\phi}{r}\right) + \frac{1}{r}\frac{\partial v_r}{\partial \phi} \right\} - \dot{M}_{r\phi}, \tag{14}$$

$$\frac{\partial \sigma_{\phi z}}{\partial t} = \mu\left(\frac{1}{r}\frac{\partial v_z}{\partial \phi} + \frac{\partial v_\phi}{\partial z}\right) - \dot{M}_{\phi z}, \tag{15}$$

$$\frac{\partial \sigma_{rz}}{\partial t} = \mu\left(\frac{\partial v_z}{\partial r} + \frac{\partial v_r}{\partial z}\right) - \dot{M}_{rz}. \tag{16}$$

3.1.2 For axisymmetric modeling

Axisymmetric modeling in cylindrical coordinates assumes the structure to be axisymmetric with respect to the axis $r = 0$. For cases with axisymmetric seismic sources, such as explosive and torque sources (SH-wave source), the 3-D seismic wavefield is completely separated into in-plane motion in the r-z plane (P-SV waves) and anti-plane motion (SH waves). In this situation, terms including ϕ derivatives can be neglected in Eqs. (8)–(16), which gives, for example, the P-SV elastodynamic equation for axisymmetric modeling in cylindrical coordinates from axisymmetric sources as

$$\rho \frac{\partial v_r}{\partial t} = f_r + \frac{1}{r}\frac{\partial}{\partial r}\left(r\sigma_{rr}\right) + \frac{\partial \sigma_{rz}}{\partial z} - \frac{\sigma_{\phi\phi}}{r}, \tag{17}$$

$$\rho \frac{\partial v_z}{\partial t} = f_z + \frac{1}{r}\frac{\partial}{\partial r}\left(r\sigma_{rz}\right) + \frac{\partial \sigma_{zz}}{\partial z}, \tag{18}$$

$$\frac{\partial \sigma_{rr}}{\partial t} = (\lambda + 2\mu)\frac{\partial v_r}{\partial r} + \lambda \frac{\partial v_z}{\partial z} + \frac{\lambda}{r}v_r - \dot{M}_{rr}, \tag{19}$$

$$\frac{\partial \sigma_{\phi\phi}}{\partial t} = \lambda \frac{\partial v_r}{\partial r} + \lambda \frac{\partial v_z}{\partial z} + \frac{\lambda + 2\mu}{r}v_r - \dot{M}_{\phi\phi}, \tag{20}$$

$$\frac{\partial \sigma_{zz}}{\partial t} = \lambda \frac{\partial v_r}{\partial r} + (\lambda + 2\mu)\frac{\partial v_z}{\partial z} + \frac{\lambda}{r}v_r - \dot{M}_{zz}, \tag{21}$$

$$\frac{\partial \sigma_{rz}}{\partial t} = \mu \left(\frac{\partial v_z}{\partial r} + \frac{\partial v_r}{\partial z}\right) - \dot{M}_{rz}. \tag{22}$$

3.2 Spherical coordinates

Spherical coordinates (r, θ, ϕ) are defined as $0 \leq r < \infty, 0 \leq \theta \leq \pi, 0 \leq \phi \leq 2\pi$, which replace notations in Eqs. (4)–(6) as follows.

$$c_1 = r, \quad c_2 = \theta, \quad c_3 = \phi, \quad h_1 = 1, \quad h_2 = r, \quad h_3 = r\sin\theta. \tag{23}$$

3.2.1 For 3-D modeling

Substituting Eq. (23) into Eqs. (4)–(6), we can get the 3-D elastodynamic equation in spherical coordinates:

$$\rho \frac{\partial v_r}{\partial t} = f_r + \frac{\partial \sigma_{rr}}{\partial r} + \frac{1}{r}\frac{\partial \sigma_{r\theta}}{\partial \theta} + \frac{1}{r\sin\theta}\frac{\partial \sigma_{r\phi}}{\partial \phi} + \frac{1}{r}\left(2\sigma_{rr} - \sigma_{\theta\theta} - \sigma_{\phi\phi} + \sigma_{r\theta}\cot\theta\right), \tag{24}$$

$$\rho \frac{\partial v_\theta}{\partial t} = f_\theta + \frac{\partial \sigma_{r\theta}}{\partial r} + \frac{1}{r}\frac{\partial \sigma_{\theta\theta}}{\partial \theta} + \frac{1}{r\sin\theta}\frac{\partial \sigma_{\theta\phi}}{\partial \phi} + \frac{1}{r}\left\{3\sigma_{r\theta} + (\sigma_{\theta\theta} - \sigma_{\phi\phi})\cot\theta\right\}, \tag{25}$$

$$\rho \frac{\partial v_\phi}{\partial t} = f_\phi + \frac{\partial \sigma_{r\phi}}{\partial r} + \frac{1}{r}\frac{\partial \sigma_{\theta\phi}}{\partial \theta} + \frac{1}{r\sin\theta}\frac{\partial \sigma_{\phi\phi}}{\partial \phi} + \frac{1}{r}\left(3\sigma_{r\phi} + 2\sigma_{\theta\phi}\cot\theta\right), \tag{26}$$

$$\frac{\partial \sigma_{rr}}{\partial t} = (\lambda + 2\mu)\frac{\partial v_r}{\partial r} + \frac{\lambda}{r}\frac{\partial v_\theta}{\partial \theta} + \frac{\lambda}{r\sin\theta}\frac{\partial v_\phi}{\partial \phi} + \frac{\lambda}{r}\left(2v_r + v_\theta\cot\theta\right) - \dot{M}_{rr}, \tag{27}$$

$$\frac{\partial \sigma_{\theta\theta}}{\partial t} = \lambda \frac{\partial v_r}{\partial r} + \frac{\lambda + 2\mu}{r}\frac{\partial v_\theta}{\partial \theta} + \frac{\lambda}{r\sin\theta}\frac{\partial v_\phi}{\partial \phi} + \frac{2(\lambda + \mu)}{r}v_r + \frac{\lambda}{r}v_\theta\cot\theta - \dot{M}_{\theta\theta}, \tag{28}$$

$$\frac{\partial \sigma_{\phi\phi}}{\partial t} = \lambda \frac{\partial v_r}{\partial r} + \frac{\lambda}{r}\frac{\partial v_\theta}{\partial \theta} + \frac{\lambda + 2\mu}{r\sin\theta}\frac{\partial v_\phi}{\partial \phi} + \frac{2(\lambda + \mu)}{r}v_r + \frac{\lambda + 2\mu}{r}v_\theta\cot\theta - \dot{M}_{\phi\phi}, \tag{29}$$

$$\frac{\partial \sigma_{r\theta}}{\partial t} = \mu \left(\frac{\partial v_\theta}{\partial r} + \frac{1}{r}\frac{\partial v_r}{\partial \theta} - \frac{1}{r}v_\theta\right) - \dot{M}_{r\theta}, \tag{30}$$

$$\frac{\partial \sigma_{\theta\phi}}{\partial t} = \frac{\mu}{r}\left(\frac{\partial v_\phi}{\partial \theta} + \frac{1}{\sin\theta}\frac{\partial v_\theta}{\partial \phi} - v_\phi\cot\theta\right) - \dot{M}_{\theta\phi}, \tag{31}$$

$$\frac{\partial \sigma_{r\phi}}{\partial t} = \mu \left(\frac{\partial v_\phi}{\partial r} + \frac{1}{r\sin\theta}\frac{\partial v_r}{\partial \phi} - \frac{1}{r}v_\phi\right) - \dot{M}_{r\phi}. \tag{32}$$

Quasi-Axisymmetric Finite-Difference Method for Realistic Modeling of Regional and Global Seismic Wavefield
— Review and Application

91

3.2.2 For axisymmetric modeling

Axisymmetric modeling in spherical coordinates assumes the structure to be axisymmetric about the axis $\theta = 0$. As we have already seen in Section 3.1.2, the in-plane P-SV motion in the r-θ plane and the anti-plane SH motion can completely been separated for axisymmetric sources. Consequently, the P-SV elastodynamic equation for axisymmetric modeling in spherical coordinates from axisymmetric sources becomes

$$\rho \frac{\partial v_r}{\partial t} = f_r + \frac{\partial \sigma_{rr}}{\partial r} + \frac{1}{r} \frac{\partial \sigma_{r\theta}}{\partial \theta} + \frac{1}{r} \left(2\sigma_{rr} - \sigma_{\theta\theta} - \sigma_{\phi\phi} + \sigma_{r\theta} \cot \theta \right), \tag{33}$$

$$\rho \frac{\partial v_\theta}{\partial t} = f_\theta + \frac{\partial \sigma_{r\theta}}{\partial r} + \frac{1}{r} \frac{\partial \sigma_{\theta\theta}}{\partial \theta} + \frac{1}{r} \left\{ 3\sigma_{r\theta} + \left(\sigma_{\theta\theta} - \sigma_{\phi\phi} \right) \cot \theta \right\}, \tag{34}$$

$$\frac{\partial \sigma_{rr}}{\partial t} = (\lambda + 2\mu) \frac{\partial v_r}{\partial r} + \frac{\lambda}{r} \frac{\partial v_\theta}{\partial \theta} + \frac{\lambda}{r} \left(2v_r + v_\theta \cot \theta \right) - \dot{M}_{rr}, \tag{35}$$

$$\frac{\partial \sigma_{\theta\theta}}{\partial t} = \lambda \frac{\partial v_r}{\partial r} + \frac{\lambda + 2\mu}{r} \frac{\partial v_\theta}{\partial \theta} + \frac{2(\lambda + \mu)}{r} v_r + \frac{\lambda}{r} v_\theta \cot \theta - \dot{M}_{\theta\theta}, \tag{36}$$

$$\frac{\partial \sigma_{\phi\phi}}{\partial t} = \lambda \frac{\partial v_r}{\partial r} + \frac{\lambda}{r} \frac{\partial v_\theta}{\partial \theta} + \frac{2(\lambda + \mu)}{r} v_r + \frac{\lambda + 2\mu}{r} v_\theta \cot \theta - \dot{M}_{\phi\phi}, \tag{37}$$

$$\frac{\partial \sigma_{r\theta}}{\partial t} = \mu \left(\frac{\partial v_\theta}{\partial r} + \frac{1}{r} \frac{\partial v_r}{\partial \theta} - \frac{1}{r} v_\theta \right) - \dot{M}_{r\theta}. \tag{38}$$

Similarly, the SH elastodynamic equation in spherical coordinates for axisymmetric sources becomes the following.

$$\rho \frac{\partial v_\phi}{\partial t} = f_\phi + \frac{\partial \sigma_{r\phi}}{\partial r} + \frac{1}{r} \frac{\partial \sigma_{\theta\phi}}{\partial \theta} + \frac{1}{r} \left(3\sigma_{r\phi} + 2\sigma_{\theta\phi} \cot \theta \right), \tag{39}$$

$$\frac{\partial \sigma_{\theta\phi}}{\partial t} = \frac{\mu}{r} \left(\frac{\partial v_\phi}{\partial \theta} - v_\phi \cot \theta \right) - \dot{M}_{\theta\phi}, \tag{40}$$

$$\frac{\partial \sigma_{r\phi}}{\partial t} = \mu \left(\frac{\partial v_\phi}{\partial r} - \frac{1}{r} v_\phi \right) - \dot{M}_{r\phi}. \tag{41}$$

This decoupling between P-SV and SH waves only holds for axisymmetric sources. Toyokuni & Takenaka (2006a) implemented arbitrary moment-tensor point sources, including shear dislocation sources, into the axisymmetric computation in spherical coordinates, using the Fourier transform of all field variables in the ϕ direction, which can be written as

$$a(t, r, \theta, \phi) = \hat{a}^0(t, r, \theta) + \sum_{m=1}^{2} \left\{ \hat{a}_C^m(t, r, \theta) \cos m\phi + \hat{a}_S^m(t, r, \theta) \sin m\phi \right\}, \tag{42}$$

where a is a variable that can be replaced by any component of the particle velocity vector, body force vector, stress tensor, and moment tensor; m is the expansion order and $\{\hat{a}^0, \hat{a}_C^m, \hat{a}_S^m\}$ are expansion coefficients. Subscripts C and S have been added to indicate coefficients for cosine and sine terms, respectively. It is sufficient to take the expansion order up to $m = 2$ with consideration of radiation patterns of moment tensor sources. Substitution of Eq. (42) into the 3-D elastodynamic equation in spherical coordinates Eqs. (24)–(32), followed by rearrangement, gives five closed systems of the partial differential equations of the expansion coefficients. The equations for $m = 0$ have the same form as Eqs. (33)–(41), whereas those for

$m = 1, 2$ become the following:

$$\rho \frac{\partial \hat{v}_{rA}^m}{\partial t} = \hat{f}_{rA}^m + \frac{\partial \hat{\sigma}_{rrA}^m}{\partial r} + \frac{1}{r} \frac{\partial \hat{\sigma}_{r\theta A}^m}{\partial \theta} + \frac{\zeta m}{r \sin \theta} \hat{\sigma}_{r\phi B}^m + \frac{1}{r} \left(2 \hat{\sigma}_{rrA}^m - \hat{\sigma}_{\theta\theta A}^m - \hat{\sigma}_{\phi\phi A}^m + \hat{\sigma}_{r\theta A}^m \cot \theta \right), \quad (43)$$

$$\rho \frac{\partial \hat{v}_{\theta A}^m}{\partial t} = \hat{f}_{\theta A}^m + \frac{\partial \hat{\sigma}_{r\theta A}^m}{\partial r} + \frac{1}{r} \frac{\partial \hat{\sigma}_{\theta\theta A}^m}{\partial \theta} + \frac{\zeta m}{r \sin \theta} \hat{\sigma}_{\theta\phi B}^m + \frac{1}{r} \left\{ 3 \hat{\sigma}_{r\theta A}^m + \left(\hat{\sigma}_{\theta\theta A}^m - \hat{\sigma}_{\phi\phi A}^m \right) \cot \theta \right\}, \quad (44)$$

$$\rho \frac{\partial \hat{v}_{\phi B}^m}{\partial t} = \hat{f}_{\phi B}^m + \frac{\partial \hat{\sigma}_{r\phi B}^m}{\partial r} + \frac{1}{r} \frac{\partial \hat{\sigma}_{\theta\phi B}^m}{\partial \theta} - \frac{\zeta m}{r \sin \theta} \hat{\sigma}_{\phi\phi A}^m + \frac{1}{r} \left(3 \hat{\sigma}_{r\phi B}^m + 2 \hat{\sigma}_{\theta\phi B}^m \cot \theta \right), \quad (45)$$

$$\frac{\partial \hat{\sigma}_{rrA}^m}{\partial t} = (\lambda + 2\mu) \frac{\partial \hat{v}_{rA}^m}{\partial r} + \frac{\lambda}{r} \frac{\partial \hat{v}_{\theta A}^m}{\partial \theta} + \frac{\lambda \zeta m}{r \sin \theta} \hat{v}_{\phi B}^m + \frac{\lambda}{r} \left(2 \hat{v}_{rA}^m + \hat{v}_{\theta A}^m \cot \theta \right) - \hat{M}_{rrA}^m, \quad (46)$$

$$\frac{\partial \hat{\sigma}_{\theta\theta A}^m}{\partial t} = \lambda \frac{\partial \hat{v}_{rA}^m}{\partial r} + \frac{\lambda + 2\mu}{r} \frac{\partial \hat{v}_{\theta A}^m}{\partial \theta} + \frac{\lambda \zeta m}{r \sin \theta} \hat{v}_{\phi B}^m + \frac{2(\lambda + \mu)}{r} \hat{v}_{rA}^m + \frac{\lambda}{r} \hat{v}_{\theta A}^m \cot \theta - \hat{M}_{\theta\theta A}^m, \quad (47)$$

$$\frac{\partial \hat{\sigma}_{\phi\phi A}^m}{\partial t} = \lambda \frac{\partial \hat{v}_{rA}^m}{\partial r} + \frac{\lambda}{r} \frac{\partial \hat{v}_{\theta A}^m}{\partial \theta} + \frac{(\lambda + 2\mu)\zeta m}{r \sin \theta} \hat{v}_{\phi B}^m + \frac{2(\lambda + \mu)}{r} \hat{v}_{rA}^m + \frac{\lambda + 2\mu}{r} \hat{v}_{\theta A}^m \cot \theta - \hat{M}_{\phi\phi A}^m, \quad (48)$$

$$\frac{\partial \hat{\sigma}_{r\theta A}^m}{\partial t} = \mu \left(\frac{\partial \hat{v}_{\theta A}^m}{\partial r} + \frac{1}{r} \frac{\partial \hat{v}_{rA}^m}{\partial \theta} - \frac{1}{r} \hat{v}_{\theta A}^m \right) - \hat{M}_{r\theta A}^m, \quad (49)$$

$$\frac{\partial \hat{\sigma}_{\theta\phi B}^m}{\partial t} = \frac{\mu}{r} \left(\frac{\partial \hat{v}_{\phi B}^m}{\partial \theta} - \frac{\zeta m}{\sin \theta} \hat{v}_{\theta A}^m - \hat{v}_{\phi B}^m \cot \theta \right) - \hat{M}_{\theta\phi B}^m, \quad (50)$$

$$\frac{\partial \hat{\sigma}_{r\phi B}^m}{\partial t} = \mu \left(\frac{\partial \hat{v}_{\phi B}^m}{\partial r} - \frac{\zeta m}{r \sin \theta} \hat{v}_{rA}^m - \frac{1}{r} \hat{v}_{\phi B}^m \right) - \hat{M}_{r\phi B}^m, \quad (51)$$

where $A, B \in \{C, S\}$. Note that P-SV and SH waves are coupled via coupling terms in these equations, including m. Through the Fourier expansion in the ϕ direction, an arbitrary moment tensor source is decomposed into five moment tensor elements: (1) Axisymmetric excitation for $m = 0$, and four double couple excitations for $m = 1, 2$ that are classified using the combinations of three parameters $\{m, A, B\}$ as (2) $\{1, C, S\}$, (3) $\{1, S, C\}$, (4) $\{2, C, S\}$, and (5) $\{2, S, C\}$. The elements (2) and (3) correspond to purely vertical dip-slip excitations shifted $\pi/2$ in the ϕ direction for each other, whereas (4) and (5) are purely strike-slip excitations shifted $\pi/4$ in the ϕ direction for each other, as shown in Figure 2. Computations of expansion coefficients using, for example, the FDM with respect to each moment tensor element via Eqs. (33)–(41) or Eqs. (43)–(51), followed by substitution of the results into Eq. (42), enables attainment of global elastic response by an arbitrary moment tensor source for the axisymmetric structural model. This requires only five times the computational resources of computations for purely axisymmetric sources.

3.3 Review of axisymmetric modeling

Because of the light computational requirement and correct treatment of 3-D seismic wavefields, axisymmetric modeling has frequently been used by researchers. Here, we briefly summarize their works.

Quasi-Axisymmetric Finite-Difference Method for Realistic Modeling of Regional and Global Seismic Wavefield
— Review and Application

93

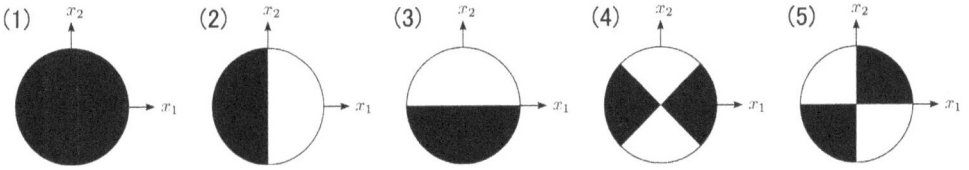

Fig. 2. Five moment tensor elements. An arbitrary moment tensor source consists of the non-double-couple component (1) for $m = 0$, and the double couple components (2)–(5) for $m = 1, 2$. We show an explosive source as representative of axisymmetric sources. The elements (2)–(5) correspond to situations with combinations of three parameters $\{m, A, B\}$ as $\{1, C, S\}, \{1, S, C\}, \{2, C, S\}$, and $\{2, S, C\}$, respectively.

Axisymmetric modeling in cylindrical coordinates has often been used to efficiently calculate realistic 3-D seismic wavefields, especially for target areas of seismic exploration. For flat-lying media, the solution on a r-z cross section with a source and receivers will be correct for full 3-D modeling for a point source. When using Cartesian 2-D modeling for the same target, the seismic source becomes a line in 3-D (point in 2-D) along a direction of structural invariance. This causes fatal errors on waveforms and makes it impossible for direct comparison between real and synthetic data, even when the real data are converted to 2-D. However, in axisymmetric modeling in cylindrical coordinates, any lateral variations on the r-z plane become physically unrealistic rings in 3-D, except in very special cases. Alterman & Karal (1968) introduced the technique for FDM computation of seismic waves in elastic layered half-space with a buried point compressional source. They applied the scheme to various investigations, e.g., of the effect of different mesh sizes on synthetic seismograms, development of Rayleigh waves on the surface, change of Rayleigh waves with depth and pulse width, and so on. Details of their FDM scheme are also in Alterman & Loewenthal (1972). Stephen (1983) adopted the cylindrical approach to compare the FDM and reflectivity synthetic seismograms for a compressional point source, using laterally homogeneous seafloor models with step and ramp discontinuities between liquid and solid, showing the two methods to be in excellent agreement. Stephen (1988) expanded the work of Stephen (1983), testing various FDM formulations to determine which ones produce acceptable solutions for seafloor problems. They used models with horizontal liquid-solid interfaces, and those with rough shape. Igel et al. (1996) performed waveform inversion of marine reflection seismograms to determine P impedance and Poisson's ratio structures in the Gulf of Mexico, through iterative calculation of synthetic seismograms by axisymmetric modeling in cylindrical coordinates.

Axisymmetric modeling in spherical coordinates is a powerful technique to obtain the realistic 3-D global seismic wavefield. Therefore, it has long been used, in spite of the restriction of structural models in rotational symmetry with respect to the axis through the seismic source. Alterman et al. (1970) pioneered the application of this method to the FDM computations of elastic wavefield radiated by an impulsive point source, for radially and laterally heterogeneous, purely mathematical sphere models. The first application of this approach to the Earth model was the work of Igel & Weber (1995). They simulated SH-wave propagation in frequency bands up to 0.1 Hz in the whole mantle model from the PREM (Dziewonski & Anderson, 1981), using displacement-stress FD schemes with an eight-point operator in space. Igel & Weber (1996) extended this high-order FDM scheme to P-SV waves. They investigated the effects of heterogeneities in the D" layer on long-period P-waves, with a dominant period of 15 s excited by an explosive source. Their computational domain was restricted to a region with angular range of 105° and maximum depth 4600 km, because of high computational requirements and perturbation of the FD stability criterion near the Earth center. Nevertheless, they successfully examined three lower mantle models, in addition to the isotropic PREM. Chaljub & Tarantola (1997) also used the displacement-stress FDM scheme for SH waves, to test sensitivity of SS precursors to the presence of topography on the 660-km discontinuity. They adopted models with topography on the discontinuity, as well as with a penetrating slab toward it at various scales, to examine its apparent depth deduced by bottomside reflection $S660S$. Igel & Gudmundsson (1997) applied a multi-domain, i.e., the FDM grid configuration with vertically-varying lateral grid spacing, to the SH algorithm. This was done to investigate frequency-dependent effects on S and SS waveforms and travel times through random upper-mantle models with pre-assumed spectral properties. Thomas et al. (2000) solved the acoustic wave equation by axisymmetric modeling in spherical coordinates, using the multi-domain including the Earth center. They used the scheme to study the influence of velocity contrasts, location, and orientation of various scatterers imposed near the CMB on precursors to $PKPdf$. Although axisymmetric modeling itself can treat an arbitrary moment-tensor point source, all works listed above concentrated on using axisymmetric sources, such as explosive and torque sources. Toyokuni & Takenaka (2006a) therefore developed a scheme to implement a non-axisymmetric source in the FDM scheme based on axisymmetric modeling in spherical coordinates, using the Fourier expansion of wavefield variables in the ϕ direction, as in Section 3.2.2. As a numerical example, they simulated which seismic phases can be related to a stagnant slab located far from a point source, with the mechanism of the 1994 deep Bolivia earthquake. Jahnke et al. (2008) extended the SH scheme of Igel & Weber (1995), for use on parallel computers with distributed memory architecture. They calculated synthetic seismograms at dominant periods down to 2.5 s for global mantle models, using high performance computers and PC networks. This scheme was used by Thorne et al. (2007) to model SH-wave propagation through cross sections of laterally varying, lower mantle models under the Cocos Plate.

4. Quasi-axisymmetric modeling

As stated in the previous section, axisymmetric modeling remains a powerful tool to obtain the 3-D seismic wavefield, because its economical calculation focuses only on a cross section, including the source and receivers. Especially in global modeling, axisymmetric modeling in spherical coordinates is the best way for iterative computation of synthetic seismograms for inverting data to image the Earth's inner structures by waveform inversion. Purely axisymmetric approximation is difficult in practice, however, because the structure along the measurement line of the seismic survey is rarely symmetric with respect to the source location. In other words, the approach cannot model seismic wave propagation on both sides of the symmetric axis through the seismic source on the measurement line. Furthermore, when one assigns lateral heterogeneity on one side of the cross section, a structural ghost appears on the opposite side because of axisymmetry, such that synthetic seismograms on the side defined as a computation target are contaminated by artificial waves reflected from the ghost that travel through the symmetry axis. In recent years, several efforts have continued to bring the synthetics of axisymmetric modeling closer to the real seismic wavefield.

4.1 Cylindrical coordinates

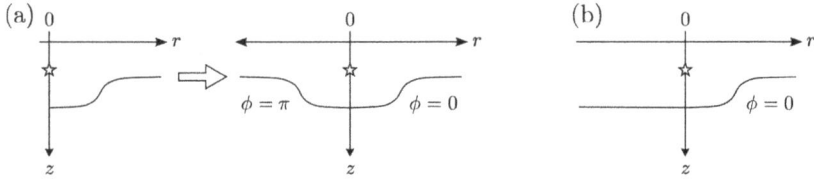

Fig. 3. Cross sections of structure models for both conventional axisymmetric modeling in cylindrical coordinates, and quasi-cylindrical approach. (a) In a conventional cylindrical domain $(0 \leq r < \infty, -\pi \leq \phi \leq \pi, -\infty < z < \infty)$, the cross section is represented by two planes located at $\phi = 0$ and $\phi = \pi$. (b) In the quasi-cylindrical domain $(-\infty < r < \infty, -\pi/2 \leq \phi \leq \pi/2, -\infty < z < \infty)$, the section is represented by a single plane at $\phi = 0$. Stars indicate seismic sources (Modified from Takenaka et al., 2003c).

In seismic exploration, treatment of an arbitrary heterogeneous structure model about the axis through a seismic source is crucial for precise comparison between synthetic and observed seismograms, since lateral heterogeneities are close to the axis, and the waveforms calculated by axisymmetric modeling are easily contaminated by artificial reflections from the structural ghost in such a situation. To overcome this difficulty, Takenaka et al. (2003a) proposed a "quasi-cylindrical approach" for seismic exploration, using a nearly linear survey with measurement lines including the source and receiver. In contrast to the conventional axisymmetric approach in cylindrical coordinates using the usual cylindrical domain $(0 \leq r < \infty, -\pi \leq \phi \leq \pi, -\infty < z < \infty)$, the quasi-cylindrical approach uses a newly defined "quasi-cylindrical domain" $(-\infty < r < \infty, -\pi/2 \leq \phi \leq \pi/2, -\infty < z < \infty)$. Although both approaches calculate the 3-D seismic wavefield on a cross section with a source and receivers, assuming a structure that is invariant in the transverse (ϕ) direction, the cross section representations are different. In a conventional cylindrical domain, we first have a rectangular half plane with infinite sides formed by movement inside an area specified by ranges $0 \leq r < \infty$ and $-\infty < z < \infty$, then rotation of this plane in the ϕ direction through 2π for coverage of the entire spatial domain. Thus, a cross section along the linear survey line of a 3-D target structure is described by two rectangular half planes located at $\phi = 0$ and $\phi = \pi$. When we assign a 2-D structure model on the $\phi = 0$ plane, the structure on the $\phi = \pi$ plane becomes symmetric, because of the calculation based on axisymmetric modeling. In this situation, the r direction becomes opposite when crossing over the symmetry axis, which makes it impossible to calculate r derivatives in the elastodynamic equation, and therefore the waves cannot travel through the symmetry axis. In fact, conventional axisymmetric modeling produces artificial reflection at the axis, because the line acts rigidly. Nevertheless, such reflection can be regarded as waves coming from the opposite side through the axis in so far as we treat them as axisymmetric wavefields. This is the reason why conventional axisymmetric modeling in cylindrical coordinates cannot treat asymmetric structures with respect to the source axis. On the other hand, the quasi-cylindrical domain first has a rectangular plane with infinite sides formed by $-\infty < r < \infty$ and $-\infty < z < \infty$, and then rotates this plane in the ϕ direction through π to cover the whole domain. In this domain, a cross section of the structure model along the survey line is described by only one plane for $\phi = 0$, and the direction of the horizontal coordinate (r) is unchanged across the vertical axis $r = 0$ (Figure 3). Hence, we can assign an arbitrary structural model on this plane, followed by reproduction of seismic wavefields propagating through the axis, calculating the r derivatives. The quasi-cylindrical approach, therefore, can calculate realistic 3-D seismic wavefields for an arbitrary cross section of a 3-D structural model with lateral heterogeneity, maintaining the efficiency of conventional axisymmetric modeling. If the structure is defined in a 2-D Cartesian domain (x, z) with shot position $x = x_0$, equations for cylindrical coordinates are solved by setting $r = x - x_0$. When we need synthetic seismograms for another shot position in the same structure, we only shift the source grid position in the numerical code, without remaking the computational structure model. Takenaka et al. (2003a) applied the method to a realistic structure model of the Nankai trough off Japan, producing possible observed seismograms by onshore-offshore seismic experimentation in the area.

4.2 Spherical coordinates

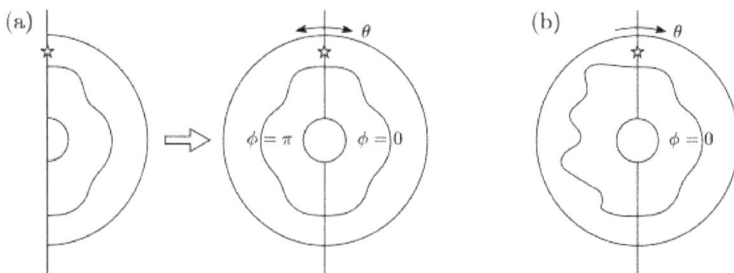

Fig. 4. Cross sections along a great circle of the Earth, for both conventional axisymmetric modeling in spherical coordinates and quasi-spherical approach.(a) For a conventional spherical domain ($0 \leq r < \infty, 0 \leq \theta \leq \pi, -\pi \leq \phi \leq \pi$), cross section is represented by two semi-circles located at $\phi = 0$ and $\phi = \pi$. (b) For the quasi-spherical domain ($0 \leq r < \infty, -\pi \leq \theta \leq \pi, -\pi/2 \leq \phi \leq \pi/2$), the section is represented by a single circle at $\phi = 0$. Stars indicate seismic sources.

Toyokuni et al. (2005) applied the quasi-cylindrical approach to spherical coordinates. The elastodynamic equation in spherical coordinates is usually solved in the conventional spherical domain ($0 \leq r < \infty, 0 \leq \theta \leq \pi, -\pi \leq \phi \leq \pi$). However, they introduced a new domain, designated a "quasi-spherical domain" ($0 \leq r < \infty, -\pi \leq \theta \leq \pi, -\pi/2 \leq \phi \leq \pi/2$), which maps the sphere in an alternate way. In a conventional spherical domain, we first have a semi-circle with infinite radius formed by rotation from $\theta = 0$ to $\theta = \pi$, then rotation of this semi-circle in the ϕ direction through 2π to cover the entire spatial domain. Thus, a cross section along a great circle of the Earth is described by two semi-circles located at $\phi = 0$ and $\phi = \pi$. When we assign a 2-D structure model on the $\phi = 0$ plane, the structure on a $\phi = \pi$ plane becomes symmetric because of axisymmetry. Similar to the cylindrical case, we cannot take θ derivatives on the source axis, which makes it impossible to propagate waves across the line. On the other hand, in the quasi-spherical domain, we first have a circle with an infinite radius formed by rotation from $\theta = -\pi$ to $\theta = \pi$, and then rotate this circle in the ϕ direction through π to cover the entire domain. In this new domain, a cross section along a great circle of the Earth is described by only one circle for $\phi = 0$, and the θ direction is unchanged across the source axis $\theta = 0$ (Figure 4). Hence, we can apply an arbitrary structure model on this plane. We further explain the concept of the quasi-spherical approach in an intuitive way. Figure 4 appears as though the quasi-spherical domain is made by gluing two hemispheres together, although such a joint does not really exist. The reality of the quasi-spherical approach is that it calculates seismic wavefields in two axisymmetric spherical structures, connecting the wavefields only on the axes with $\theta = 0$ and $\theta = \pi$, which makes us approximately treat wave propagation through an asymmetric structure. In Figure 5, we define a blue semi-circle as structure A and a red semi-circle as structure B. Wavefields propagating in structures A and B are solutions of the elastodynamic equation for axisymmetric structures A and B, respectively. However, computation of θ derivatives at the source axis connects and exchanges wavefields for both structures, which results in apparent treatment of realistic wavefields propagating in an arbitrary asymmetric structure made by combining two semi-circles. This concept is easy to understand with reference to the Riemann surface. When we consider a double-valued function, a two-sheeted Riemann surface should be defined. This surface is made by joining the two sheets crosswise along the "branch cuts", so that values can move from the upper to lower images. Although each sheet is continuous through the branch cuts and the function could have values even on the lower sheet, from the top view of the surface it appears like two sheets glued together.

We call the method of solving the elastodynamic equation in spherical coordinates in the quasi-spherical domain the "quasi-spherical approach". This approach enables modeling of seismic wave propagation in a 2-D slice of a global Earth model of arbitrary lateral heterogeneity, with similar computation time and storage as 2-D modeling, but with full consideration of 3-D wave propagation. Using a method to implement arbitrary moment tensor point sources for conventional axisymmetric modeling (Toyokuni & Takenaka,

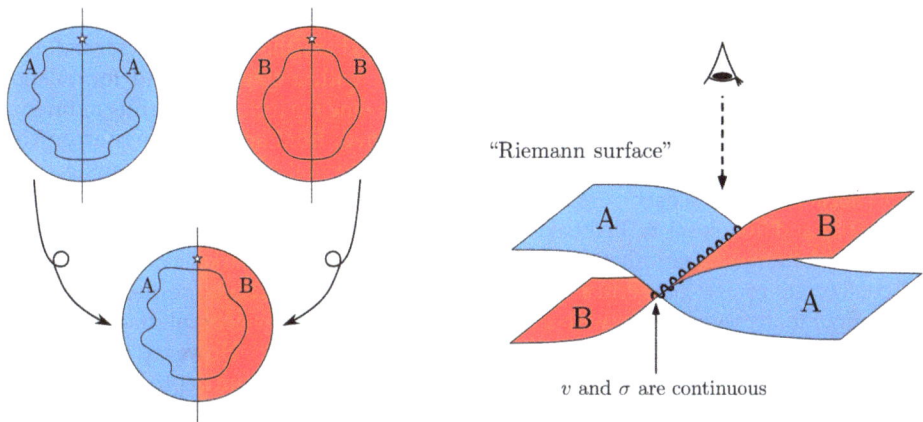

Fig. 5. Schematic drawings of the concept of quasi-spherical approach. (Left) The elastodynamic equation in spherical coordinates is solved separately for both blue and red axisymmetric structures. However, when both wavefields are connected at the source axis, the resulting wavefield appears to propagate through an asymmetric Earth model. (Right) The concept is similar to the Riemann surface. Viewing from the top, the structure looks as if it is made up by gluing two structures, although these structures are continuous across the joint.

2006a), Toyokuni & Takenaka (2006b) simulated the seismic wavefield from the 1994 deep Fiji earthquake. This was done to investigate waveform characteristics observed at Antarctica, propagated through an asymmetric structure with anomalous density and seismic wavespeeds below New Zealand. Toyokuni & Takenaka (2011) extended the quasi-spherical FDM scheme to treat attenuative structures and the Earth's center.

5. FDM implementation

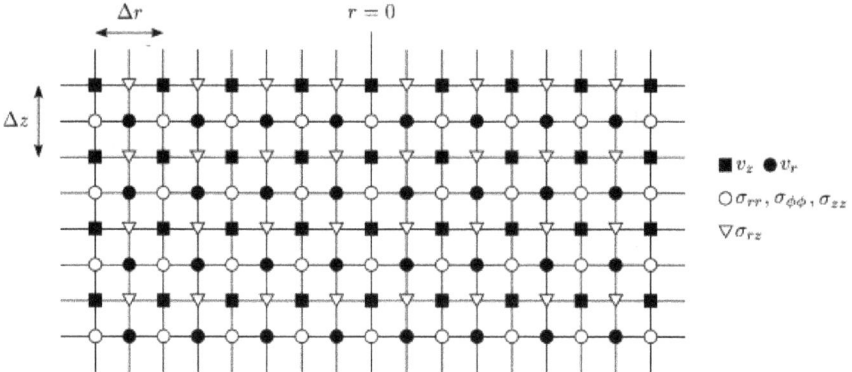

Fig. 6. Staggered-grid distribution used in quasi-cylindrical FDM computations of Takenaka et al. (2003a). The grids for the vertical component of particle velocity v_z and the normal stress components $\sigma_{rr}, \sigma_{\phi\phi}, \sigma_{zz}$ are located on the source axis $r = 0$. Δr and Δz are grid spacings in the r and z directions, respectively.

Although quasi-axisymmetric modeling can be applied to variety of numerical methods, all previous works developed numerical schemes based on the FDM with second-order accuracy in time and fourth-order accuracy in space, with a staggered-grid formulation. Takenaka et al. (2003a) constructed a staggered-grid scheme for rectangular grids of uniform spacing, for quasi-cylindrical computations of P-SV waves from an explosive source. They used a grid configuration with grid points for v_z and normal stress components σ_{rr}, $\sigma_{\phi\phi}$, and σ_{zz} located on the axis $r = 0$, as shown in Figure 6. On the other hand, Toyokuni et al. (2005) and Toyokuni & Takenaka (2006b) used a staggered-grid scheme in spherical coordinates for quasi-spherical computations using nonuniform (Pitarka, 1999) and uniform grid configurations for the vertical (r) and the angular (θ) directions, respectively. Such grid configurations were chosen with smaller vertical grid spacings near interfaces with high contrast of material parameters, e.g., the free surface and the CMB. However, the structural models in these computations were defined over an area with maximum depth 5321 km, so the computations did not treat waves propagating through the Earth center because of problems in this region. The FDM computations of seismic wavefields in spherical coordinates with uniform gridding in the θ direction fail near the Earth center, because of two reasons: (1) The extremely small lateral grid spacings near the center perturb the FDM stability criterion, and (2) the singularity of the elastodynamic equation at the center $r = 0$. To solve the first problem, Toyokuni & Takenaka (2011) applied the so-called multi-domain technique (e.g., Aoi & Fujiwara, 1999; Thomas et al., 2000; Wang & Takenaka, 2001; 2010), in which several domains consisting of FD grids with different lateral grid spacings are connected in the r

direction, with coarser lateral grids around the center. The second problem was solved using linear interpolation of wavefield variables in the r direction, giving values of particle velocity and stress at the center. Further, Toyokuni & Takenaka (2011) introduced anelastic attenuation into the quasi-spherical FDM. The anelastic behavior of Earth material can be approximated by viscoelastic models, in which the stress-strain relations contain the convolution integral in the time domain, so that time-domain computation such as the FDM had difficulty treating the integral. However, a method using so-called memory variables, which replace the convolution integral with ordinary differential equations for additional internal variables, was invented in 1980s following improvements in Cartesian coordinates (e.g., Carcione et al., 1988a;b; Emmerich & Korn, 1987; JafarGandomi & Takenaka, 2007). Toyokuni & Takenaka (2011) applied the scheme for the first time to the FDM computations in spherical coordinates. The studies with the quasi-spherical FDM used a grid configuration with grid points for v_r, σ_{rr}, $\sigma_{\theta\theta}$, $\sigma_{\phi\phi}$, and $\sigma_{\theta\phi}$ located on the axis $\theta = 0$, as shown in Figure 7.

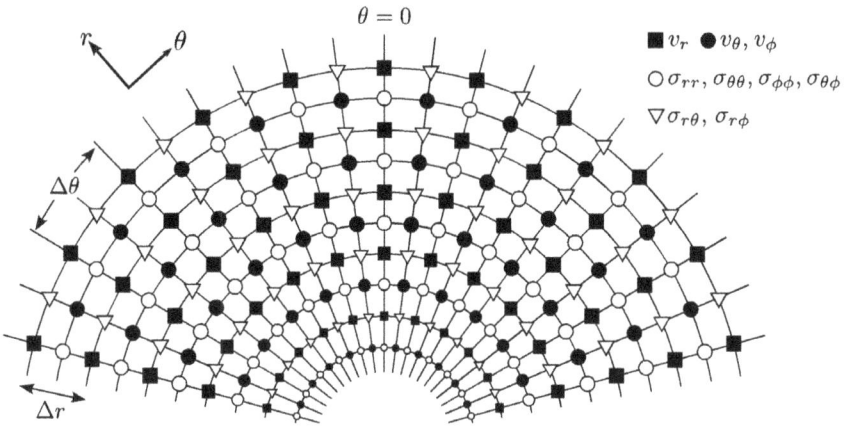

Fig. 7. Staggered-grid distribution used in quasi-spherical FDM computations of, for example, Toyokuni et al. (2005). The grids for the vertical component of particle velocity v_r, the normal stress components $\sigma_{rr}, \sigma_{\theta\theta}, \sigma_{\phi\phi}$, and the $\{\theta\phi\}$-component of the stress tensor $\sigma_{\theta\phi}$ are located on the source axis $\theta = 0$. Δr and $\Delta \theta$ are grid spacings in the r and θ directions, respectively.

As mentioned in the previous section, in the staggered-grid scheme, the derivatives of a field quantity are naturally defined halfway between the grid points where the field quantity is defined. Thus, terms on the right-hand side of the elastodynamic equation, including spatial derivatives, are consistently evaluated at the same grid position where the field quantity on the left-hand side is defined. However, this is not the case for terms that do not include spatial derivatives, so these terms have sometimes been evaluated using linear interpolation, despite a decline in accuracy of these terms to second order. To retain fourth-order computation in space at nearly all grid points except along and near the source axis and several computational boundaries, the quasi-axisymmetric schemes prepare the elastodynamic equation that has been rewritten through identities, such as

$$\frac{\sigma_{rr}}{r} = \frac{1}{r}\frac{\partial}{\partial r}(r\sigma_{rr}) - \frac{\partial \sigma_{rr}}{\partial r}, \tag{52}$$

followed by discretization (e.g., Takenaka et al., 2003a; Toyokuni et al., 2005). Finally and for example, equations for the *P-SV* waves corresponding to Eqs. (17)–(22) used by the quasi-cylindrical computations become

$$\rho \frac{\partial v_r}{\partial t} = f_r + \frac{1}{r} \frac{\partial}{\partial r} \left[r(\sigma_{rr} - \sigma_{\phi\phi}) \right] + \frac{\partial \sigma_{\phi\phi}}{\partial r} + \frac{\partial \sigma_{rz}}{\partial z}, \tag{53}$$

$$\rho \frac{\partial v_z}{\partial t} = f_z + \frac{1}{r} \frac{\partial}{\partial r} (r\sigma_{rz}) + \frac{\partial \sigma_{zz}}{\partial z}, \tag{54}$$

$$\frac{\partial \sigma_{rr}}{\partial t} = \lambda \frac{1}{r} \frac{\partial}{\partial r} (rv_r) + 2\mu \frac{\partial v_r}{\partial r} + \lambda \frac{\partial v_z}{\partial z} - \dot{M}_{rr}, \tag{55}$$

$$\frac{\partial \sigma_{\phi\phi}}{\partial t} = (\lambda + 2\mu) \frac{1}{r} \frac{\partial}{\partial r} (rv_r) - 2\mu \frac{\partial v_r}{\partial r} + \lambda \frac{\partial v_z}{\partial z} - \dot{M}_{\phi\phi}, \tag{56}$$

$$\frac{\partial \sigma_{zz}}{\partial t} = \lambda \frac{1}{r} \frac{\partial}{\partial r} (rv_r) + (\lambda + 2\mu) \frac{\partial v_z}{\partial z} - \dot{M}_{zz}, \tag{57}$$

$$\frac{\partial \sigma_{rz}}{\partial t} = \mu \left(\frac{\partial v_z}{\partial r} + \frac{\partial v_r}{\partial z} \right) - \dot{M}_{rz}. \tag{58}$$

Similarly, for the quasi-spherical approach, equations for *P-SV* waves corresponding to Eqs. (33)–(38) can be rewritten as

$$\rho \frac{\partial v_r}{\partial t} = f_r + \frac{1}{r} \frac{\partial}{\partial r} \left[r(2\sigma_{rr} - \sigma_{\theta\theta} - \sigma_{\phi\phi}) \right] - \frac{\partial}{\partial r} (\sigma_{rr} - \sigma_{\theta\theta} - \sigma_{\phi\phi}) + \frac{1}{r \sin\theta} \frac{\partial}{\partial \theta} (\sin\theta \sigma_{r\theta}), \tag{59}$$

$$\rho \frac{\partial v_\theta}{\partial t} = f_\theta + \frac{3}{r} \frac{\partial}{\partial r} (r\sigma_{r\theta}) - 2 \frac{\partial \sigma_{r\theta}}{\partial r} + \frac{1}{r \sin\theta} \frac{\partial}{\partial \theta} \left[\sin\theta(\sigma_{\theta\theta} - \sigma_{\phi\phi}) \right] + \frac{1}{r} \frac{\partial \sigma_{\phi\phi}}{\partial \theta}, \tag{60}$$

$$\frac{\partial \sigma_{rr}}{\partial t} = \frac{2\lambda}{r} \frac{\partial}{\partial r} (rv_r) + (-\lambda + 2\mu) \frac{\partial v_r}{\partial r} + \frac{\lambda}{r \sin\theta} \frac{\partial}{\partial \theta} (\sin\theta v_\theta) - \dot{M}_{rr}, \tag{61}$$

$$\frac{\partial \sigma_{\theta\theta}}{\partial t} = \frac{2(\lambda + \mu)}{r} \frac{\partial}{\partial r} (rv_r) - (\lambda + 2\mu) \frac{\partial v_r}{\partial r} + \frac{\lambda}{r \sin\theta} \frac{\partial}{\partial \theta} (\sin\theta v_\theta) + \frac{2\mu}{r} \frac{\partial v_\theta}{\partial \theta} - \dot{M}_{\theta\theta}, \tag{62}$$

$$\frac{\partial \sigma_{\phi\phi}}{\partial t} = \frac{2(\lambda + \mu)}{r} \frac{\partial}{\partial r} (rv_r) - (\lambda + 2\mu) \frac{\partial v_r}{\partial r} + \frac{\lambda + 2\mu}{r \sin\theta} \frac{\partial}{\partial \theta} (\sin\theta v_\theta) - \frac{2\mu}{r} \frac{\partial v_\theta}{\partial \theta} - \dot{M}_{\phi\phi}, \tag{63}$$

$$\frac{\partial \sigma_{r\theta}}{\partial t} = 2\mu \frac{\partial v_\theta}{\partial r} - \frac{\mu}{r} \frac{\partial}{\partial r} (rv_\theta) + \frac{\mu}{r} \frac{\partial v_r}{\partial \theta} - \dot{M}_{r\theta}, \tag{64}$$

and for *SH* waves corresponding to Eqs. (39)–(41) become

$$\rho \frac{\partial v_\phi}{\partial t} = f_\phi + \frac{3}{r} \frac{\partial}{\partial r} (r\sigma_{r\phi}) - 2 \frac{\partial \sigma_{r\phi}}{\partial r} + \frac{2}{r \sin\theta} \frac{\partial}{\partial \theta} \left(\sin\theta \sigma_{\theta\phi} \right) - \frac{1}{r} \frac{\partial \sigma_{\theta\phi}}{\partial \theta}, \tag{65}$$

$$\frac{\partial \sigma_{\theta\phi}}{\partial t} = \frac{2\mu}{r} \frac{\partial v_\phi}{\partial \theta} - \frac{\mu}{r \sin\theta} \frac{\partial}{\partial \theta} (\sin\theta v_\phi) - \dot{M}_{\theta\phi}, \tag{66}$$

$$\frac{\partial \sigma_{r\phi}}{\partial t} = 2\mu \frac{\partial v_\phi}{\partial r} - \frac{\mu}{r} \frac{\partial}{\partial r} (rv_\phi) - \dot{M}_{r\phi}. \tag{67}$$

As mentioned above, a characteristic of quasi-axisymmetric modeling is the computation of seismic wavefields even on the source axis. This permits waves to propagate across the axis, from a structure assigned on the right half, to that on the left half of the cross section, and vice versa. For direct computation of the elastodynamic equation on the source axis, we must also solve singularity problems associated with the axis. The elastodynamic equation in cylindrical coordinates has terms containing $\sigma_{\theta\theta}/r$ and v_r/r, which cannot be directly calculated on the axis $r = 0$. Takenaka et al. (2003a) exploited the formulae derived from limiting operations using the l'Hospital rule, which is also used in Toyokuni et al. (2005) and associated works. For example, formulae for evaluation of wavefield variables on the $\theta = 0$ and $\theta = \pm\pi$ axes in quasi-spherical computations are

$$a\cot\theta \to \frac{\partial a}{\partial\theta} \quad (\theta \to 0, \pm\pi), \tag{68}$$

where the variable a can be replaced by $\sigma_{r\theta}$, $\sigma_{\theta\theta}$, $\sigma_{\phi\phi}$, $\sigma_{\theta\phi}$, v_θ, or v_ϕ.

Since the FDM calculates seismic wavefields only on grid points distributed across computation space, accurate treatment of material discontinuities inside the grid cells has been a serious problem. One possible solution to this problem is the introduction of so-called effective parameters for the density and elastic moduli, calculated by volume arithmetic averaging of densities and volume harmonic averaging of elastic moduli in the cells. The effective parameters scheme enables us to place a material discontinuity at an arbitrary position inside a grid cell (e.g, Boore, 1972; Moczo et al., 2002). Toyokuni & Takenaka (2009) extended the scheme to spherical coordinates and developed a FORTRAN subroutine ACE that calculates the effective parameters analytically for an arbitrary spatial grid distribution within the four major, standard Earth models.

6. Applications

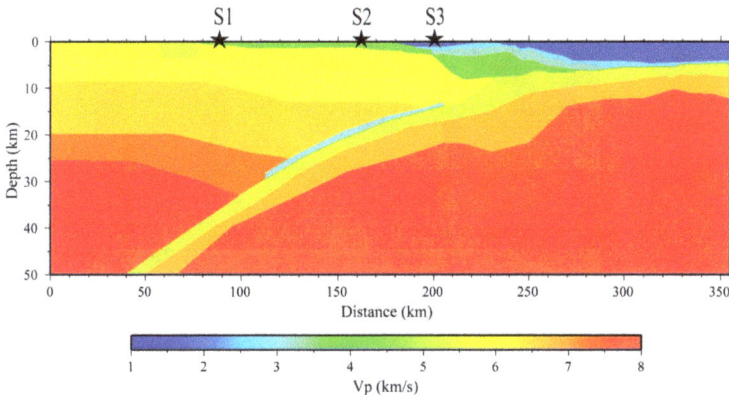

Fig. 8. *P*-wave velocity model used for simulation of onshore-offshore seismic experiment. Stars indicate shot locations (after Takenaka et al., 2003b).

This section shows examples of wavefield computation using quasi-axisymmetric modeling. First, we display an application of the quasi-cylindrical FDM to a realistic structure model

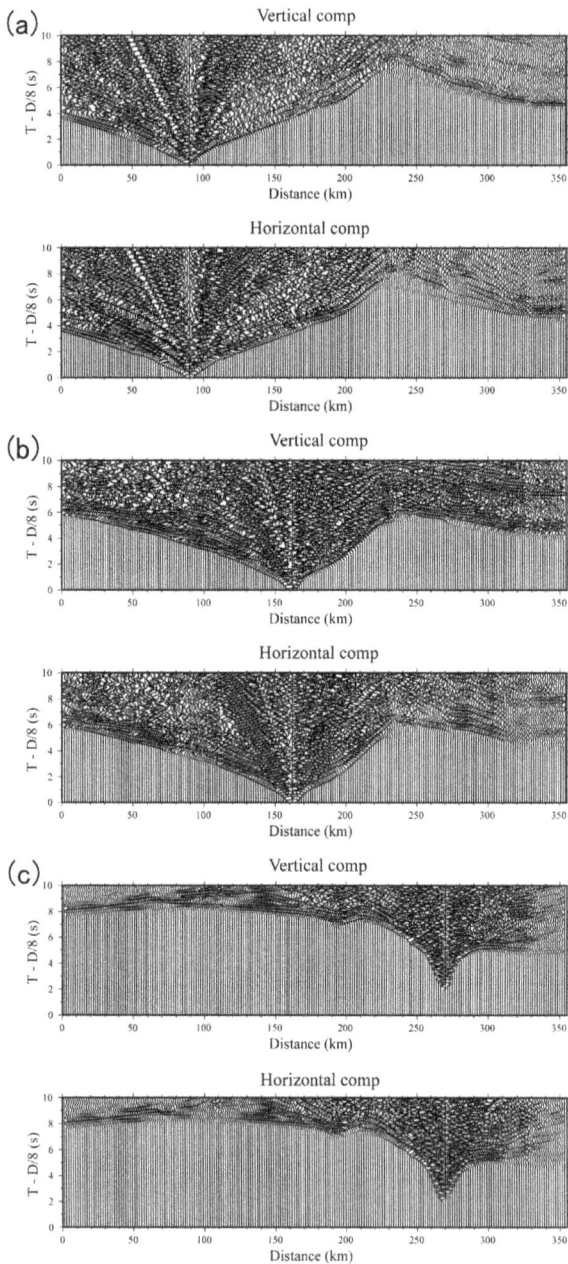

Fig. 9. Synthetic velocity seismograms calculated for three shot positions (a) S_1; (b) S_2; and (c) S_3 in Figure 8, using Nankai trough model. Both vertical and horizontal components are shown for all cases. Amplitudes of seismograms are scaled linearly with offset. A lowpass filter (< 3 Hz) has been applied (after Takenaka et al., 2003b).

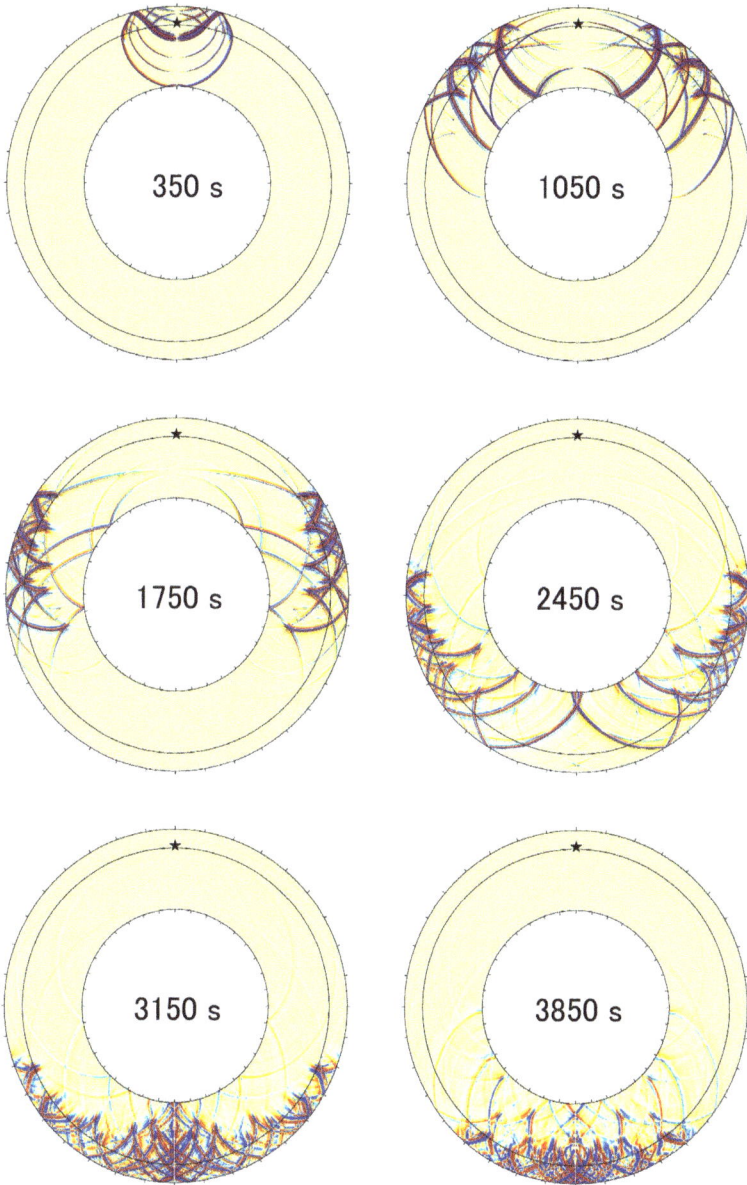

Fig. 10. *SH*-wave snapshots up to a period of 4 s at six time steps, showing generation and propagation of various seismic phases in the spherically symmetric Earth model PREM. Each frame uses the same color scale: red and blue indicate plus and minus amplitudes, respectively. Solid circles are the free surface, the 670-km discontinuity, and the CMB. Seismic source is a 600-km deep shear point source, indicated by a star.

with subducting slab (Takenaka et al., 2003b;c). Figure 8 indicates the P-wave velocity model for the Nankai trough region, Japan, where the Philippine Sea Plate is subducting toward the Eurasian Plate (Kodaira et al., 2002). Each layer in the model has a constant P-wave velocity, corresponding to the color scale. The V_P/V_S was assumed to be 1.73, except sea water where V_P and V_S were set to be 1.5 km/s and zero, respectively. Densities for the solid layers were evaluated using the formula of Darbyshire at al. (2000). The model was defined on a 7100 × 1000 grid of spacing 50 m in both horizontal and vertical directions. The time increment was 2.5×10^{-3} s. We calculated synthetic seismograms for three horizontal positions (S_1, S_2, and S_3) of 100 m deep seismic sources, as shown in Figure 8. Note that S_1 and S_3 are land and sea shots, respectively, and S_2 is located at the land-sea boundary. The source time function was a phaseless, bell-shaped pulse of width 0.5 s. Figure 9 shows synthetic seismograms for both vertical and horizontal components of particle velocity on the land surface and sea bottom for the three shots. The FDM computations simulated all possible seismic phases in the computation time window. Because of the completeness of the FDM seismograms, we can perform a direct comparison with observed seismograms, which is very important for testing and improving the structural models obtained by seismic surveys.

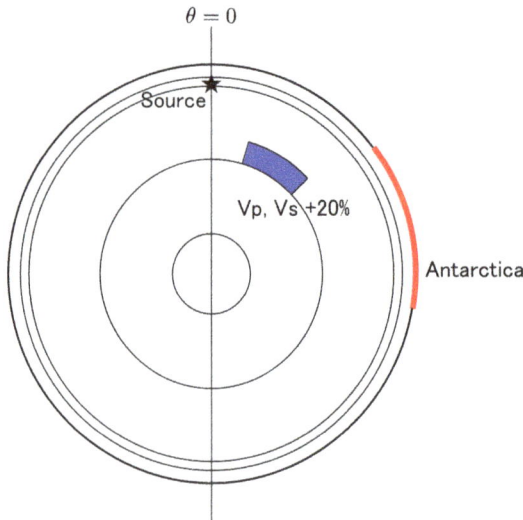

Fig. 11. Cross section of structural model for computation of synthetic seismograms. Circles indicate the free surface, 400-km and 670-km discontinuities, the CMB and ICB. Seismic source is located at depth 651 km, underneath northern Bolivia. Blue area is an anomaly of seismic wave speed, placed just above the CMB within a range of 3480km $\leq r \leq$ 4180km and $16.18° \leq \theta \leq 46.18°$, having a +20% velocity increase of P- and S-wavespeeds from the PREM basis. A red line indicates angular range of Antarctica ($52.78° \leq \theta \leq 99.58°$).

Next, we apply the quasi-spherical FDM for the spherically symmetric Earth model PREM (Dziewonski & Anderson, 1981), to show SH wave propagation for a shear source. Simulation of SH waves is useful for extracting effects related to S waves, since such sources do not exist in nature. It also enables computations for higher frequencies, through a reduction of computational requirements compared to P-SV or 3-D wave simulations. These attributes are why many authors have been working with global SH-wave computations (e.g., Chaljub &

Tarantola, 1997; Igel & Weber, 1995; Igel & Gudmundsson, 1997; Jahnke et al., 2008; Thorne et al., 2007; Wang & Takenaka, 2011; Wysession & Shore, 1994). We use a shear point source $M_{21} = -M_{12}$ at depth 600 km, with source time function described as a phaseless, bell-shaped pulse of width 4 s. The computational model is defined on a $2843(r) \times 27000(\theta)$ grid with maximum depth 2891 km (CMB), since SH waves cannot propagate into the outer core. The free surface condition has been applied to the top and bottom boundaries of the computational domain. Uniform grid spacing is used in both vertical and angular directions. The time increment is 7.0×10^{-3} s. Figure 10 shows sequential snapshots at six time steps, which allows us to confirm fundamental properties of SH-waves reverberating inside the crust and mantle.

Finally, to investigate characteristics of observable waveforms in the intra-Antarctic region, we calculate global synthetic seismograms with the quasi-spherical FDM for an asymmetric model with a simply-shaped, high seismic wavespeed anomaly superimposed on the attenuative PREM. Numerous temporal broadband seismic stations have been recently installed in this region, in association with the International Polar Year (IPY) 2007–2008 (Kanao et al., 2009). The seismic source is located in northern Bolivia at depth 651 km, the same location as the 1994 deep Bolivia earthquake. However, the mechanism is a simple dip-slip source with nonzero moment-tensor components of $M_{13} = M_{31}$. The source time function is a phaseless, bell-shaped pulse with duration 60 s. The anomaly is expressed as a region containing perturbations on P- and S-wavespeeds set at $+20.0$ % above the PREM basis, within vertical and angular ranges of $3480\text{km} \leq r \leq 4180\text{km}$ and $16.18° \leq \theta \leq 46.18°$. This is representative of a high velocity anomaly beneath southern South America, deduced by seismic tomography. We calculate wavefields on a longitudinal cross section, including the source and the anomaly, to see how the observed seismograms in Antarctica are affected by the anomaly. As shown in Figure 11, the angular range of Antarctica for this situation is $52.78° \leq \theta \leq 99.58°$. We use the ACE subroutine (Toyokuni & Takenaka, 2009) to generate the effective parameters for the PREM, with respect to the given grid distribution in the radial direction. The results are presented by synthetic seismograms along the Earth surface, and sequential snapshots of wave propagation. Figures 12 and 13 indicate respectively the angular (θ) and transverse (ϕ) components of synthetics at various angular ranges in Antarctica, calculated for (a) the PREM and (b) the model, with a high velocity anomaly. Differential seismograms in panel (c) are obtained by subtracting the PREM results from those of the asymmetric model, which clearly illustrate various phases affected by the anomaly region. Since the anomaly is located just above the CMB, we see that the core reflection such as ScS, $sScS$, and their multiple reflections, have been strongly affected by the region, as expected. These results suggest probable characteristics of observed seismograms in the intra-Antarctic region. Figure 14 shows sequential snapshots of the vertical (r) component of the seismic wavefield propagating on a cross section at every 300 s, from 300 s to 3900 s after excitation. We can see the asymmetric wavefield about the source axis, caused by the anomaly. The computation required 2.4 Gbytes of memory in a single precision calculation, with computation time of 27.3 hours on eight CPUs with IBM POWER6 architecture (4.7 GHz clock speed), for a total duration of 5000 s after excitation.

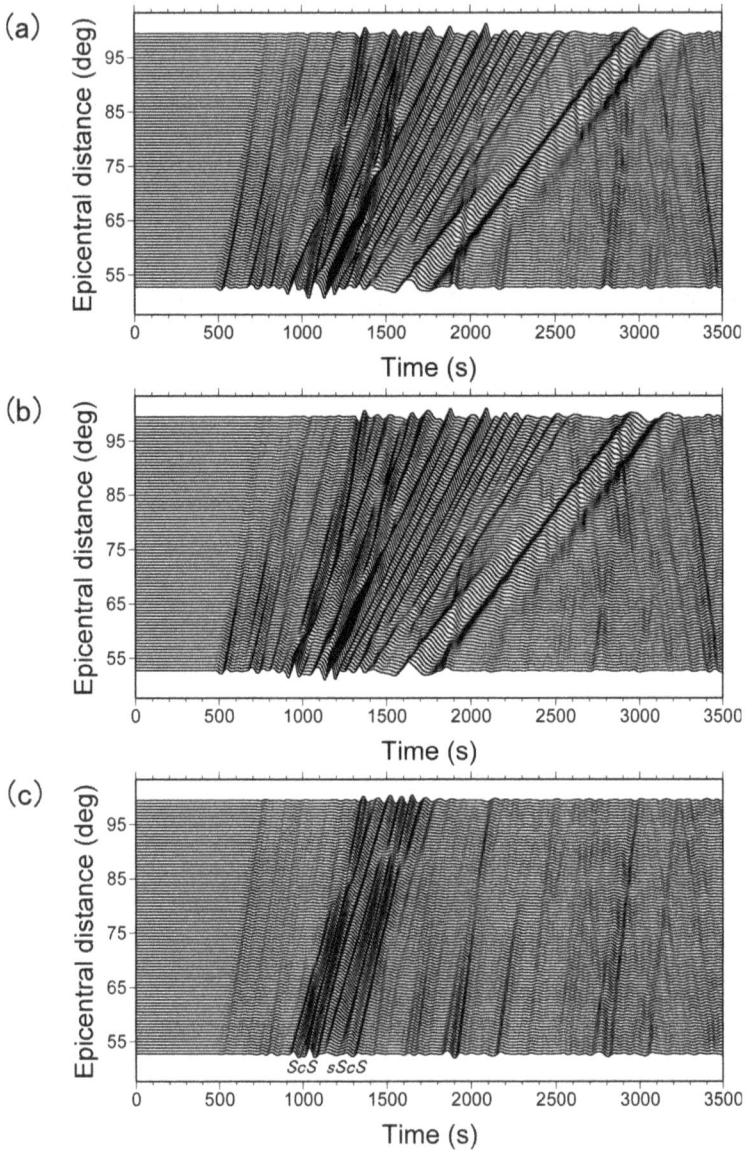

Fig. 12. Synthetic seismograms of v_θ at the Earth surface within the angular range of Antarctica, calculated for (a) the PREM, and (b) the model, including a high velocity anomaly. Differential seismograms (c) are calculated by subtracting (a) from (b), which indicate various phases affected by the anomaly. All traces were low-pass filtered with cutoff period 60 s, and are shown at the same scale.

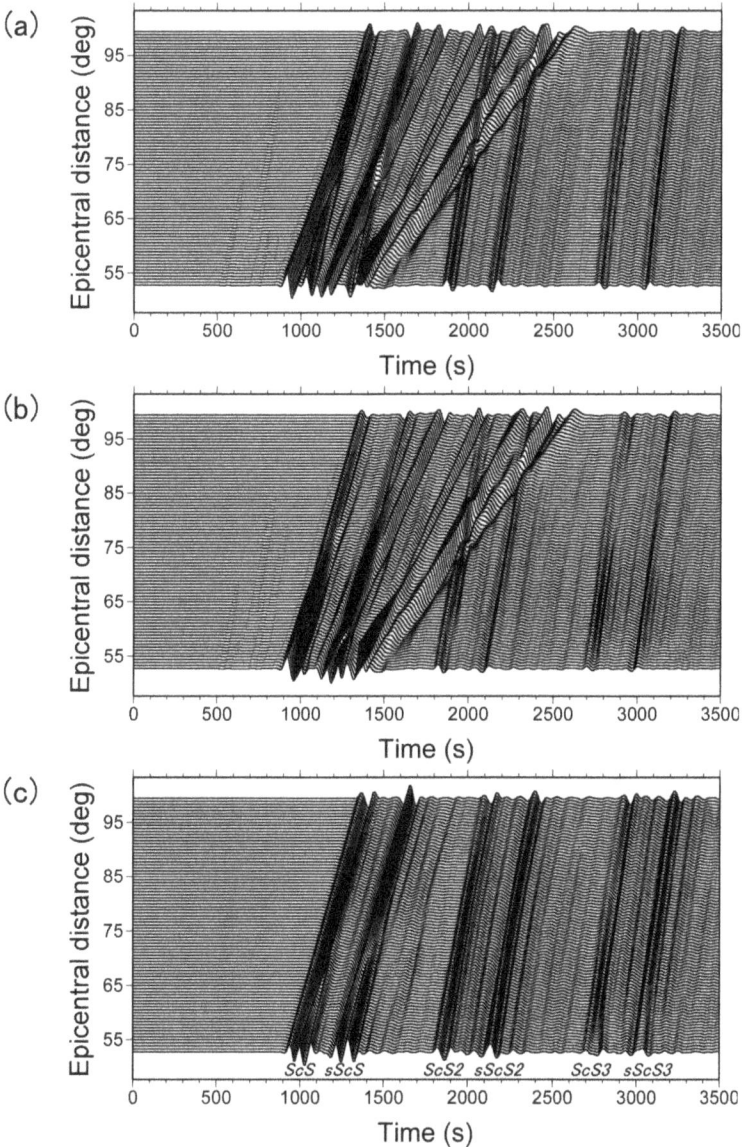

Fig. 13. Synthetic seismograms of v_ϕ at the Earth surface within the angular range of Antarctica, calculated for (a) the PREM, and (b) the model, including a high velocity anomaly. Differential seismograms (c) are calculated by subtracting (a) from (b), which indicate various phases affected by the anomaly. All traces were low-pass filtered with cutoff period 60 s, and are shown in the same scale.

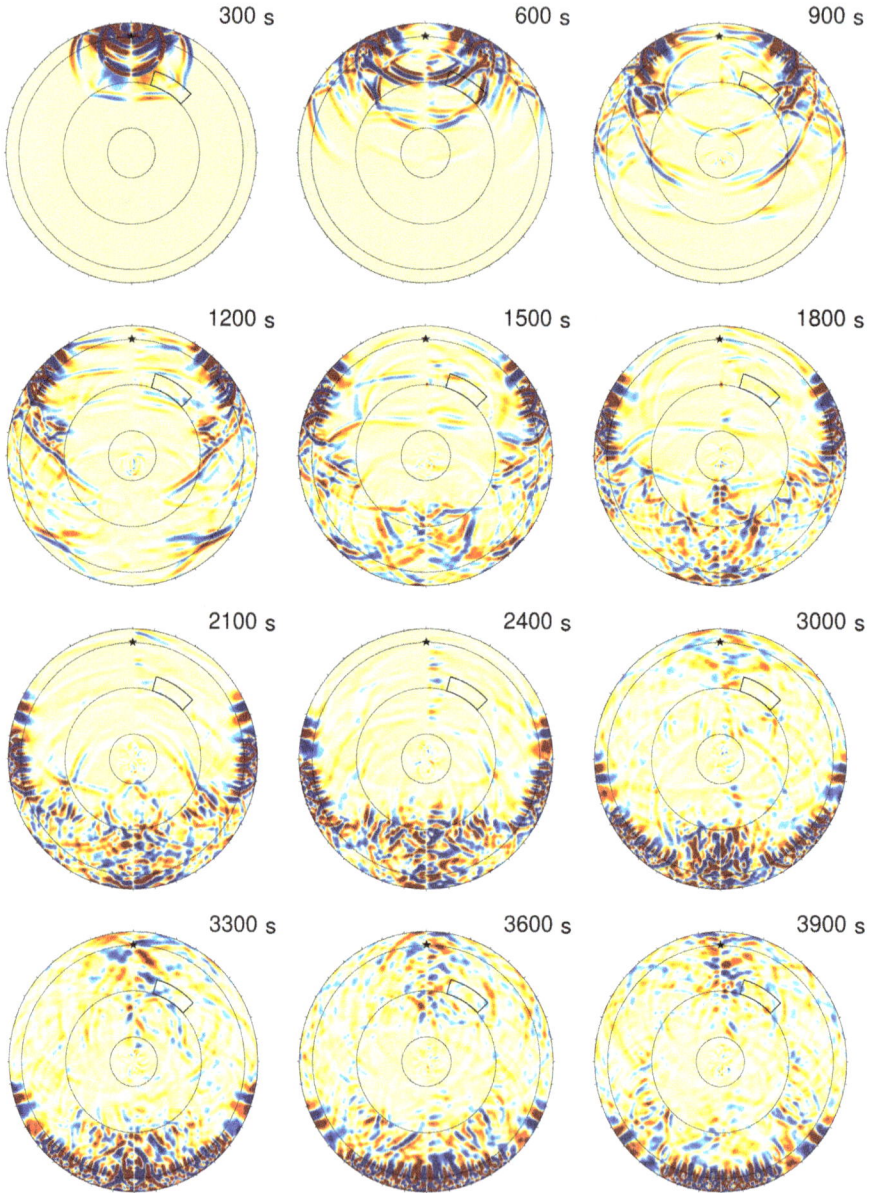

Fig. 14. Sequential v_r snapshots at 12 time steps, calculated for the model with a high velocity anomaly. Each frame uses the same color scale: red and blue indicate plus and minus amplitudes, respectively. Solid circles are the free surface, 670-km discontinuity, the CMB and ICB. Solid box represents location of the anomaly. Seismic source is a 651-km deep vertical dip-slip source, indicated by a star.

7. Conclusions

We have reviewed recent developments of numerical computation methods that have accuracy and computational efficiency for realistic seismic wavefields, using the FDM. Traditional axisymmetric modeling solves the 3-D seismic wave propagation only on a 2-D cross section of a structure model including a seismic source and receivers, under the assumption that the structure is invariant in the transverse direction about the axis through the source. However, realistic structures with asymmetry cannot be treated in principle. Quasi-axisymmetric modeling represents methods solving the seismic wave equation in newly defined quasi-cylindrical / spherical coordinates, rather than the usual cylindrical / spherical coordinates. This type of modeling retains the efficiency of axisymmetric modeling but can treat an arbitrary asymmetric structure, thereby providing a breakthrough for the problem of traditional axisymmetric strategies.

8. References

Aki, K. & Richards, P.G. (2002). *Quantitative Seismology Second Edition*, University Science Books, ISBN 0-935702-96-2, Sausalito.

Alterman, Z. & Karal, Jr., F.C. (1968). Propagation of elastic waves in layered media by finite difference methods. *Bull. Seism. Soc. Am.*, Vol. 58, No. 1, 367–398.

Alterman, Z. & Loewenthal, D. (1972). Computer generated seismograms, In: *Methods in Computational Physics, Vol. 12*, Bolt, B.A., (Ed.), 35–164, Academic Press, New York.

Alterman, Z.S., Aboudi, J. & Karal, F.C. (1970). Pulse propagation in a laterally heterogeneous solid elastic sphere. *Geophys. J. R. astr. Soc.*, Vol. 21, No. 3, 243–260.

Aoi, S. & Fujiwara, H. (1999). 3D finite-difference method using discontinuous grids. *Bull. Seism. Soc. Am.*, Vol. 89, No. 4, 918–930.

Boore, D.M. (1972). Finite difference methods for seismic wave propagation in heterogeneous materials, In: *Methods in Computational Physics, Vol. 11*, Bolt, B.A., (Ed.), 1–37, Academic Press, New York.

Carcione, J.M., Kosloff, D. & Kosloff, R. (1988a). Wave propagation simulation in a linear viscoelastic medium. *Geophys. J.*, Vol. 95, No. 3, 597–611.

Carcione, J.M., Kosloff, D. & Kosloff, R. (1988b). Visco-acoustic wave propagation simulation in the earth. *Geophysics*, Vol. 53, No. 6, 769–777.

Chaljub, E. & Tarantola, A. (1997). Sensitivity of *SS* precursors to topography on the upper-mantle 660-km discontinuity. *Geophys. Res. Lett.*, Vol. 24, No. 21, 2613–2616.

Darbyshire, F.A., White, R.S. & Priestley, K.F. (2000). Structure of the crust and uppermost mantle of Iceland from a combined seismic and gravity study, *Earth Planet. Sci. Lett.*, Vol. 181, No. 3, 409–428.

Dziewonski, A.M. & Anderson, D.L. (1981). Preliminary reference Earth model. *Phys. Earth Planet. Int.*, Vol. 25, No. 4, 297–356.

Emmerich, H. & Korn, M. (1987). Incorporation of attenuation into time-domain computations of seismic wave fields. *Geophysics*, Vol. 52, No. 9, 1252–1264.

Igel, H. & Weber, M. (1995). SH-wave propagation in the whole mantle using high-order finite differences. *Geophys. Res. Lett.*, Vol. 22, No. 6, 731–734.

Igel, H. & Weber, M. (1996). P-SV wave propagation in the Earth's mantle using finite differences : Application to heterogeneous lowermost mantle structure. *Geophys. Res. Lett.*, Vol. 23, No. 5, 415–418.

Igel, H., Djikpéssé, H. & Tarantola, A. (1996). Waveform inversion of marine reflection seismograms for P impedance and Poisson's ratio. *Geophys. J. Int.*, Vol. 124, No. 2, 363–371.

Igel, H. & Gudmundsson, O. (1997). Frequency-dependent effects on travel times and waveforms of long-period S and SS waves. *Phys. Earth Planet. Int.*, Vol. 104, No. 1 − 3, 229–246.

JafarGandomi, A. & Takenaka, H. (2007). Efficient FDTD algorithm for plane-wave simulation for vertically heterogeneous attenuative media. *Geophysics*, Vol. 72, No. 4, H43–H53, doi:10.1190/1.2732555.

Jahnke, G., Thorne, M.S., Cochard, A. & Igel, H. (2008). Global SH-wave propagation using a parallel axisymmetric spherical finite-difference scheme: application to whole mantle scattering. *Geophys. J. Int.*, Vol. 173, No. 3, 815–826, doi:10.1111/j.1365-246X.2008.03744.x.

Kanao, M., Wiens, D., Tanaka, S., Nyblade, A. & Tsuboi, S. (2009). Broadband seismic deployments in east Antarctica: IPY contribution to understanding Earth's deep interior -AGAP/GAMSEIS-, *The 16th KOPRI International Symposium on Polar Sciences, June 10-12, Incheon, Korea, Proceedings*, 90–94.

Kodaira, S., Kurashimo, E., Park, J.-O., Takahashi, N., Nakanishi, A., Miura, S., Iwasaki, T., Hirata, N., Ito, K. & Kaneda, Y. (2002). Structural factors controlling the rupture process of a megathrust earthquake at the Nankai trough seismogenic zone, *Geophys. J. Int.*, Vol. 149, No. 3, 815–835.

Moczo, P., Kristek, J., Vavryčuk, V., Archuleta, R.J. & Halada, L. (2002). 3D heterogeneous staggered-grid finite-difference modeling of seismic motion with volume harmonic and arithmetic averaging of elastic moduli and densities. *Bull. Seism. Soc. Am.*, Vol. 92, No. 8, 3042–3066.

Moczo, P., Robertsson, J.O.A. & Eisner, L. (2007). The finite-difference time-domain method for modeling of seismic wave propagation, In: *Advances in Wave Propagation in Heterogenous Earth, Advances in Geophysics, Vol. 48*, Wu, R.-S.; Maupin, V. & Dmowska, R., (Eds.), 421–516, Academic Press, ISBN 978-0-12-018850-5, New York.

Pitarka, A. (1999). 3D elastic finite-difference modeling of seismic motion using staggered grids with nonuniform spacing. *Bull. Seism. Soc. Am.*, Vol. 89, No. 1, 54–68.

Shearer, P.M. (1999). *Introduction to Seismology*, Cambridge University Press, ISBN 0-521-66953-7, Cambridge.

Stephen, R.A. (1983). A comparison of finite difference and reflectivity seismograms for marine models. *Geophys. J. R. astr. Soc.*, Vol. 72, No. 1, 39–57.

Stephen, R.A. (1988). A review of finite difference methods for seismo-acoustics problems at the seafloor. *Rev. Geophys.*, Vol. 26, No. 3, 445–458.

Takenaka, H. (1993). Computational methods for seismic wave propagation in complex subsurface structures. *Zisin (J. Seis. Soc. Japan)*, Vol. 46, 191–205 (in Japanese with English abstract).
http://www.journalarchive.jst.go.jp/english/jnltop_en.php?cdjournal=zisin1948

Takenaka, H. (1995). Modeling seismic wave propagation in complex media. *J. Phys. Earth*, Vol. 43, No. 3, 351–368.
http://www.journalarchive.jst.go.jp/english/jnltop_en.php?cdjournal=jpe1952

Takenaka, H., Furumura, T. & Fujiwara, H. (1998). Recent developments in numerical methods for ground motion simulation, In: *The Effects of Surface Geology on Seismic Motion*, Irikura, K.; Kudo, K.; Okada, H. & Sasatani, T., (Eds.), 91–101, Balkema, ISBN 90-5809-030-2, Rotterdam.

Takenaka, H., Tanaka, H., Okamoto, T. & Kennett, B.L.N. (2003a). Quasi-cylindrical 2.5D wave modeling for large-scale seismic surveys. *Geophys. Res. Lett.*, Vol. 30, No. 21, 2086, doi:10.1029/2003GL018068.

Takenaka, H., Tanaka, H., Okamoto, T. & Kennett, B.L.N. (2003b). Quasi-cylindrical approach of 2.5-D wave modelling for explosion sesimic experiments. *Abstracts of The XXIII General Assembly of the International Union of Geodesy and Geophysics (IUGG)*, SS01/03A/D-093, A.467, Sapporo, Japan, 30 June – 11 July 2003.

Takenaka, H., Tanaka, H., Okamoto, T. & Kennett, B.L.N. (2003c). Efficient 2.5-D wave modeling for explosion sesimic experiments by quasi-cylindrical approach, *Programme and Abstracts, The Seismological Society of Japan 2003, Fall Meeting*, A007, Kyoto, Japan, October 2003.

Tarantola, A. (2005). *Inverse Problem Theory and Methods for Model Parameter Estimation*, Society for Industrial and Applied Mathematics, ISBN 0-89871-572-5, Philadelphia.

Thomas, Ch., Igel, H., Weber, M. & Scherbaum, F. (2000). Acoustic simulation of P-wave propagation in a heterogeneous spherical earth: numerical method and application to precursor waves to *PKPdf. Geophys. J. Int.*, Vol. 141, No. 2, 307–320.

Thorne, M.S., Lay, T., Garnero, E.J., Jahnke, G. & Igel, H. (2007). Seismic imaging of the laterally varying D″ region beneath the Cocos Plate. *Geophys. J. Int.*, Vol. 170, No. 2, 635–648, doi:10.1111/j.1365-246X.2006.03279.x

Toyokuni, G., Takenaka, H., Wang, Y. & Kennett, B.L.N. (2005). Quasi-spherical approach for seismic wave modeling in a 2-D slice of a global Earth model with lateral heterogeneity. *Geophys. Res. Lett.*, Vol. 32, L09305, doi:10.1029/2004GL022180.

Toyokuni, G. & Takenaka, H. (2006a). FDM computation of seismic wavefield for an axisymmetric earth with a moment tensor point source. *Earth Planets Space*, Vol. 58, No. 8, e29–e32.

Toyokuni, G. & Takenaka, H. (2006b). Efficient method to model seismic wave propagation through deep inside the earth: Quasi-spherical approach. *Chikyu Monthly*, Vol. 28, No. 9, 607–611 (in Japanese).

Toyokuni, G. & Takenaka, H. (2009). ACE–A FORTRAN subroutine for analytical computation of effective grid parameters for finite-difference seismic waveform modeling with standard Earth models. *Computers & Geosciences*, Vol. 35, No. 3, 635–643, doi:10.1016/j.cageo.2008.05.005.

Toyokuni, G. & Takenaka, H. (2011). Accurate and efficient modeling of global seismic wave propagation for an attenuative Earth model including the center. *Phys. Earth Planet. Int.*, under review.

Wang, Y. & Takenaka, H. (2001). A multidomain approach of the Fourier pseudospectral method using discontinuous grid for elastic wave modeling. *Earth Planets Space*, Vol. 53, No. 3, 149–158.

Wang, Y. & Takenaka, H. (2010). A scheme to treat the singularity in global seismic wavefield simulation using pseudospectral method with staggered grids. *Earthquake Science*, Vol. 23, No. 2, 121–127, doi:10.1007/s11589-010-0001-x.

Wang, Y. & Takenaka, H. (2011). SH-wavefield simulation for laterally heterogeneous whole-Earth model using the pseudospectral method. *Sci China Earth Sci*, doi:10.1007/s11430-011-4244-8.

Wysession, M.E. & Shore, P.J. (1994). Visualization of whole mantle propagation of seismic shear energy using normal mode summation. *Pure Appl. Geophys.*, Vol. 142, No. 2, 295–310.

6

The Latest Mathematical
Models of Earthquake Ground Motion

Snezana Gjorgji Stamatovska

*'Ss. Cyril and Methodius' University-Institute of Earthquake Engineering and
Engineering Seismology (UKIM-IZIIS), Skopje
Republic of Macedonia*

1. Introduction

Strong motion instrument networks have enabled creation of a large number of databanks ranging from small to regional and world ones. This data is of a great importance for the investigations aimed at prediction of strong earthquake ground motion parameters by application of empirical mathematical models fitted to the databanks. These mathematical models are referred to as ground motion models or attenuation laws. They define the relationships between ground motion parameters and factors that affect the amplitudes of ground motion as are the released energy, the regional characteristics, the local soil characteristics, the type of fault, the radiation pattern, etc.

Ground motion models are defined by application of the regression analysis method. Regression coefficients and standard deviation are obtained as a result of the regression analysis. Standard deviation is the measure for the dispersion of the data around the computed medium or median value for which a distribution function defined by the probability density function is assumed.

Regression coefficients and standard deviation are the input parameters for the probabilistic seismic hazard analyses (Cornell 1968). Despite the evident results of the progress made in the use of the seismic hazard methodology, there are still uncertainties by which the hazard curves are computed. The mathematical models of ground motion have a big influence upon the results obtained from the seismic hazard analyses that are applied in practice. This justifies the efforts made by a large number of researchers worldwide toward development of mathematical models that will best fit the available databanks obtained from occurred strong earthquakes. As a result, there is a big number of different mathematical models of ground motion.

The presented investigations refer to the latest mathematical models of ground motion during earthquakes. These are: the azimuth dependent mathematical model and the mathematical model based on radius vectors.

2. Azimuth dependent mathematical model

Based on data from records on earthquakes that occurred from the Vranchea focus in Romania, the author has developed an azimuth dependent mathematical model of ground motion. It includes the focal mechanism, the size of the seismic field represented by an ellipse with a shape dependent on the relative relationship of its semi-axes and with a longitudinal axis in the direction of the projection of the fault plain upon the surface as well as the position of the instrument location (Stamatovska, 1996). Presented for this mathematical model are the idea used in defining the mathematical equation for a single earthquake, the general procedure of definition of the azimuth dependent mathematical model for any selected azimuth and its application in the seismic hazard analyses. The detailed description of the procedure of its development is aimed at its easier understanding and use by other researchers. This also contributes to easier understanding of the procedure by which the author has developed a new mathematical model based on radius vectors.

2.1 Mathematical equation

The starting point is a general empirical ground motion model in which ground motion parameter- Y depends on magnitude- M, distance- R and local soil conditions- S. It is given in Equation 1

$$\ln Y = b + b_M M + b_R \ln(R_h + C) + b_S S + \sigma_{\ln Y} P \tag{1}$$

where,

Y -peak ground acceleration- PGA, or peak ground velocity- PGV or peak ground displacement- PGD; parameter of dynamic response of a linear or nonlinear model of a single degree of freedom system– SDOF, as well as Fourier Amplitude Spectrum- FS
M -magnitude
R_h -hypocentral distance in km
S -parameter that includes the effect of local soil conditions and has values, for example, 0 for rock, 1 for alluvium, 2 for deep alluvium
C -constant by which is defined the shape of the attenuation in the epicentral zone expressed in km
b, b_M, b_R, b_S -regression coefficients
$\sigma_{\ln Y}$ -standard deviation
P -binary variable, which has the value of 0 and 1 for median and median plus one standard deviation, respectively.

The model is based on the following theoretical assumptions: term $e^{b_M M}$ involves the relationship between energy and magnitude; coefficient b_R has a negative value and accounts for the spherical spreading of the seismic wave energy, while term $b_S S$ includes the effect of local soil conditions.

The ground motion model given in Equation 1 is simplified by use of records of occurred strong earthquakes obtained on rock soil type or referent soil with $V_S \geq 700 m / s$, by which the parameter defining the effect of the local soil conditions is omitted. With this, the parameters of ground motion under strong earthquake effect are only a function of distance and magnitude.

2.2 Mathematical equation for a single earthquake

The solution of the mathematical equation of a single earthquake came from the analyses of the records of an earthquake obtained at two locations, i.e., by instruments situated at equal epicentral distance from the earthquake epicenter. For each of the two locations, the epicentral distance and the focal depth are equal. The difference is in their position in respect to the projection of the fault upon the surface, i.e., the angle between the direction of the fault plane and the direction toward the instrument location. Hence, the differences in the recorded amplitudes at these two locations result from the position of the location in respect to the projection of the fault plane and the characteristics of the region in the direction of that location. If the recorded amplitudes, for example, amplitudes of *PGA* with equal value are connected by an isoseismal, then it is clear that, although the two considered locations are at equal epicentral distances, due to the different recorded amplitudes, the two locations will not lie on the same isoseismal. This means that the characteristics of the focus and the region in the direction toward the location perform faster or slower attenuation of the energy of the seismic waves by which they define the form of the isoseismals of equal *PGA*. Since the earthquake depth is the same for both locations, it is clear that the regional characteristics perform correction through the epicentral distances wherefore the form of the seismic field on the surface is not a circle. Therefore, the model of ground motion for each individual earthquake is a function of corrected epicentral distance or epicentral distance divided by a single function, the so called ρ, whose value depends on the form of the isoseismal of equal amplitudes of *PGA* and the angle between the fault plane and the direction of the location, i.e., the radiation pattern.

During mathematical modelling, particular importance is given to idealization of the form of the seismic field on the surface. For the azimuth dependent mathematical model developed by the author, it is assumed that this form may range from a circle to any shape of an ellipse with a longitudinal axis in the direction of the projection of the fault plane upon the surface (Figure 1). The shape of the ellipse is defined by the ratio of the semi-axes $a:b$, whereas the position of any two points M and M_i lying on it, is defined by radius vectors $\vec{\rho}$ and $\vec{\rho_i}$, whose moduli are equal to ρ and ρ_i.

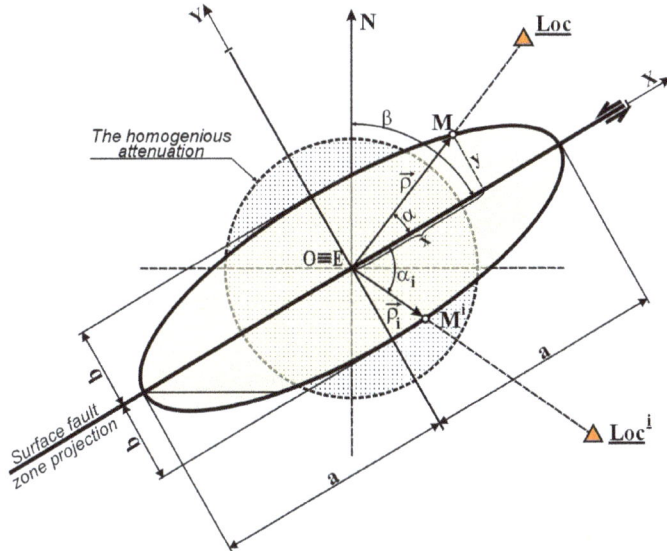

Fig. 1. Function ρ

$$\rho = |\vec{\rho}| = \sqrt{x^2 + y^2} \tag{2}$$

$$tg\alpha = y/x \tag{3}$$

$$x^2 / a^2 + y^2 / b^2 = 1 \tag{4}$$

$$\rho = \sqrt{\frac{1 + tg^2\alpha}{a^{-2} + tg^2\alpha}} \tag{5}$$

$$\beta_i^L \pm \alpha_i = \beta \tag{6}$$

So the mathematical equation for the PGA of an earthquake acquires a form dependent on the corrected epicentral distance $\dfrac{R_e}{\rho}$:

$$PGA = b_0 (\frac{R_e}{\rho})^{b_1} e^{\sigma_{\ln PGA}} \tag{7}$$

where,

b_0 and b_1 are regression coefficients

$\dfrac{R_e}{\rho}$ - corrected epicentral distance, and

$\sigma_{\ln PGA}$ - standard deviation

2.3 Regression analysis method

The exploration through analysis of a large number of published ground motion models (Joyner & Boore, 1981; 1988; Boore & Joyner 1982; Ambraseys & Bommer, 1992; Ambraseys et al., 1996; Boore et al., 1993; Sabetta & Pugliese, 1987, 1996; Idriss, 1991; Sadigh, 1993; Sadigh at al., 1993; Campbell, 1981) has pointed out the primary importance of the empirical model developed by application of the double regression method. This method (Joyner & Boore, 1981) involves the mode in which earthquakes occur in nature, one at a time, which is encompassed by the first step. Their connection is the objective of the second step. Accordingly, the regression analysis method is carried out in two steps as follows:

First step: Definition of ground motion models for each occurred earthquake taken separately, and,

Second step: Connection of all occurred earthquakes, i.e., different magnitudes and focal depths.

2.3.1 First step of regression analysis

The first step of the regression analysis involves definition of regression coefficients b_0 and b_1 , and standard deviation $\sigma_{\ln PGA}$. To carry out the first step, it is necessary to perform parametric analysis in which the value of the parameters affecting function ρ will vary.

These are: the azimuth of the projection of the fault plane upon the surface β and the ratio of the semi-axes of the ellipse of the seismic field $a : b$.

The procedure itself is reduced to the following:

1. An initial value for the azimuth of the projection of the fault plane on the surface- β (Figure 2a) is selected;
2. The $a : b$ ratio is defined for value of $b = 1.$, by which the relative ratio of the semi-axes of the seismic field is $a : 1 = a$ (Figure 2a)
3. An initial value of the relative ratio $a = 1$. (Figure 2a) is defined;
4. The values of function ρ for all instrument locations and the values of the corrected epicentral distances $\dfrac{R_e}{\rho}$ are computed;
5. Linear regression is carried out for dependent random variable PGA and independent random variable a $\dfrac{R_e}{\rho}$. Then, the regression coefficients b_0 and b_1 and the standard deviation $\sigma_{\ln PGA}$ from the first step is computed.
6. The value of the relative ratio a is changed for an increase of Δa and the procedure from item 4. (Figure 2b) is repeated;
7. A new value of azimuth β with an increase $\Delta \beta$ is selected and the procedure pursuant to 1 (Figure 2c) is repeated.

A number of solutions is obtained. Out of these, the one for which the standard deviation has the least value is selected. With this, the ground motion model due to an earthquake is defined. In the same way, the ground motion models are defined for all occurred earthquakes originating from a single focus.

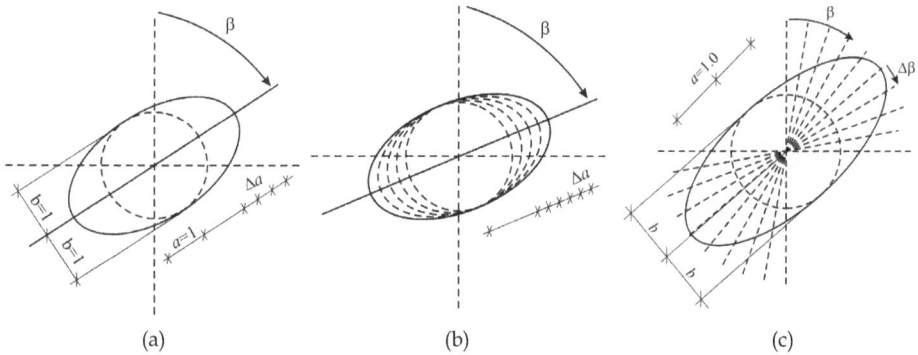

(a) (b) (c)

Fig. 2. Procedure referring to the first step of the regression analysis

2.3.2 Second step of regression analysis

In the second step of the regression analysis, all the occurred earthquakes originating from the same focus are connected and regression coefficients b, b_R and b_M and the standard deviation $\sigma_{\ln PGA}$ are computed. The data used in the second step of the regression analysis are: earthquake magnitude- M and hypocentral distance- R_h as independent variables and

PGA as dependent variable (Equation 1). Hypocentral distance is computed according to the following formula:

$$R_h{}^2 = \left(\frac{R_e}{\rho}\right)^2 + h^2 \tag{8}$$

while value $\dfrac{R_e}{\rho}$ is computed separately for each occurred earthquake and for all the instrument locations on which the records from that earthquake are obtained.

A key issue in the second step of the regression analysis is the connection of all the earthquakes (Figure 3) and definition of the ground motion model given by Equation 1.

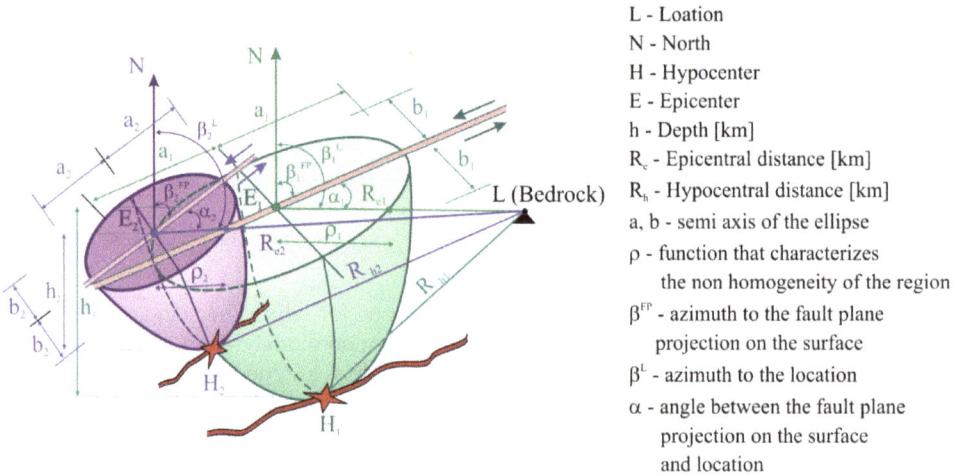

L - Loation
N - North
H - Hypocenter
E - Epicenter
h - Depth [km]
R_e - Epicentral distance [km]
R_h - Hypocentral distance [km]
a, b - semi axis of the ellipse
ρ - function that characterizes
     the non homogeneity of the region
β^{FP} - azimuth to the fault plane
     projection on the surface
β^L - azimuth to the location
α - angle between the fault plane
     projection on the surface
     and location

Fig. 3. Connection of earthquakes – second step of regression analysis

The solution is possible only if a ground motion model is defined for a direction toward a location, in which case it is necessary to perform normalization of value $\dfrac{R_e}{\rho}$. The normalization is performed separately for each occurred earthquake with value ρ_i defined for the direction toward the selected location by use of the ground motion model computed in the first step of the regression analysis performed for that earthquake (Figure 4). All the normalized values are used in the second step of the regression analysis.

It is possible to compute ground motion models for different directions (azimuths according to locations) in which case it is necessary to perform normalization of $\dfrac{R_e}{\rho}$ for each selected direction, separately.

The value of constant C is defined by its variation (for example, from 0 km to 200 km, by a step of 1, or 2, or more km) and execution of the second step of the regression analysis for each of its values. A number of solutions is obtained out of which the one for which the standard deviation in the second step of the regression analysis is minimal, is selected.

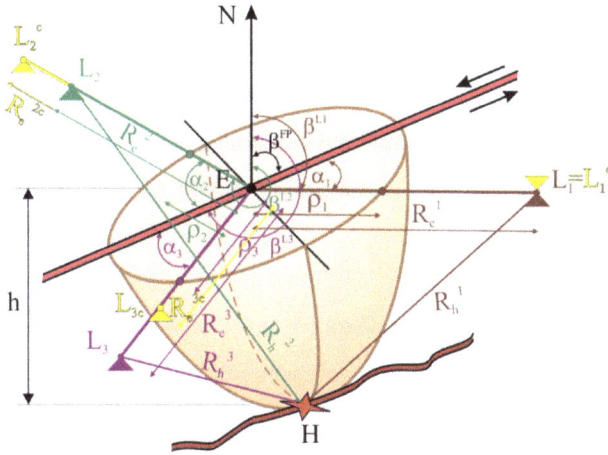

Normalization according to the azimuth's location

$$\text{Location } L_1: \quad \frac{R_e^1}{\rho_1}\rho_1; \quad \frac{R_e^2}{\rho_2}\rho_1; \quad \frac{R_e^3}{\rho_3}\rho_1;$$

$$\text{Location } L_2: \quad \frac{R_e^1}{\rho_1}\rho_2; \quad \frac{R_e^2}{\rho_2}\rho_2; \quad \frac{R_e^3}{\rho_3}\rho_2;$$

$$\text{Location } L_3: \quad \frac{R_e^1}{\rho_1}\rho_3; \quad \frac{R_e^2}{\rho_2}\rho_3; \quad \frac{R_e^3}{\rho_3}\rho_3;$$

Fig. 4. Normalization over selected azimuth

2.4 Advantages

The advantages of the azimuth dependent ground motion model are:

- Definition of separate ground motion models for different directions
- The mathematical form of the azimuth dependent ground motion model (Equation 1) is applicable in seismic hazard methodology;
- Application in definition of ground motion models for spectral characteristics of ground motion expressed by response spectra and the Fourier Amplitude Spectrum.
  In this case, the results from the first step of the regression analysis (Stamatovska, 2008) (β, a, b_0, b_1 and $\sigma_{\ln PGA}$ from the first step) defined for PGA are used, and it is only in the second step that the PGA value is replaced by the value of the spectral characteristic of the earthquake, as for example, the spectrum of the linear model of SDOF (absolute acceleration– SA, relative velocity– SV, relative displacement SD), the Fourier Amplitude Spectrum– FS and the spectrum of the nonlinear model of SDOF (acceleration spectrum, displacement spectrum, ductility factor and alike);
- In case of a new earthquake, only the ground motion model for the new earthquake is defined in the first step. All the previous results from the first step obtained for the preceding earthquakes are used (preceding earthquakes + the new earthquake) and the second step of the regression analysis is carried out;

- Improvement of the azimuth dependent ground motion model is possible through idealization of the seismic field upon the surface via including irregular forms defined by radius vectors.

2.5 Application in probabilistic seismic hazard analyses - PSHA

The application of the azimuth dependent ground motion model in PSHA is based on the following two steps:

- Definition of azimuth dependent ground motion models for different azimuth directions;
- Definition of sub-sources in a seismic source.

To define the ground motion model for any azimuth direction of a seismic source, it is necessary to pre-define ground motion models for each occurred earthquake from that source by application of the first step of the regression analysis of the azimuth-dependent empirical mathematical model (Stamatovska, 1996, 2002, 2006, 2008; Stamatovska & Petrovski, 1996, 1997) presented by Equation 1.

Important parameters from the first step of the regression analysis for each occurred earthquake are: the azimuth of the projection of the fault upon the surface- β and the value of the relative ratio a. By using these parameters, the value of function ρ_i can be computed for each selected direction i defined by azimuth- β_i. In doing so, angle- α_i, as an angle between the azimuth of the projection of the fault plane upon the surface- β and the selected azimuth- β_i is defined by using Equation 6.

With the value of function ρ_i normalization for the selected azimuth is performed. Each corrected epicentral distance $\dfrac{R_e}{\rho}$ in which ρ is the value computed for the azimuth of the instrument location, is multiplied by ρ_i.

This procedure is iterated separately for each occurred earthquake originating from the investigated seismic focus (for example, if four strong earthquakes took place, it is iterated 4 times). All the normalized values are used in the second step of the regression analysis and the regression coefficients b, b_M and b_R as well as the standard deviation $\sigma_{\ln Y}$ are computed. With this, the ground motion model for that azimuth is defined. By selection of a new azimuth (new location) and iteration of the entire procedure described in this part, ground motion models for different azimuth directions are obtained. This step is schematically presented in Figure 5.

The computed ground motion models can directly be applied in analyses of seismic hazard for all the software packages in which the ground motion model is assigned or reduced to the mathematical form presented in Equation 1 in the case of a point seismic source. In all other cases of seismic sources, it is necessary to model sub-sources.

2.5.1 Definition of sub-sources in seismic source

In the methods for computation of seismic hazard (Cornell, 1968), the seismic source is modelled as point, line or area source. Each point of the seismic source, defined by

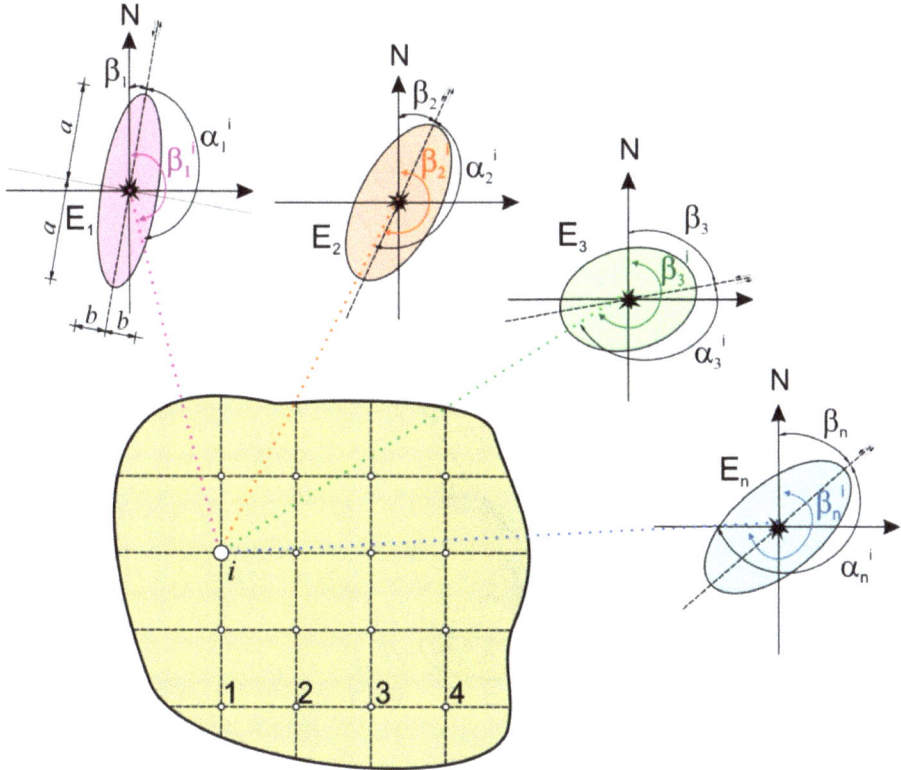

Fig. 5. Application of the results obtained in the first step of the regression analysis for definition of the model of ground motion at a selected location

coordinates (x, y) where x is east longitude, while y – north latitude, is a potential epicenter of a future earthquake from that focus. The possibility that the model of the seismic source be represented by a point (in the case of a point seismic source), or a number of points (in the case of a linear or an area model of seismic source) facilitates the procedure to be applied if a software package is developed for the purpose of avoiding a large number of computations. Then, the area of the seismic source is modelled by sub-sources with very small areas $\Delta S = \Delta x \Delta y$, to be harmonized with the computed ground motion models for different azimuths (Figure 6).

The above means that the azimuths of the end points of the small seismic sub-source computed in respect to a single point in region-i for which the seismic hazard is computed should tend to a single azimuth value. This is possible in all cases where the seismic hazard is computed for a point in the region that is sufficiently distant to reach an azimuth (Figure 6, point 1). However, particular attention should be paid to a point of the region that is very close to the seismic source (Figure 6, point 2) when the azimuth of the end points of the small seismic sub-source do not tend to an azimuth but there is a considerable difference among them. It is further necessary to reduce the area of the seismic sub-source $\Delta S_1 \langle \Delta S$, or $\Delta \beta_1 \langle \Delta \beta$ (Figure 6).

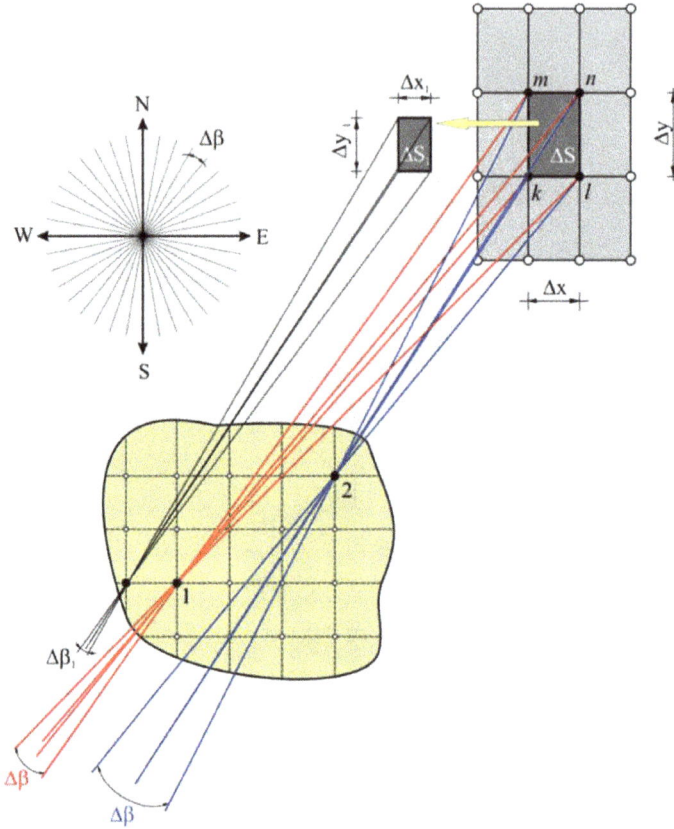

Fig. 6. Effect of modelling of seismic source and epicentral distance upon the extent of deviation from an azimuth

3. Mathematical model based on radius vectors

The mathematical model based on radius vectors represents an advanced azimuth dependent mathematical model. It is developed as an azimuth dependent model of a random shape of a seismic field defined by radius vectors in different azimuth directions.

3.1 Theoretical background

The ground motion model defined on the basis of radius vectors has the same mathematical form as the azimuth dependent model, or,

$$\ln Y = b + b_M M + b_R \ln(R_h + C) + \varepsilon \tag{9}$$

$$R_h{}^2 = (R_e^c)^2 + h^2 \quad R_e^c = R_e(\frac{\rho_L}{\rho_i}) \quad \frac{\rho_L}{\rho_i} = \frac{\left|\overrightarrow{\rho_L}\right|}{\left|\overrightarrow{\rho_i}\right|} \tag{10}$$

where: Y is the ground motion parameter (peak acceleration, velocity, displacement, horizontal vector, spectral amplitude, etc.), ρ_i is the modulus of the radius vector in respect to any instrument location, whereas ρ_L is the modulus of the radius vector in respect to the location/or the direction for which the ground motion model is defined. The effect of the local soil conditions is not included in this mathematical model due to usage of records obtained on one type of local soil conditions (for example, rock with $V_s \geq 700m / s$).

3.2 Method

The method for definition of this model consists of two parts. The first part involves preparation of data to be used in the regression analysis. In this part, the shape of the recorded seismic field defined by radius vectors (Fig. 7) is established. Each radius vector begins at the earthquake epicentre and runs in the direction from the epicentre to the instrument location. Its modulus is equal to the absolute value of peak acceleration /or velocity/ or displacement/ of ground or vector defined for horizontal direction under the earthquake effect. Applying the normalized seismic field for a selected azimuth/ or direction toward a selected location, the value of the relative relationship of $\dfrac{\rho_L}{\rho_i}$ or $\dfrac{\rho_i}{\rho_L}$ moduli (Fig. 8) is defined. This relationship is a dimensionless number and enables obtaining the regional characteristics in different directions. It is used to correct the epicentral distances. This is carried out separately for each earthquake that has occurred from a single seismic focus.

In the second part, the multi linear regression analysis method is used. The data for the regression analysis are: PGA - dependent variable, M and R_h - independent variables. Each regression analysis results in regression coefficients b, b_M, b_R and standard deviation - $\sigma_{\ln Y}$. The number of regression analyses depends on the number of variations of constant C (for example, 27 analyses with variable C ranging from 0 to 130 km, with a step of 5 km). From the multitude solutions, the one for which the standard deviation is minimal is selected.

The second part is equal to the second step of the regression analysis applied in the azimuth dependent model. In this way, the simplest mathematical model for prediction of characteristics of future earthquakes from a single seismic focus is obtained. According to the author, this model is the closest to the physical model since it includes a realistically occurred seismic field recorded by strong motion instruments.

The described procedure is based on the idea that the amplitudes of ground motion obtained for different epicentral distances and different azimuths result from the effect of the amount of the energy released by the earthquake, the focal mechanism and the regional characteristics at different azimuths from the earthquake hypocenter.

3.3 Method verification

The method verification has been performed on the basis of the created data bank of available three-component records of strong earthquakes that occurred on March 4, 1977 (epicenter 45.8N and 26.8E, M=7.2, h=109 km), August 30, 1986 (epicenter 45.52N and 26.49E, M=7.0, h=131 km), May 30, 1990 (epicenter 45.872N and 26.885E, M=6.7, h=99.1 km) and

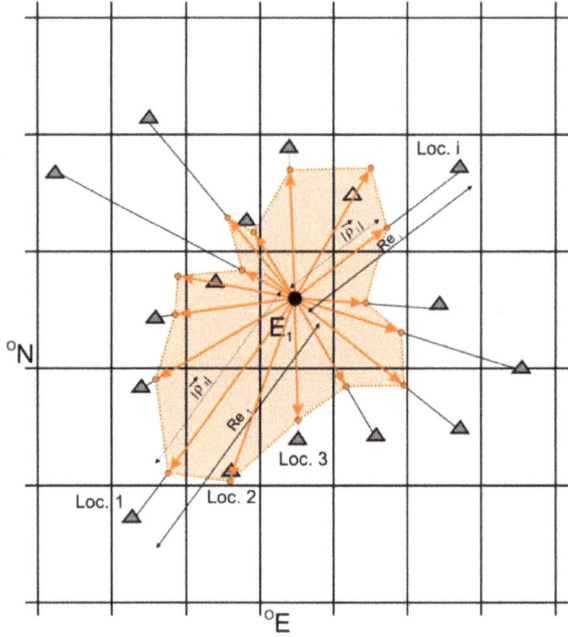

Fig. 7. Recorded seismic field of PGA at rock

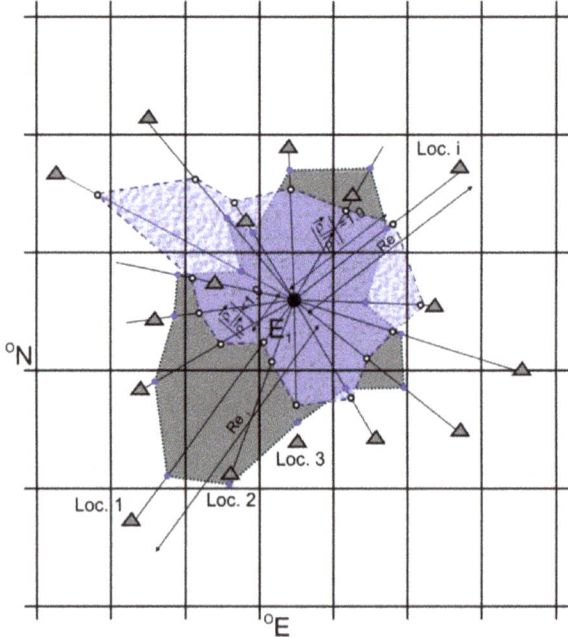

Fig. 8. Normalized seismic field for the azimuth toward location i

May 31, 1990 (epicenter 45.852N, 26.882E, M=6.1, h=89.1 km). The data bank includes data from records of occurred deep earthquakes at the Vranchea focus (Romania) obtained by the instruments of the Romanian, Bulgarian and Former Yugoslav strong motion networks.

The isoseismals of the recorded PGA seismic field (in cm/s^2 for $V_S \geq 700m/s$) referring to the earthquakes that occurred at the Vranchea focus are given in figures 9, 10 and 11.

Two separate investigations have been performed. In the first one, the ground motion parameter are the peak ground accelerations from the two horizontal components, while in the second investigation, the ground motion parameter is the higher value of the two horizontal components of the peak ground acceleration. Mathematical models of ground motion have been defined for seven azimuths toward the following instrument locations: BUC (Bucharest), CFR (Carcaliu), CVD (Chernavoda), IASI (Iasi), VLM (Valeni de Munte) and VRI (Vrincioaia). For all these, the regression coefficients and standard deviations are given (Tables 1 and 2). The results shown in Table 1 refer to two horizontal components, whereas those in Table 2 refer to the larger component of the two horizontal components. The March 4, 1977 earthquake is included only for an azimuth toward the INC (INCERC-Bucharest) location.

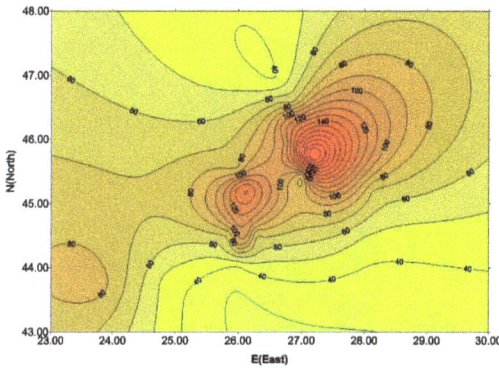

Fig. 9. The earthquake of 30th August 1986 – recorded PGA seismic field

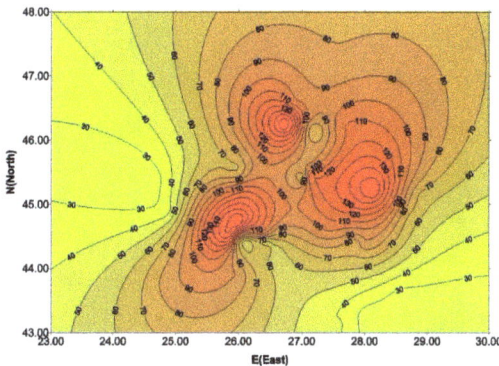

Fig. 10. The earthquake of 30th May 1990 – recorded PGA seismic field

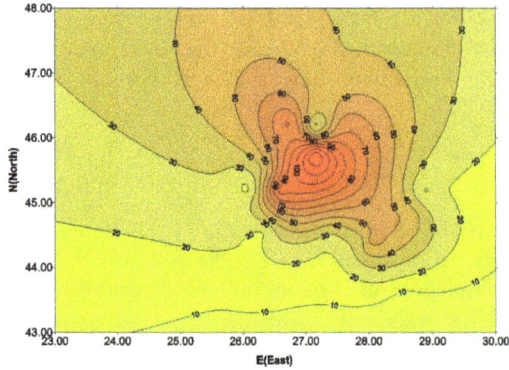

Fig. 11. The earthquake of 31st May 1990 – Recorded PGA seismic field

The data used for definition of the mathematical model based on radius vectors for the MLR azimuth based on the larger of the two horizontal components (a total of 95 PGA) are given in Table A1 (Appendix A). The isoseismals of the normalized seismic field $\left|\rho_{VLM} / \rho_i\right|$ for the VLM azimuth are given in figures 12, 13 and 14.

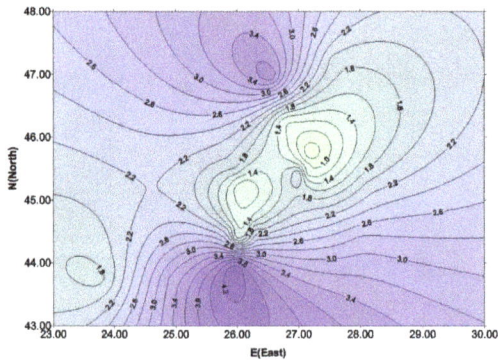

Fig. 12. The earthquake of 30th August 1986 - Normalized seismic field for the VLM azimuth

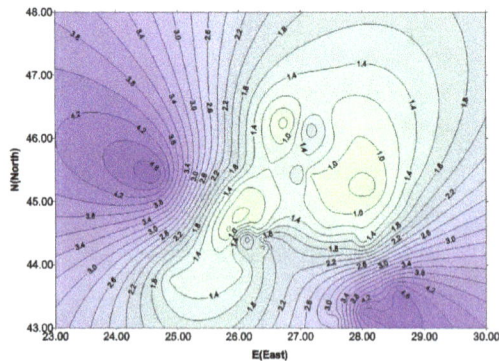

Fig. 13. The earthquake of 30th May 1990 – Normalized seismic field for the VLM azimuth

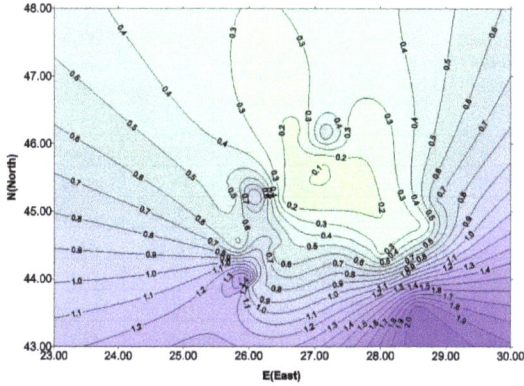

Fig. 14. The earthquake of 31st May 1990 – Normalized seismic field for the VLM azimuth

Mathematical Model: $\ln PGA = b + b_M M + b_R \ln Rh + \sigma_{\ln PGA}$

Azimuth	Regression coefficients			Standard deviation $\sigma_{\ln PGA}$
	b	b_M	b_R	
INC	-1.84230	1.50539	-0.79342	0.37103
BUC	-2.08125	1.61035	-0.87901	0.33432
CFR	0.52772	0.98049	-0.53216	0.40309
CVD	2.53490	0.77706	-0.67739	0.35774
IASI	1.19074	1.03637	-0.75200	0.32129
VLM	-4.33168	1.78635	-0.64281	0.40225
VRI	2.13673	0.82625	-0.63389	0.31867

Table 1. Regression coefficients and standard deviations based on two horizontal components

Mathematical Model: $\ln PGA = b + b_M M + b_R \ln Rh + \sigma_{\ln PGA}$

Azimuth	Regression coefficients			Standard deviation $\sigma_{\ln PGA}$
	b	b_M	b_R	
INC	-1.40590	1.49455	-0.84663	0.35791
BUC	- 1.60526	1.59385	-0.93390	0.32036
CFR	0.94361	0.96645	-0.57296	0.38277
CVD	2.95699	0.76408	-0.72328	0.33394
IASI	1.60496	1.02434	-0.79915	0.29758
VLM	-3.91229	1.76977	-0.68350	0.39286
VRI	2.58231	0.80355	-0.67176	0.29063

Table 2. Regression coefficients and standard deviations based on the larger component out of the two horizontal components

Azimuth	Magnitude M	Hypocentral Distance R_h (km)	Predicted PGA-L (cm/s^2)		Recorded PGA (two horizontal components) (cm/s^2)		Predicted PGA (cm/s^2)	
			50% non-exceedance	84% non-exceedance			50% non-exceedance	84% non-exceedance
INC	7.2	187.80	137.34	196.44	137.81	115.30	124.59	180.31
BUC	7.0	188.32	105.613	145.495	-95.77	-81.06	98.18	137.16
CFR	7.0	188.19	110.820	162.500	-70.04	-69.62	99.88	149.47
CVD	7.0	221.72	81.354	113.608	40.69	-51.13	74.85	107.04
IASI	7.0	241.85	80.589	108.520	51.27	76.36	75.05	103.48
VLM	7.0	139.56	164.125	243.104	-123.02	-146.71	148.15	221.52
VRI	7.0	137.87	134.006	179.202	-107.90	63.11	121.24	166.74
BUC	6.7	207.47	59.812	82.399	-63.34	-61.58	55.62	77.71
CFR	6.7	159.36	91.218	133.756	164.01	88.83	81.32	121.68
CVD	6.7	217.22	65.656	91.686	77.27	93.26	60.11	85.97
IASI	6.7	184.53	73.568	99.066	73.44	81.56	67.40	92.94
VLM	6.7	139.17	96.703	143.238	-118.19	91.52	86.85	129.86
VRI	6.7	99.87	130.764	174.867	91.66	-120.47	116.08	159.64
BUC	6.1	194.51	24.413	33.632	15.66	-16.56	22.40	31.29
CFR	6.1	152.41	52.401	76.838	-59.01	-46.55	46.24	69.19
IASI	6.1	181.26	40.364	54.354	38.02	-40.51	36.68	50.58
VLM	6.1	130.89	34.871	51.652	13.91	-13.85	30.93	46.25
VRI	6.1	89.95	86.623	115.839	-33.53	78.47	75.55	103.91

Table 3. Comparison between recorded and predicted values of PGA

Applying the regression coefficients and standard deviations from Tables 1 and 2, the PGA values have been computed with a non-exceedance of 50% and 84%, or as median and median + 1 standard deviation (Table 3).

The obtained results point to good fitting of the data from the mathematical model based on radius vectors, particularly in the case of use of the higher component from the two horizontal components. This is confirmed by the small values of the computed standard deviations ($\sigma_{\ln}Y \leq 0.4$) as well as the values of the median and median+1 standard deviation for the predicted PGA (PGA-L in Table 3).

The obtained PGA values depend on the instrument type, its transmission characteristics, maintenance, knowledge of the characteristics of the local profile of the instrument location, the procedures for processing of records, etc. The effect of the mathematical operations is reduced to minimum since only one multi linear regression analysis is performed.

3.4 Advantages and disadvantages

The advantages and disadvantages of the ground motion model based on radius vectors are:

- The advantage of the mathematical model based on radius vectors is that it uses a recorded seismic field. In this case, the uncertainties that are incorporated in the computation of the mathematical model of the earthquake ground motion result from the accuracy of the records.

- The disadvantage of this model is the case of use of a small number of records of occurred earthquakes and their non-uniform distribution in respect to the different azimuths. In such a case of a small number of records, the irregular closed polygon of the seismic field upon the surface will represent a polygonal figure with longer sides. This is not a deficiency of the method itself but a deficiency related to the available number of records and position of instruments. As such, it will be overcome by gradual increase of the number of instruments and records.

4. Conclusions and recommendations

The conclusions and recommendations referring to the presented ground motion models are the following:

- The azimuth dependent ground motion model defined by application of the double regression analysis contains all the specificities of the occurred individual earthquakes originating from a single seismic source;
- In an indirect way, by application of a parametric analysis, it includes in itself the characteristics of the seismic focus and the position of the location in respect to the projection of the fault plane upon the surface, or radiation pattern;
- The results obtained in the first step of the regression analysis can be controlled by the results computed by use of seismological data– seismograms. An example for this is the azimuth of the projection of the fault plane on the surface - β ;
- It is possible to develop a method for computation of azimuth dependent ground motion model by use of results from seismological investigations, or taking the direction of the projection of the fault plane on the surface from the seismological investigations. This will extensively simplify the computation of the azimuth dependent ground motion model since the first step of the regression analysis will involve only parametric analysis of the relative ratio of the semi-axes of the ellipse of the seismic field $a : 1 = a$;
- Two models are applicable in seismic hazard analyses;
- The ground motion model based on radius vectors will yield even better results if the position of the instrument within an observation network is permanent, if it is regularly maintained and calibrated, if there are as many as possible instruments within the network and if the triggering thresholds are such that records of a number of occurred earthquakes are obtained from as many as possible instruments. So, the more exactly the recorded seismic field is defined, the more reduced will be the values of the standard deviations in the mathematical model of ground motion based on radius vectors.

The author believes that, in future, advantage will be given to the model based on radius vectors particularly due to the increasing number of recording instruments and number of records of occurred strong earthquakes generated from single seismic foci.

5. Acknowledgement

The author wishes to extend her gratitude to the Ministry of Education and Science of R. Macedonia and to UKIM–IZIIS for permanent moral and financial support of her investigations.

Appendix A

No.	Data source code *)	Comp. **)	Instrument location			Mag. M	Depth h (km)	Corrected epicentral distance (km)	Hypocentral distance (km)	Peak ground acceleration PGA (cm/s²)	Normalized seismic field $\|\rho_{VLM}/\rho_\perp\|$
			Code	N (rad)	E (rad)						
1	1	2	FOC	0.798	0.474	7	131	36.897	136.097	227.7609	0.6441
2	1	2	VRI	0.801	0.466	7	131	58.413	143.433	-107.904	1.3596
3	1	1	DOC	0.819	0.463	7	131	580.25	594.854	-38.9911	3.7626
4	1	1	CFR	0.789	0.491	7	131	283.008	311.856	-70.039	2.0947
5	1	1	MLR	0.794	0.453	7	131	79.065	153.011	-79.122	1.8542
6	1	1	ISR	0.788	0.463	7	131	57.393	143.021	109.075	1.345
7	1	2	IAS	0.824	0.481	7	131	390.608	411.99	76.3557	1.9214
8	1	1	BAC	0.813	0.469	7	131	261.127	292.144	67.7456	2.1656
9	1	1	BUC	0.774	0.454	7	131	207.264	245.193	-95.7646	1.532
10	1	2	CVD	0.774	0.489	7	131	513.447	529.895	-51.1277	2.8694
11	2	2	BLV	0.776	0.451	7	131	277.749	307.092	67.2604	2.1812
12	2	1	BRN	0.777	0.46	7	131	221.264	257.136	-75.7762	1.9361
13	2	2	CVD	0.774	0.489	7	131	574.741	589.482	45.6613	3.213
14	2	2	EXP	0.776	0.456	7	131	158.18	205.382	113.8977	1.2881
15	2	1	FOC	0.798	0.474	7	131	38.055	136.416	220.8287	0.6644
16	2	2	GRG	0.767	0.453	7	131	798.046	808.727	33.5727	4.3699
17	2	1	INC	0.776	0.457	7	131	259.74	290.905	67.3488	2.1783
18	2	1	ONS	0.807	0.467	7	131	99.906	164.749	-119.651	1.2261
19	2	2	PRS	0.78	0.454	7	131	123.727	180.193	117.0445	1.2534
20	2	1	RMS	0.792	0.473	7	131	56.003	142.469	-126.626	1.1586
21	2	2	RMS	0.792	0.472	7	131	95.582	162.163	-70.9702	2.0672
22	2	1	TRM	0.764	0.434	7	131	743.82	755.267	46.128	3.1805
23	2	2	VLM	0.789	0.455	7	131	48.131	139.562	-146.708	1
24	3	2	KOZ	0.763	0.415	7	131	506.124	522.802	84.76	1.7309
25	1	1	ARR	0.792	0.43	6.7	99.1	881.87	887.421	-24.632	4.7984
26	1	1	BAC	0.813	0.469	6.7	99.1	90.24	134.03	-101.178	1.1682
27	1	2	BIR	0.807	0.482	6.7	99.1	74.799	124.16	113.7463	1.0391
28	1	1	BUC	0.774	0.454	6.7	99.1	340.121	354.264	-63.3387	1.8661
29	1	1	CFR	0.789	0.491	6.7	99.1	89.936	133.826	164.013	0.7206
30	1	1	CVD	0.774	0.489	6.7	99.1	257.233	275.662	-88.4745	1.3359
31	1	1	ARM	0.775	0.455	6.7	99.1	396.126	408.334	-52.2339	2.2628
32	1	1	MLR	0.794	0.453	6.7	99.1	151.993	181.446	-65.624	1.8011
33	1	2	SDR	0.794	0.46	6.7	99.1	70.538	121.641	-97.237	1.2155
34	1	2	VRI	0.801	0.466	6.7	99.1	12.149	99.842	-120.474	0.9811
35	1	2	IAS	0.824	0.481	6.7	99.1	225.596	246.403	81.5571	1.4492
36	2	1	ADJ	0.805	0.474	6.7	99.1	60.617	116.169	-66.3789	1.7806
37	2	2	BAA	0.781	0.5	6.7	99.1	319.905	334.903	-69.6289	1.6975
38	2	1	BIR	0.807	0.482	6.7	99.1	77.785	125.981	109.3795	1.0806
39	2	1	BLV	0.776	0.451	6.7	99.1	130.099	163.544	-159.892	0.7392
40	2	1	BRN	0.777	0.46	6.7	99.1	161.769	189.71	-115.588	1.0225
41	2	1	DRS	0.774	0.461	6.7	99.1	246.684	265.845	-82.9311	1.4252
42	2	2	FOC	0.798	0.474	6.7	99.1	43.077	108.057	83.2419	1.4199
43	2	2	FTS	0.775	0.486	6.7	99.1	276.548	293.768	76.9566	1.5359
44	2	2	GRG	0.767	0.453	6.7	99.1	309.034	324.535	-87.4576	1.3514
45	2	1	INC	0.776	0.457	6.7	99.1	279.72	296.756	69.8092	1.6931
46	2	1	MET	0.773	0.463	6.7	99.1	392.448	404.767	53.9582	2.1905
47	2	2	MLT	0.775	0.46	6.7	99.1	298.264	314.297	67.4054	1.7535
48	2	2	MTR	0.775	0.454	6.7	99.1	322.933	337.796	-65.0369	1.8173

*) Source of data: 1 INFP – Romania; 2 INCERC – Romania; 3 Bulgaria; 4 Former Yugoslavia; 5 GEOTEC – Romania **) Components: 1 N-S; 2 E-W

Table A1. (continues on next page) Data used for definition of mathematical model based on radius vectors for the VLM azimuth

No.	Data source code *)	Comp. **)	Instrument location Code	N (rad)	E (rad)	Mag. M	Depth h (km)	Corrected epicentral distance (km)	Hypocentral distance (km)	Peak ground acceleration PGA (cm/s²)	Normalized seismic field $\|\rho_{VLM}/\rho\|$
49	2	1	ONS	0.807	0.467	6.7	99.1	26.624	102.614	177.9046	0.6644
50	2	2	PIT	0.783	0.434	6.7	99.1	651.248	658.745	-35.0827	3.369
51	2	2	PND	0.774	0.461	6.7	99.1	209.246	231.527	96.5762	1.2238
52	2	2	PRS	0.78	0.454	6.7	99.1	101.328	141.733	171.5427	0.689
53	2	1	RMS	0.792	0.473	6.7	99.1	55.295	113.483	121.5669	0.9723
54	2	2	RMS	0.792	0.472	6.7	99.1	89.55	133.566	73.699	1.6037
55	2	2	SLB	0.778	0.478	6.7	99.1	172.285	198.753	102.0212	1.1585
56	2	2	TIT	0.775	0.461	6.7	99.1	387.054	399.539	50.5505	2.3381
57	2	2	TLC	0.788	0.503	6.7	99.1	278.851	295.937	-71.7137	1.6481
58	2	2	TRM	0.764	0.434	6.7	99.1	389.337	401.751	86.0013	1.3743
59	2	1	VLM	0.789	0.455	6.7	99.1	97.704	139.165	-118.194	1
60	2	2	CVD	0.774	0.489	6.7	99.1	244.985	264.269	93.2554	1.2674
61	3	1	VRN	0.755	0.489	6.7	99.1	1442.841	1446.24	25.0339	4.7214
62	3	2	KVR	0.758	0.495	6.7	99.1	1280.872	1284.7	27.1648	4.351
63	3	1	SHB	0.76	0.498	6.7	99.1	1377.353	1380.913	24.9266	4.7417
64	3	2	RUS	0.766	0.454	6.7	99.1	314.193	329.452	87.8256	1.3458
65	3	1	BZV	0.752	0.48	6.7	99.1	807.732	813.788	45.5224	2.5964
66	3	2	PRV	0.753	0.479	6.7	99.1	942.059	947.257	38.6623	3.0571
67	1	2	ARM	0.775	0.455	6.1	89.1	145.281	170.427	-16.5572	0.8402
68	1	1	BIR	0.807	0.482	6.1	89.1	15.557	90.448	-65.7703	0.2115
69	1	1	CFR	0.789	0.491	6.1	89.1	29.153	93.748	-59.009	0.2358
70	1	1	CVD	0.774	0.489	6.1	89.1	48.303	101.351	-54.929	0.2533
71	1	1	ISR	0.788	0.463	6.1	89.1	12.582	89.984	92.414	0.1505
72	1	1	SDR	0.794	0.46	6.1	89.1	17.664	90.834	44.312	0.314
73	1	2	VRI	0.801	0.466	6.1	89.1	2.183	89.127	78.4674	0.1773
74	1	2	IAS	0.824	0.481	6.1	89.1	54.208	104.295	-40.5088	0.3434
75	2	1	ADJ	0.805	0.474	6.1	89.1	22.207	91.826	-22.4738	0.619
76	2	2	BAA	0.781	0.5	6.1	89.1	54.169	104.274	-48.0672	0.2894
77	2	2	BIR	0.807	0.482	6.1	89.1	14.95	90.345	-68.4408	0.2033
78	2	1	BLV	0.776	0.451	6.1	89.1	81.847	120.987	-29.5639	0.4706
79	2	2	BRN	0.777	0.46	6.1	89.1	114.489	145.075	18.9555	0.7339
80	2	1	CLS	0.772	0.477	6.1	89.1	131.325	158.698	-19.5534	0.7115
81	2	1	CMN	0.788	0.45	6.1	89.1	49.684	102.016	32.8062	0.4241
82	2	1	CMN	0.788	0.449	6.1	89.1	55.098	104.76	29.6712	0.4689
83	2	2	CVD	0.774	0.489	6.1	89.1	49.446	101.901	-53.6487	0.2593
84	2	1	DRS	0.774	0.461	6.1	89.1	90.552	127.037	26.2509	0.53
85	2	1	FOC	0.798	0.474	6.1	89.1	3.057	89.152	-132.605	0.1049
86	2	2	FTS	0.775	0.486	6.1	89.1	69.84	113.21	35.4889	0.392
87	2	1	GRG	0.767	0.453	6.1	89.1	392.623	402.606	-8.0253	1.7335
88	2	1	MTR	0.768	0.454	6.1	89.1	276.282	290.294	-10.9373	1.272
89	2	1	ONS	0.807	0.467	6.1	89.1	7.08	89.381	82.9353	0.1677
90	2	2	PND	0.774	0.461	6.1	89.1	85.804	123.698	27.3609	0.5085
91	2	2	SLB	0.778	0.478	6.1	89.1	61.524	108.278	-33.1588	0.4196
92	2	2	TLC	0.788	0.503	6.1	89.1	116.48	146.65	-20.113	0.6917
93	2	1	VLM	0.789	0.455	6.1	89.1	95.886	130.893	13.912	1
94	3	1	SHB	0.76	0.498	6.1	89.1	599.346	605.933	6.6992	2.0767
95	3	2	RUS	0.766	0.454	6.1	89.1	191.412	211.133	16.8093	0.8276

*) Source of data: 1 INFP – Romania; 2 INCERC – Romania; 3 Bulgaria; 4 Former Yugoslavia; 5 GEOTEC – Romania   **) Components: 1 N-S; 2 E-W

Table A1. (continued) Data used for definition of mathematical model based on radius vectors for the VLM azimuth

6. References

Ambraseys, N.N & Bommer, J.J (1992). On the Attenuation of Ground Acceleration in Europe. *Proc. of the 10th World Conference on Earthquake Engineering*, Vol.1, pp. 675-678.

Ambraseys et al., (1996). Prediction of Horizontal Response Spectra in Europe. Earthquake Engineering and Structural Dynamics, Vol.25, 371-400.

Boore, D.M. & Joyner, W.B (1982). The Empirical Prediction of Ground Motion. BSSA 72, S43-S60.

Boore et al., (1993). A Summary of Recent Results Connecting for Prediction of Strong Motion in Western North America. Proc. of the International Workshop on Strong Motion Data,Vol.2, December 13-17, Menlo Park, California.

Campbell, W.K. (1981). Near–Source Attenuation of Peak Horizontal Acceleration. BSSA 71, No.6, pp. 2039-2070. December 1981.

Cornell, C.A. Engineering Seismic Risk Analysis, Bulletin of the Seismological Society of America, Vol. 58, No. 58, 1968, pp.1503-1606.

Idriss, M. I. "Selection of Earthquake Ground Motion at Rock Sites". Report Prepared for the Structures Division, Building and Fire Research Laboratory, National Institute of Standards and Technology, Department of Civil Engineering. University of California, Davis, September 1991.

Joyner, W.B. & Boore, D.M. (1981). Peak Horizontal Acceleration and Velocity from Strong-Motion Records Including Records from the 1979 Imperial Valley, California Earthquake. BSSA 71, No.6, pp. 2011-2039. December, 1981.

Joyner, W.B. &. Boore, D.M. (1988). Measurement, Characteristics and Prediction of Ground Motion. Proceedings of Earthquake Engineering and Soil Dynamics Division, ASCE, 1988, Vol.11 G.T., p. 43-102

Sabetta, F. & Pugliese, A. (1987). Attenuation of Peak Horizontal Acceleration and Velocity from Italian Strong-Motion Records. BSSA 77, No.5, pp.1491-1513, October 1987.

Sabetta, F.& Pugliese, A. (1996). Estimation of Response Spectra and Simulation of Nonstationary Earthquake Ground Motions. BSSA, Vol. 86, No.2. pp. 337-352. April 1996.

Sadigh, K.R., "A Review of Attenuation Relationships for Rock Site Conditions from Shallow Crustal Earthquakes in an Interplate Environment". Proceedings of the International Workshop on Strong Motion Data, Vol. 2, Menlo Park, California, December 13-17, (1993).

Sadigh et al., "Specification of Long–Period Ground Motions: Updates Attenuation Relationships for Rock Site Conditions and Adjustments Factors for Near– Fault Effects". Proceedings of the International Workshop on Strong Motion Data, Vol. 2, Menlo Park, California, December 13-17, (1993).

Stamatovska, S. (1996). Empirical Non-Homogeneous Attenuation Acceleration Laws for Intermediate Earthquakes from Vrancea Subdustion Zone (doctoral thesis in Macedonian). Institute of Earthquake Engineering and Engineering Seismology, University "Ss. Cyril and Methodius", Skopje, Republic of Macedonia.

Stamatovska, S.G.& Petrovski, D.S. (1996). Empirical Attenuation Acceleration Laws for Vrancea Intermediate Earthquakes. 11 WCEE, paper No.14, Mexico.

Stamatovska, S.G.& Petrovski, D.S. (1997). Non-homogeneous Attenuation Acceleration Laws for Vrancea Intermediate Earthquakes, 14th International Conference on Structural Mechanics in Reactor Technology (SMiRT 14), Lyon, France, August 17-22.

Stamatovska, S.G. (2002). A New Azimuth Dependent Empirical Strong Motion Model for Vrancea Subduction Zone. 12th European Conference on Earthquake Engineering, paper Reference 324, London, United Kingdom.

Stamatovska, S.G. (2006). A New Ground Motion Model – Methodological Approach. Acta Geodaetica et Geophysica Hungarica, Vol. 41 (3-4), pp.409-423.

Stamatovska, S.G. (2008). Ground Motion Models - State of the Art. Acta Geodaetica et Geophysica Hungarica, Vol. 43 (2-3), pp.267-284.

7

Coupling Modeling and Migration for Seismic Imaging

Hervé Chauris and Daniela Donno
Centre de Géosciences, Mines Paristech, UMR Sisyphe 7619
France

1. Introduction

Seismic imaging consists of retrieving the Earth's properties, typically velocity and density models, from seismic measurements at the surface. It can be formulated as an inverse problem (Bamberger et al., 1982; Beylkin, 1985; Lailly, 1983; Tarantola, 1987). The resolution of the inverse problem involves two seismic operators: the *modeling* and the *migration* operators.

The modeling operator M applies to a given velocity model $m(\mathbf{x})$, where $\mathbf{x}$ denotes the spatial coordinates, and indicates how to generate the corresponding shot gathers at any position in the model, usually at the surface. It consists of solving the wave equation for given velocity and density parameters. Fig. 1 and 2 illustrate the acoustic wave propagation for different travel times in two different velocity models. A point source generates a roughly circular wavefront for short travel times. The wavefront is then largely distorted due to the heterogeneous aspect of the velocity model. In simple models, it is easy to derive which part of the wave energy is diffracted, reflected, transmitted or refracted (Fig. 1). In more complex models, the wave modeling is obtained by numerically solving the wave equation, here with a finite difference scheme in the time domain (Fig. 2). In our definition, the migration operator is the adjoint M' of the modeling operator. It is related to kinematic migration, in the sense that the adjoint operator does not necessarily consider proper amplitudes. Equivalently, the modeling operator is also known to be the demigration operator.

Both the modeling and the migration operators can be very complicated. They provide the link between the time/data domain (shot, receiver and time) and the space/model domain ($\mathbf{x}$ positions). For example, a homogeneous model with a local density anomaly will create a data gather containing the direct arrival and a diffraction curve. For more complex models, the corresponding data gather is complicated, even under the Born approximation. Fig. 3 illustrates the fact that a single input trace extracted from a shot gather contributes to a large portion of the migrated image, whatever the type of migration used for implementation. The same conclusion holds for elastic modeling or more sophisticated wave equations.

We analyze in this work the combination of the modeling and the migration operators, with the objective of showing that the coupling of these operators can provide a large number of benefits for seismic imaging purposes. In particular, we consider the following general operator $H = M'[m + \delta m] \, W \, M[m]$. Here, we call H the generalized Hessian. The operator W is typically a weighting or filtering matrix. The modeling and migration operators may be defined in two different models m and $m + \delta m$, with δm representing a model perturbation.

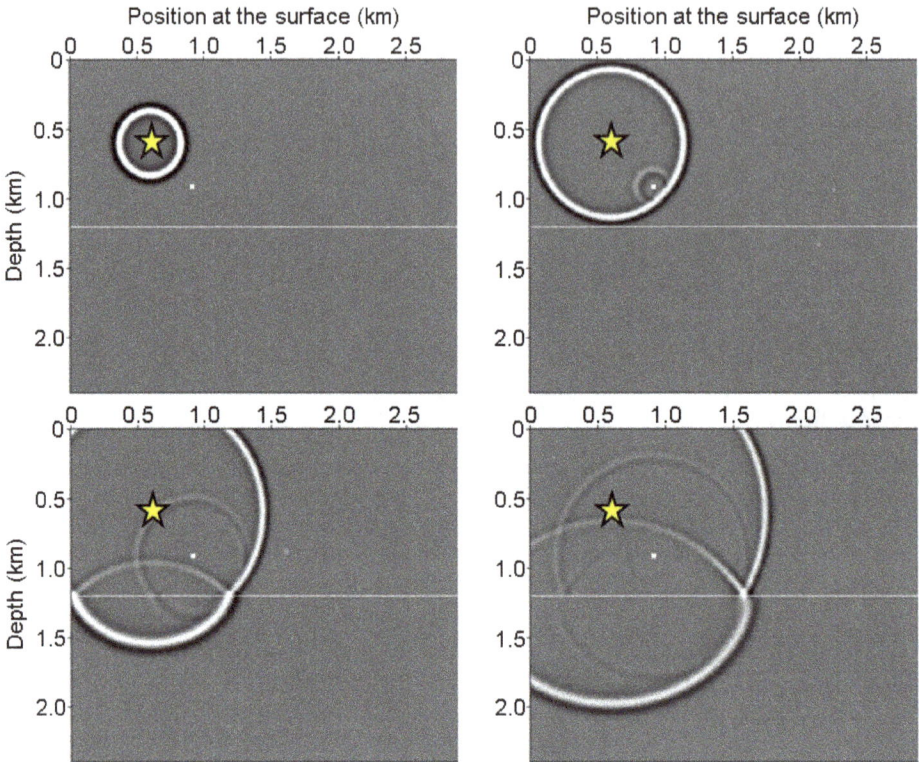

Fig. 1. Snapshots of the acoustic wave propagation in a simple model for different travel time values (from left to right and top to bottom). The velocity model consists of two different homogeneous layers and a diffraction point (white point). The source position is indicated by a star.

The exact descriptions of W and δm are given in the following sections, depending on the applications. In the strict definition of the Hessian, the model perturbation δm is equal to zero and W is an identity matrix. The classical Hessian, also known as normal operator $M'M$, naturally appears in the solution of the imaging problem as proposed by Tarantola (1987). It makes the link between images defined in the same domains, here the space domain.

Compared to the effects of M or M', we would like to demonstrate through different examples from the literature that the application of operator H has several advantages. In section 2 of this work, we present the exact expression of the Hessian and we show why this operator naturally appears in the resolution of the inverse problems as in the migration case. Then, we concentrate on three main seismic imaging tasks: pre-processing steps for reducing migration artifacts (section 3), true-amplitude imaging processes (section 4), and image sensitivity to model parameters (section 5). We also discuss the difficulties to construct the Hessian (very large matrix) and to invert for it (ill-conditioned matrix), and present several strategies to avoid its computation by sequentially applying the modeling and migration operators. We review the different approaches proposed within the geophysical community, mainly during

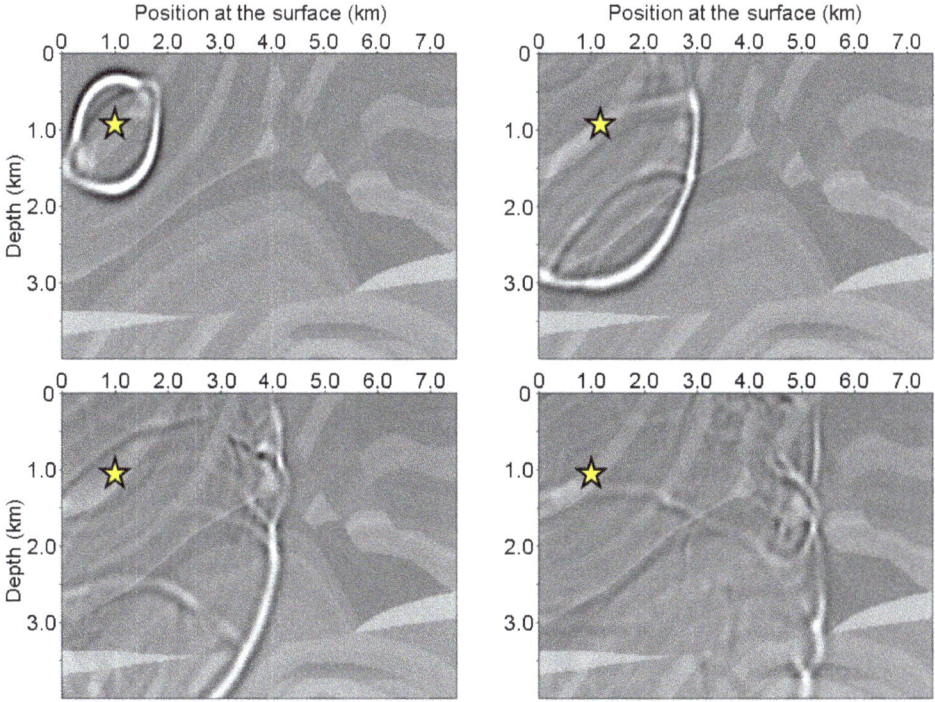

Fig. 2. Snapshots of the acoustic wave propagation in a complex model for different travel time values (from left to right and top to bottom). The velocity model is displayed in the image background. The source position is indicated by a star.

the last decade, to deal with these problems. Finally, in section 6, we conclude by suggesting new possible research directions, mainly along the estimation of unknown model parameters, where the coupling of modeling and migration could be useful.

2. Hessian and linearized migration

Non-linear seismic inversion consists of minimizing the differences between the observed data d^{obs} recorded at the surface and the computed data $d(m)$ generated in a given velocity model m, such that the objective function in the least-squares sense (Tarantola, 1987) is written as

$$J(m) = \frac{1}{2}||d(m) - d^{\mathrm{obs}}||^2. \tag{1}$$

The definition of the Hessian is given by the second derivative of the objective function with respect to the velocity model

$$H(\mathbf{x}, \mathbf{y}) = \frac{\partial^2 J(m)}{\partial m^2}, \tag{2}$$

where $\mathbf{x}$ and $\mathbf{y}$ denote two spatial positions.

Position at the surface (km)

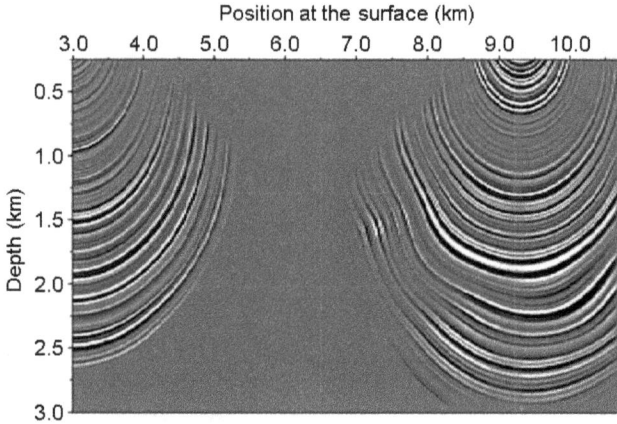

Fig. 3. Migration of two traces in a heterogeneous model.

In the case of *linear* least-squares inversion, where $d(m) = M\,m$, then the Hessian is:

$$H = M'M, \tag{3}$$

where M and M' represent the modeling and the migration operators, respectively. Under the Born approximation (single scattering), the velocity model is decomposed into two parts: $m = m_0 + \delta m$, where m_0 is referred to the background model and δm to a velocity perturbation associated to the reflectivity (Fig. 4). The background model m_0 should contain the large wavelengths (low frequencies) of the velocity model and is classically obtained by travel time tomography (Bishop et al., 1985) or by migration velocity analysis techniques (Chauris et al., 2002; Mulder & ten Kroode, 2002; Shen & Symes, 2008; Symes, 2008b). Migration aims at finding the reflectivity model δm, assuming a known smooth background model. In the linear case, the solution of equation 1 using the Hessian gives us the migration image as

$$\delta m = -H^{-1}M'\left(d - d^{\mathrm{obs}}\right). \tag{4}$$

If migration is only obtained as the gradient of equation 1 with respect to the model m ($\nabla J(m) = \frac{\partial J(m)}{\partial m}$), then only the kinematic part of the migration is retrieved (Lailly, 1983; Tarantola, 1987):

$$\delta m = -K\nabla J(m) = -KM'\left(d - d^{\mathrm{obs}}\right), \tag{5}$$

where K is a positive matrix. The application of the inverse of the Hessian yields better migration estimates, by getting a balance between amplitudes at shallow and deeper depths. As it will be discussed in section 4, the Hessian matrix is mainly diagonally banded. For simple models, its scaling properties are contained in the diagonal terms of the Hessian, while the non-diagonal terms take into account the limited-bandwidth of the data. The application of the Hessian in the inversion can be seen as a deconvolution process. Moreover, as indicated by Pratt et al. (1988), the Hessian can also potentially deal with multiscattering effects, such as multiples.

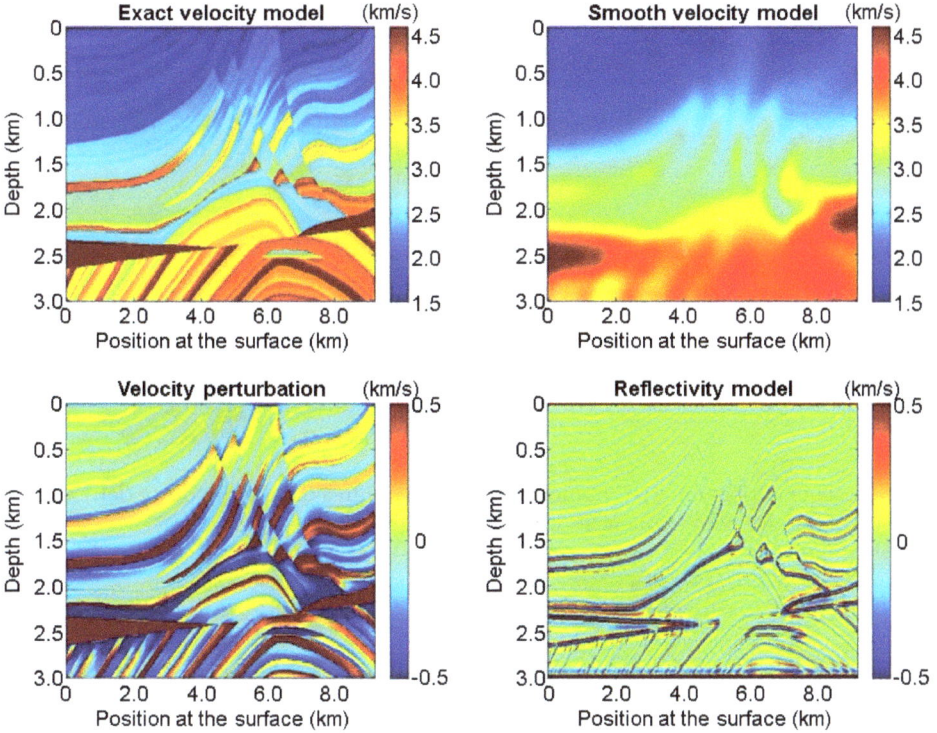

Fig. 4. Exact velocity model (top left), smoothed velocity model (top right), difference between the exact and the smooth velocity models (bottom left) and filtered version of the model difference to get a reflectivity model by taking into account the finite-frequency behavior of the migration result (bottom right).

The exact expression of the Hessian in the linear inversion case can be written using the Green's functions (Plessix & Mulder, 2004; Pratt et al., 1988)

$$H(\mathbf{x},\mathbf{y}) = \sum_{\omega} \omega^4 |S(\omega)|^2 \sum_{s} G^*(\mathbf{s},\mathbf{y},\omega)G(\mathbf{s},\mathbf{x},\omega) \cdot \sum_{r} G^*(\mathbf{r},\mathbf{y},\omega)G(\mathbf{r},\mathbf{x},\omega), \tag{6}$$

where $\mathbf{s}$ and $\mathbf{r}$ correspond to the source and receiver coordinates, S to the source term and ω to the angular frequency. The star symbol denotes the complex conjugate. The diagonal term is

$$H(\mathbf{x},\mathbf{x}) = \sum_{\omega} \omega^4 |S(\omega)|^2 \sum_{s} |G(\mathbf{s},\mathbf{x},\omega)|^2 \cdot \sum_{r} |G(\mathbf{r},\mathbf{x},\omega)|^2. \tag{7}$$

The physical meaning of the Hessian is presented in (Pratt et al., 1988; Ravaut et al., 2004; Virieux & Operto, 2009). Applied to a Dirac velocity perturbation, it provides the resolution operator. Fig. 5 displays the Hessian in a 1-D homogeneous model. For a delta-type source, the Hessian is diagonal and constant along the diagonal as there is not decay in amplitudes in a 1-D propagation case. The band-limited source, here a Ricker with a maximum frequency of 30 Hz, introduces non-zero terms off the main diagonal (Fig. 5, right).

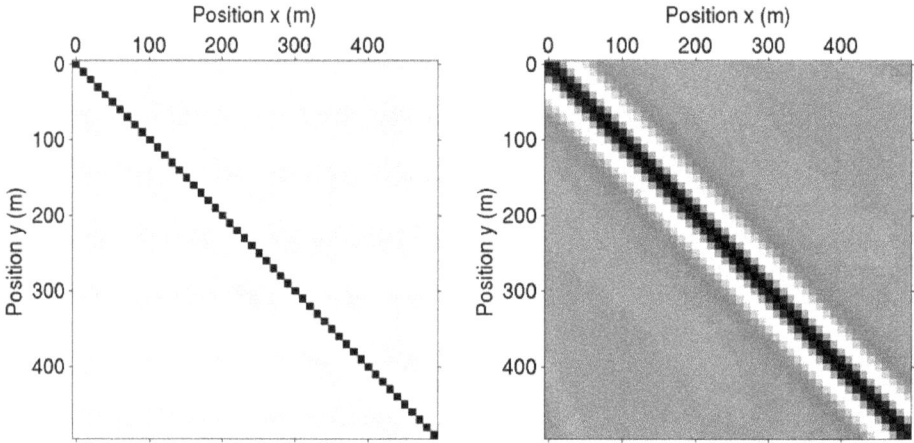

Fig. 5. Hessian matrices for a 1-D homogeneous model with a delta-type source (left) and with a Ricker source with frequencies up to 30 Hz (right).

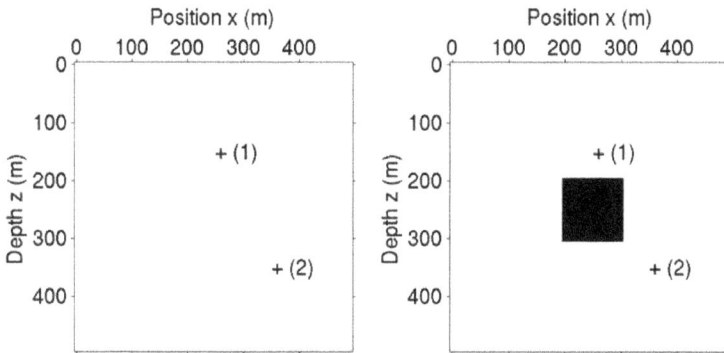

Fig. 6. Homogeneous (left) and heterogeneous (right) models used for the Hessian operator computations. Two points are selected and marked in the models.

We study the Hessian in 2-D in two different models, an homogeneous model at 1.9 km/s and the same model with a velocity perturbation of 1 km/s in the central part (Fig. 6). The maximum frequency of the data is 30 Hz. The Hessian remains mainly diagonal (Fig. 7), with an amplitude decaying with depth due to the geometrical spreading of energy and to the acquisition at the surface. The same structure is also observed in (Pratt et al., 1988; Ravaut et al., 2004; Virieux & Operto, 2009). Non-diagonal terms are present due to the band-limited data (up to 30 Hz) and to the heterogeneity of the model.

Finally, $H(\mathbf{x}, \mathbf{x}_0)$ is represented for fixed $\mathbf{x}_0$ at either positions $(250, 150)$ or $(350, 350)$ meters (Fig. 6). The resolution degrades with depth and is function of the velocity model used to compute the Green's functions (Fig. 8). These results are consistent with those obtained by Ren et al. (2011). From these illustrations, it appears that a good approximation of the Hessian, as for example proposed by Plessix & Mulder (2004), should take into account

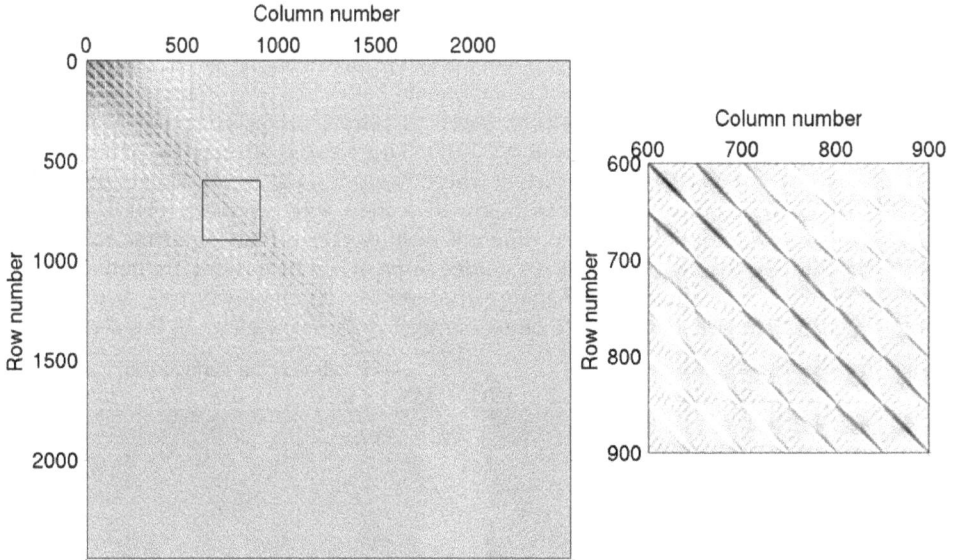

Fig. 7. Hessian matrix for the heterogeneous model of Fig. 6, and a zoom of the area delineated by the black square.

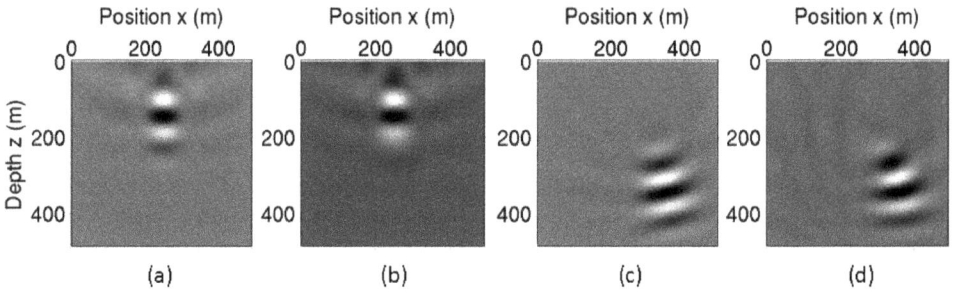

Fig. 8. Hessian responses for the selected points in Fig. 6: (a) point no. 1 of the homogeneous model, (b) point no. 1 of the heterogeneous model, (c) point no. 2 of the homogeneous model and (d) point no. 2 of the heterogeneous model.

three different aspects: the limited acquisition geometry, the geometrical spreading and the maximum frequency of the data.

3. Pre-processing for reducing migration artifacts

The quality of a migrated image is strongly influenced by uneven or partial illumination of the subsurface, which creates distortions in the migrated image. Such partial illumination can be caused by the complexity of the velocity model in the overburden, as well as by limited or irregular acquisition geometries. In the case of complex overburdens, the partial illumination is due to strong velocity variations that prevent the seismic energy either from reaching the reflectors or from propagating back to the surface, where it is recorded. The irregularity of the data spatial sampling is instead mainly due to practical acquisition constraints, such as truncated recording aperture, coarse source-receiver distributions, holes due to surface obstacles or cable feathering in marine acquisitions. In both cases, with complex overburden and with poorly sampled data, the resulting effect is that strong artifacts degrade the migrated images (Nemeth et al., 1999; Salomons et al., 2009). Fig. 9 shows an example of acquisition artifacts in a common-offset migrated section where 50 input traces are missing in the central part (Fig. 9, right). Compared with the migrated section with all traces (Fig. 9, left), we note that artifacts are localized around different positions, as a function of the reflectivity and the model used for the migration. In milder cases, when distortions are limited to the image amplitudes, true-amplitude imaging processes can be employed (see section 4 for more details). Otherwise, data need to be regularized prior to imaging. In this section, we

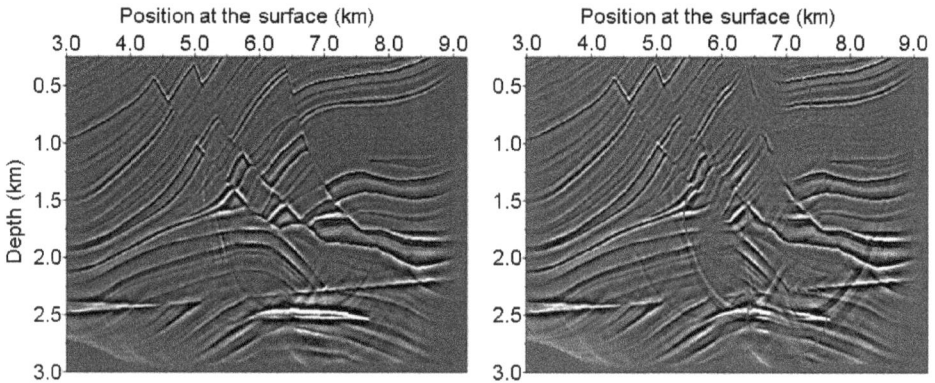

Fig. 9. Example of a migrated section with all input traces (left) and with 50 missing traces in the central part (right). The main differences are underlined with red circles.

consider the data interpolation methods that combine the migration M' and demigration M operators. Seismic reflection data acquired on an irregular grid are migrated (using a velocity model as accurate as possible) and then demigrated with the same model back into the data space onto a regular grid. In this case, the application of MWM' is required, where W is a (diagonal) weighting matrix with zero weights for dead traces and non-zero weights for live traces according to their noise level, i.e. the inverse of the standard deviation of the noise in the data (Kühl & Sacchi, 2003; Trad, 2003). Note that this combined operator, defined in the data domain, can also be interpreted as the Hessian of an alternative objective function (Ferguson, 2006). Although expensive, this process provides a good data interpolation (and even extrapolation) technique because it accounts more correctly for the propagation effects in the reflector overburden (Santos et al., 2000). With the use of the same model m for the modeling and migration parts, the kinematic aspect of the wave propagation is preserved (Bleistein, 1987). Moreover, interpolation using migration followed by demigration allows to model only those events that the migration operator can image: the demigration result is thus free of multiples (Duquet et al., 2000).

Several approaches have been proposed in the literature for implementing these techniques. We can distinguish between methods that rely on a direct inversion of the combination of migration and demigration (Ferguson, 2006; Stolt, 2002) and methods that consecutively apply the migration and modeling operators to reconstruct the data at the new locations. The latter methods can be separated into algorithms based on partial prestack migration (Chemingui & Biondi, 2002; Ronen, 1987) and those based on full prestack migration, either with a Kirchhoff operator (Duquet et al., 2000; Nemeth et al., 1999; Santos et al., 2000) or with a wavefield-continuation operator (Kaplan et al., 2010; Kühl & Sacchi, 2003; Trad, 2003).

4. True-amplitude migration schemes

The application of the inverse of the classical Hessian can be seen as a deconvolution step applied to the migration result (Aoki & Schuster, 2009). It corrects for uneven subsurface illumination (due to energy spreading and heterogeneous velocity models), takes into account the limited and non-regular acquisition geometry, and potentially increases the resolution. An extremely rich literature is available on this subject. We cite in this section the key references and detail some of them.

We distinguish between approaches based on the high frequency approximation (Beylkin & Burridge, 1990; Lecomte, 2008; Operto et al., 2000), and on the wave-equation approximation (Ayeni & Biondi, 2010; Gherasim et al., 2010; Valenciano et al., 2006; Wang & Yang, 2010; Zhang et al., 2007). Recent extensions have been proposed for sub-surface offsets (Valenciano et al., 2009).

An interesting approach has been developed in Jin et al. (1992); Operto et al. (2000), where the authors have proposed to modify the original objective function

$$J_Q(m) = \frac{1}{2}||Q\left(d(m) - d^{\text{obs}}\right)||^2, \tag{8}$$

such that the Hessian becomes diagonal. This is possible by choosing a correct weighting factor Q in the context of ray theory. The estimation of the Hessian reduced to a diagonal term is the way to correct for illumination, but it is valid only under the high frequency approximation and an infinite acquisition geometry (Lecomte, 2008). For band-limited data, other non-diagonal terms should be considered (Chavent & Plessix, 1999; Symes, 2008a; Virieux & Operto, 2009). Different strategies have been developed for estimating the non-diagonal terms (Kiyashcnenko et al., 2007; Operto et al., 2006; Plessix & Mulder, 2004; Pratt et al., 1988; Ren et al., 2011; Shin et al., 2001; Yu et al., 2006), among them the mass lumping technique (Chavent & Plessix, 1999) and the phase encoding (Tang, 2009). In practice, the estimation of the pseudo-inverse of the Hessian remains a difficult task, as the operator is large and ill-conditioned. Alternatives have been proposed to avoid the computation of the Hessian. A first approach consists of iteratively minimizing equation 1 using a gradient approach, as done in equation 5. An example is given in Fig. 10. Starting from a homogeneous model close to the exact model, J is minimized with a simple non-linear steepest descent algorithm. The model is laterally invariant, with a velocity perturbation around 400 m depth. A single shot with a maximum offset of 2 km was used. After a single iteration (Fig. 10, middle), the position of the top interface is correctly retrieved. This corresponds to the kinematic migration. After 100 iterations (Fig. 10, right), the velocity jump at the top interface is also well retrieved (+100 m/s). Since all frequencies up to 30 Hz were used at the same time, it is not possible to fully update the smooth part of the velocity model. For that reason, the second interface around 500 m is positioned at about 10 m above the exact location. More importantly here, the velocity jump is under-estimated, because no Hessian has been applied to correctly balance amplitudes (Fig. 10, right). A quasi-Newton approach (Pratt et al., 1988)

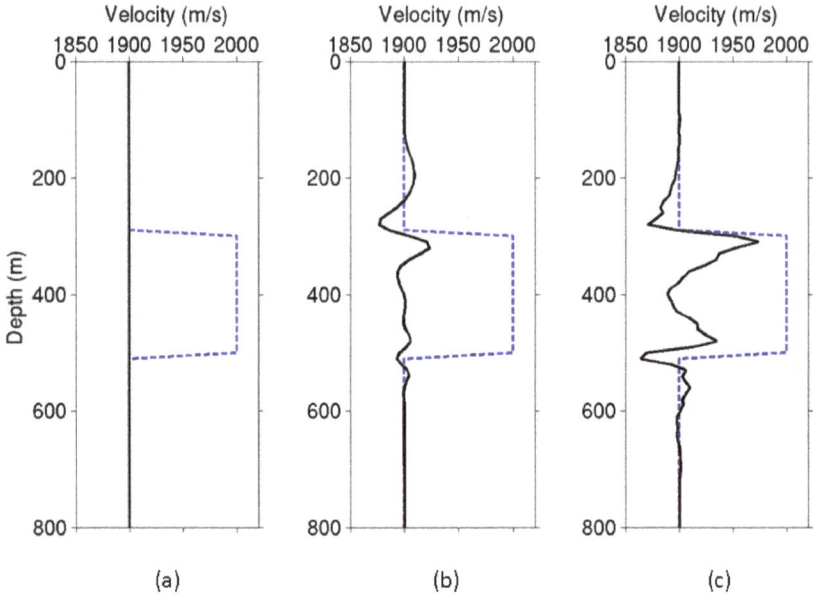

(a)   (b)   (c)

Fig. 10. Exact (dotted line) and initial (solid line) velocity models in (a), resulting model after a single iteration (solid line) in (b), and resulting model after 100 iterations (solid line) in (c).

would have been more suited in that respect, since the preconditioning of the gradient by an approximated inverse Hessian yields improved convergence rates in iterative methods.

Other approaches for the estimation of the Hessian matrix (Guitton, 2004; Herrmann et al., 2009; Nemeth et al., 1999; Rickett, 2003; Symes, 2008a; Tygel et al., 1996) consist of migrating and demigrating a result several times and of computing optimal scaling and filtering operators. This is valid in the case of single scattering. A recent article exactly shows the type of scaling and filtering to apply (Symes, 2008a). The first step consists of performing a classical prestack migration with $m_{mig} = M'd$ and, from this result, of regenerating data with the adjoint operator $d_{new} = M \, m_{mig}$. A second migration is run to obtain $m_{remig} = M' d_{new}$. Then the inverse Laplace filter $\text{Lap}^{-(n-1)/2}$ is applied $m_{filt} = \text{Lap}^{-(n-1)/2} m_{remig}$, where n is the space dimension. In that case, m_{filt} and m_{mig} are very similar except for a scaling factor S: $S \, m_{filt} = m_{mig}$. The final result is obtained as $m_{inv} = S \, \text{Lap}^{-(n-1)/2} m_{mig}$. According to Symes (2008a), this strategy is successful if the migrated result consists of nicely defined dips (see Fig 11). For this reason, curvelet or space-phase domains are well suited for these types of applications (Herrmann et al., 2009). Curvelets can be seen as an extension of wavelets to multi-dimensional spaces and are characterized by elongated shapes (Candès et al., 2006; Chauris & Nguyen, 2008; Do, 2001; Herrmann et al., 2008). All curvelets can be deduced from a reference one (Fig. 12). For true-amplitude purposes, curvelet should be understood in a broad sense as being close to the representation of local plane waves. We refer to curvelets in the next section for other applications. To summarize the approach, the effect of the inverse of the Hessian can be obtained through two migration processes and a modeling step. The scaling part only is not sufficient. A Laplace operator needs also to be applied.

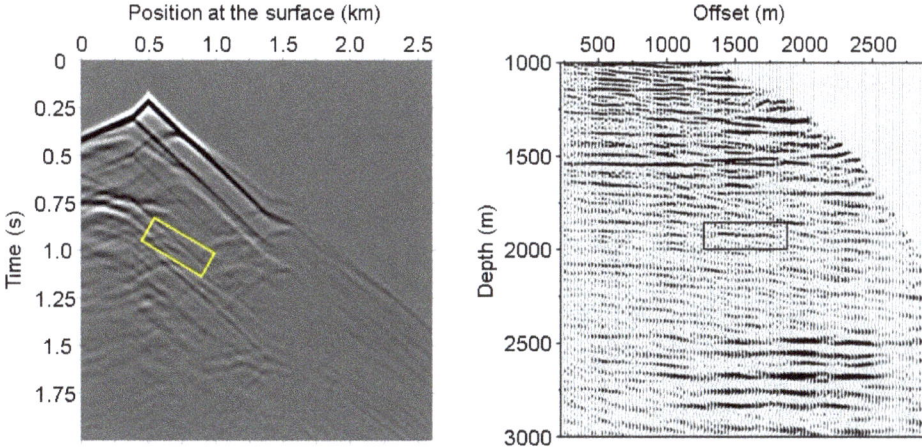

Fig. 11. Seismic data gathers can be seen as a combination of local event "curvelets", both in the unmigrated (left) and migrated (right) domains.

5. Image sensitivity

Starting from a reference migrated section, the objective of the image sensitivity techniques presented in this section is to predict the migrated section that would have been migrated with a different velocity model. For example, the migrated image in Fig. 13 (left) was obtained by using a smooth version of the exact velocity model, whereas the second migrated section (Fig. 13 right) was built with a homogeneous model. The two gathers clearly differ in terms of positioning and focusing.

In this section, we study the extended Hessian operator $H = M'[m + \delta m] M[m]$. This approach is an alternative to fully migrate the same input data for different velocity models, even though other efficient strategies have been proposed in that direction (Adler, 2002). An important aspect of the techniques based on the extended Hessian operator is that the kinematic of events remains the same through the migration/modeling operator for the same background velocity model m (Bleistein, 1987). Original ideas were first developed in the case of time migration (Fomel, 2003b). The extended Hessian H can be simplified in different ways, depending on the approximation behind the modeling and migration operators. For example, in the work of Chauris & Nguyen (2008), the operator H has a very simple shape. For that, the authors use ray tracing (high frequency approximation) and decompose the reference migrated image into curvelets (Fig. 12). The application of H to a curvelet is restricted to a shift, a rotation and a stretch of that curvelet. In practice, the model perturbation δm should be small for this method to be valid. With this strategy and for a given velocity anomaly, it is possible to predict which part of the migrated section is affected (Fig. 14). As for the approaches proposed by Symes (2008a) and Herrmann et al. (2009), a key aspect is to decompose the migrated image as a combination of local events such as curvelets. Then each curvelet is potentially distorted, if the rays connecting the curvelet to the surface penetrate the velocity perturbation. The spatial position and the orientation of the curvelets are thus important. In that context, the objective is to derive the dependency of the migrated image with respect to a given velocity anomaly.

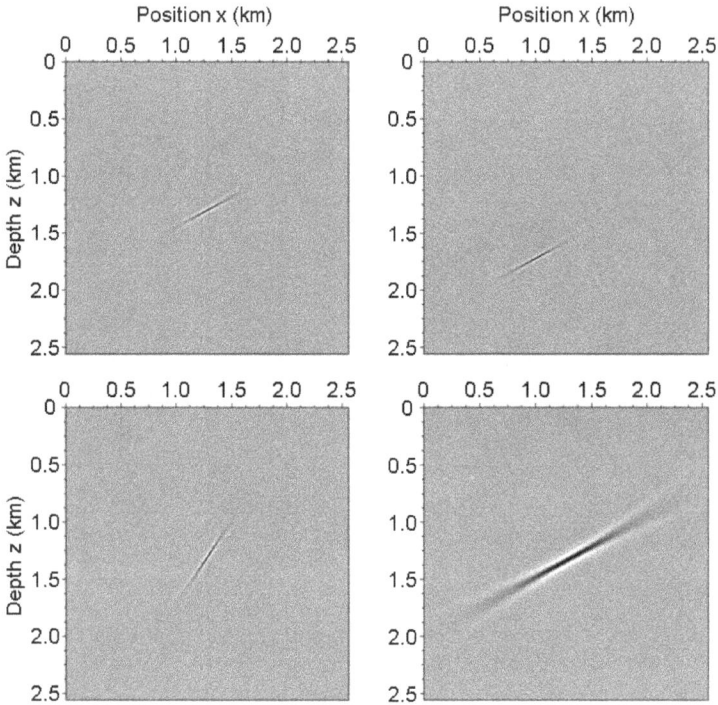

Fig. 12. Representation of different curvelets in the spatial domain. They all can be deduced from the reference curvelet (top left), either after translation/shift (top right), rotation (bottom left) or dilation/stretch (bottom right).

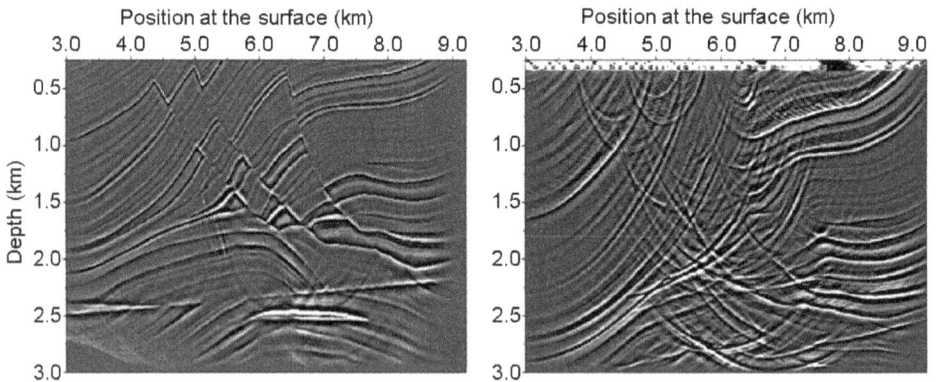

Fig. 13. Migrated images with the same input data but with two different velocity models, the correct smoothed model (left) and a constant velocity model at 3 km/s (right).

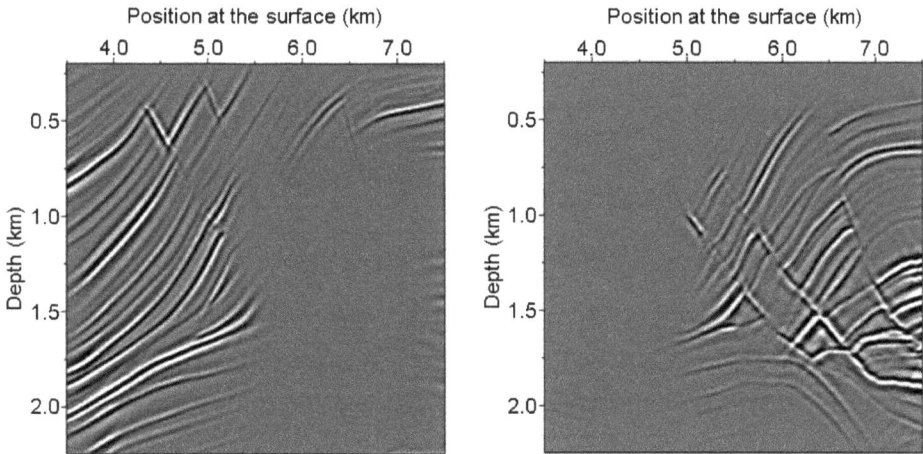

Fig. 14. Part of the migrated image unperturbed (left) and perturbed (right) by a velocity anomaly.

Finally, we refer to Chauris & Benjemaa (2010), where the authors extend the method of Chauris & Nguyen (2008) to heterogeneous models in a wave-equation approach. They propose an approximation of the Hessian that can be efficiently computed. In that case, the model perturbation δm can also be large. An example of sub-salt imaging with synthetic data is presented (Fig. 15). The first step consists of migrating the data in a given velocity model (here a smooth model that does not contain the salt body) for a series of different time-delays. A time shift is introduced during the imaging condition (Chauris & Benjemaa, 2010; Sava & Fomel, 2006). These images are considered as new input data. It is then possible to predict the new migrated section obtained in a different velocity model, at least from a kinematic point of view and with a slight frequency lost. When the new model is a smooth model with the salt body, interfaces below salt become visible (Fig. 15). In practice, the migration information is preserved on different time-delay sections, except when the exact model is used: in that case, most of the energy is concentrated around small time-delay values.

6. Discussion

In the case of single scattering, the Hessian has an explicit expression. We have reviewed different strategies to efficiently compute it or part of it, usually terms around the diagonal. However, for multiple scattering, e.g. in the case of multiples, the different approaches are not valid. Further work should be conducted along that direction. Pratt et al. (1988) indicated how to compute the Hessian without relying on the Born application. Alternatively, iterative processes for the resolution of the inverse problem potentially may deal with multiple scattering, but this should be further demonstrated.

The aim of this work is reviewing methods that combine the migration and modeling operators for seismic imaging purposes. However, it is worth noting that for seismic processing tasks several approaches exist that use a specific operator and its adjoint, particularly for data interpolation (Berkhout & Verschuur, 2006; Trad et al., 2002; van

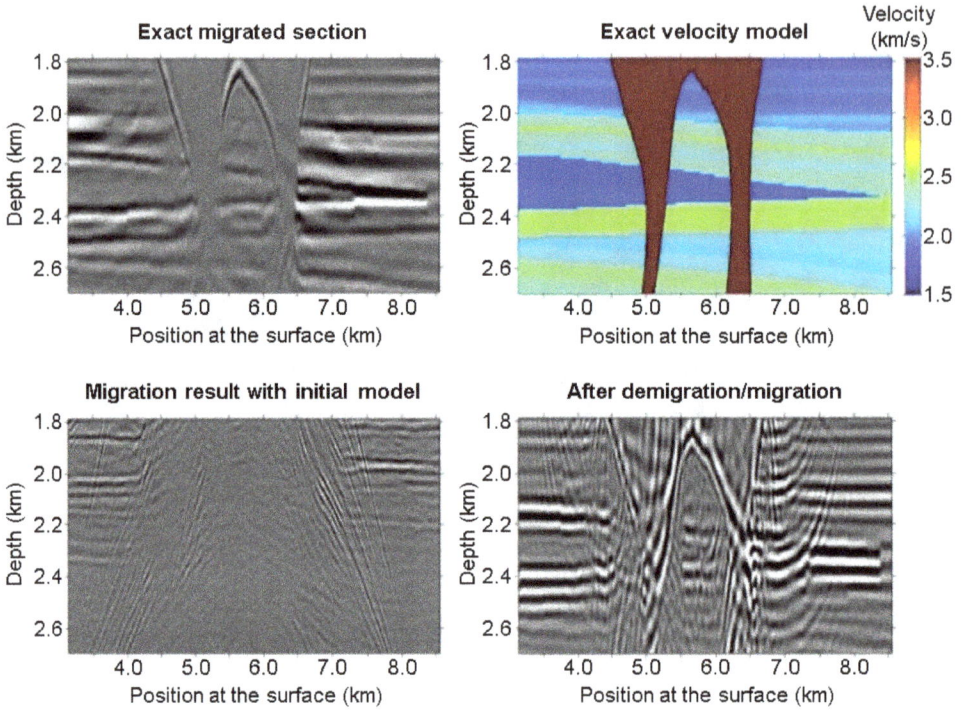

Fig. 15. Exact migrated section (top left), exact velocity model (top right), migration result with the initial velocity model (bottom left), and migration result after demigration/migration (bottom right).

Groenestijn & Verschuur, 2009), for multiple prediction (Pica et al., 2005; van Dedem & Verschuur, 2005) and for signal/noise separation (Nemeth et al., 2000). Moreover, we refer to Fomel (2003a) for other applications (stacking, redatuming, offset continuation), for which a technique is proposed to obtain a unitary modeling operator in the context of high frequency approximation.

In the developments mentioned above, the background velocity model is supposed to be known. In the context of velocity model building, we think interesting research directions should be developed along that line. For example, full waveform inversion is a general technique to retrieve the Earth's properties. However, the objective function is very oscillating and a gradient approach for the minimization leads to a local minimum. Alternative approaches have been proposed, among them Plessix et al. (1995). The Migration Based Travel Time (MBTT) method first migrates the data, and then uses the stack version to generate new data. The objective function consists of minimizing the differences between the new data and the observed data. As a benefit and compared to the classical method, it enlarges the attraction basin during the minimization process. We believe further work in that direction can deliver interesting results.

7. Conclusion

Extended research has been conducted around applying the Hessian operator in the context of pre-processing/interpolation techniques for reducing migration artifacts (section 3) and true-amplitude migration (section 4). In fact, with the use of Hessian, it is possible to correct for a limited acquisition, to provide more reliable amplitudes and to increase the resolution. However, we believe that the extended Hessian operator (refer to section 5) is a powerful tool for model estimation and that further research should be conduced along that line in the coming years.

8. Acknowledgements

The authors would like to thank a number of persons for fruitful discussions and new insights into seismic imaging. They are especially grateful to Henri Calandra (Total), Eric Dussaud (Total), Fons ten Kroode (Shell), Gilles Lambaré (CGGVeritas), Patrick Lailly (IFP), Wim Mulder (Shell), Mark Noble (Mines Paristech), Stéphane Operto (Géoazur), René-Edouard Plessix (Shell), Bill Symes (Rice university), Jean Virieux (Grenoble university) and Sheng Xu (CGGVeritas).

9. References

Adler, F. (2002). Kirchhoff image propagation, *Geophysics* 67(1): 126–134.

Aoki, N. & Schuster, G. (2009). Fast least-squares migration with a deblurring filter, *Geophysics* 74(6): WCA83–WCA93.

Ayeni, G. & Biondi, B. (2010). Target-oriented joint least-squares migration/inversion of time-lapse seismic data sets, *Geophysics* 75(3): R61–R73.

Bamberger, A., Chavent, G., Lailly, P. & Hemon, C. (1982). Inversion of normal incidence seismograms, *Geophysics* 47: 737–770.

Berkhout, A. J. & Verschuur, D. J. (2006). Focal transformation, an imaging concept for signal restoration and noise removal, *Geophysics* 71(6): A55–A59.

Beylkin, G. (1985). Imaging of discontinuities in the inverse scattering problem by inversion of a causal generalized Radon transform, *Journal of Mathematical Physics* 26: 99–108.

Beylkin, G. & Burridge, R. (1990). Linearized inverse scattering problems in acoustics and elasticity, *Wave motion* 12: 15–52.

Bishop, T. N., Bube, K. P., Cutler, R. T., Langan, R. T., Love, P. L., Resnick, J. R., Shuey, R. T. & Spinder, D. A. (1985). Tomographic determination of velocity and depth in laterally varying media, *Geophysics* 50: 903–923.

Bleistein, N. (1987). On the imaging of the reflector in the Earth, *Geophysics* 52: 931–942.

Candès, E., Demanet, L., Donoho, D. & Ying, L. (2006). Fast discrete curvelet transform, *SIAM Multiscale Modeling and Simulation* 5: 861–899.

Chauris, H. & Benjemaa, M. (2010). Seismic wave-equation demigration/migration, *Geophysics* 75(3): S111–S119.

Chauris, H. & Nguyen, T. (2008). Seismic demigration/migration in the curvelet domain, *Geophysics* 73(2): S35–S46.

Chauris, H., Noble, M., Lambaré, G. & Podvin, P. (2002). Migration velocity analysis from locally coherent events in 2-D laterally heterogeneous media, Part I: theoretical aspects, *Geophysics* 67: 1202–1212.

Chavent, G. & Plessix, R.-E. (1999). An optimal true-amplitude least-squares prestack depth-migration operator, *Geophysics* 64(2): 508–517.

Chemingui, N. & Biondi, B. (2002). Seismic data reconstruction by inversion to common offset, *Geophysics* 67: 1575–1585.

Do, M. N. (2001). *Directional multiresolution image representations*, PhD thesis, Swiss Federal Institute of Technology Lausanne.

Duquet, B., Marfurt, K. J. & Dellinger, J. A. (2000). Kirchhoff modeling, inversion for reflectivity, and subsurface illumination, *Geophysics* 65: 1195–1209.

Ferguson, R. J. (2006). Regularization and datuming of seismic data by weighted, damped least squares, *Geophysics* 71(5): U67–U76.

Fomel, S. (2003a). Asymptotic pseudounitary stacking operators, *Geophysics* 68(3): 1032–1042.

Fomel, S. (2003b). Time-migration velocity analysis by velocity continuation, *Geophysics* 68: 1662–1672.

Gherasim, M., Albertin, U., Nolte, B., Askim, O., Trout, M. & Hartman, K. (2010). Wave-equation angle-based illumination weighting for optimized subsalt imaging, *SEG, Expanded Abstracts*, pp. 3293–3297.

Guitton, A. (2004). Amplitude and kinematic corrections of migration images for nonunitary imaging operators, *Geophysics* 69(4): 1017–1024.

Herrmann, F. J., Brown, C. R., Erlangga, Y. A. & Moghaddam, P. P. (2009). Curvelet-based migration preconditioning and scaling, *Geophysics* 74(4): A41–A46.

Herrmann, F. J., Wang, D., Hennenfent, G. & Moghaddam, P. P. (2008). Curvelet-based seismic data processing: A multiscale and nonlinear approach, *Geophysics* 73(1): A1–A5.

Jin, S., Madariaga, R., Virieux, J. & Lambaré, G. (1992). Two-dimensional asymptotic iterative elastic inversion, *Geophysical Journal International* 108: 575–588.

Kaplan, S. T., Naghizadeh, M. & Sacchi, M. D. (2010). Data reconstruction with shot-profile least-squares migration, *Geophysics* 75(6): WB121–WB136.

Kiyashcnenko, D., Plessix, R.-E., Kashtan, B. & Troyan, V. (2007). A modified imaging principle for true-amplitude wave-equation migration, *Geophysical Journal International* 168: 1093–1104.

Kühl, H. & Sacchi, M. D. (2003). Least-squares wave-equation migration for AVP/AVA inversion, *Geophysics* 68(1): 262–273.

Lailly, P. (1983). *The seismic inverse problem as a sequence of before stack migrations*, Conference on Inverse Scattering, Theory and Applications, SIAM, Philadelphia.

Lecomte, I. (2008). Resolution and illumination analysis in PSDM: a ray-based approach, *The Leading Edge* 27: 650–663.

Mulder, W. A. & ten Kroode, A. P. E. (2002). Automatic velocity analysis by Differential Semblance Optimization, *Geophysics* 67(4): 1184–1191.

Nemeth, T., Sun, H. & Schuster, G. T. (2000). Separation of signal and coherent noise by migration filtering, *Geophysics* 65: 574–583.

Nemeth, T., Wu, C. & Schuster, G. (1999). Least-squares migration of incomplete reflection data, *Geophysics* 64(1): 208–221.

Operto, S., Virieux, J., Dessa, J.-X. & Pascal, G. (2006). Crustal imaging from multifold ocean bottom seismometers data by frequency-domain full-waveform tomography: application to the Eastern Nankai trough, *Journal of Geophysical Research* 111(B09306).

Operto, S., Xu, S. & Lambaré, G. (2000). Can we quantitatively image complex structures with rays?, *Geophysics* 65(4): 1223–1238.

Pica, A., Poulain, G., David, B., Magesan, M., Baldock, S., Weisser, T., Hugonnet, P. & Herrmann, P. (2005). 3D surface-related multiple modeling, *The Leading Edge* 24: 292–296.

Plessix, R.-E., Chavent, G. & De Roeck, Y. (1995). Automatic and simultaneous migration velocity analysis and waveform inversion of real data using a MBTT/WBKBJ formulation, *Expanded Abstracts*, Soc. Expl. Geophys., pp. 1099–1101.

Plessix, R.-E. & Mulder, W. (2004). Frequency-domain finite-difference amplitude-preserving migration, *Geophysical Journal International* 157: 913–935.

Pratt, G., Shin, C. & Hicks, G. (1988). Gauss-Newton and full Newton methods in frequency-space seismic waveform inversion, *Geophysical Journal International* 133: 341–362.

Ravaut, C., Operto, S., Improta, L., Virieux, J., Herrero, A. & Dell'Aversana, P. (2004). Multi-scale imaging of complex structures from multifold wide-aperture seismic data by frequency-domain full-wavefield inversions: application to a thurst belt, *Geophysical Journal International* 159: 1032–1056.

Ren, H., Wu, R.-S. & Wang, H. (2011). Wave equation least square imaging using the local angular Hessian for amplitude correction, *Geophysical Prospecting* 59: 651–661.

Rickett, J. E. (2003). Illumination-based normalization for wave-equation depth migration, *Geophysics* 68: 1371–1379.

Ronen, J. (1987). Wave-equation trace interpolation, *Geophysics* 52: 973–984.

Salomons, B., Milcik, P., Goh, V., Hamood, A. & Rynja, H. (2009). Least squares migration applied to improve top salt definition in Broek salt diapir, *Expanded Abstracts*, European Association Exploration Geophysicists, p. U007.

Santos, L. T., Schleicher, J., Tygel, M. & Hubral, P. (2000). Seismic modeling by demigration, *Geophysics* 65: 1281–1289.

Sava, P. & Fomel, S. (2006). Time-shift imaging condition in seismic migration, *Geophysics* 71(6): S209–S217.

Shen, P. & Symes, W. W. (2008). Automatic velocity analysis via shot profile migration, *Geophysics* 73(5): VE49–VE59.

Shin, C., Jang, S. & Min, D.-J. (2001). Improved amplitude preservation for prestack depth migration by inverse scattering theory, *Geophysical Prospecting* 49: 592–606.

Stolt, R. H. (2002). Seismic data mapping and reconstruction, *Geophysics* 67: 890–908.

Symes, W. W. (2008a). Approximate linearized inversion by optimal scaling of prestack depth migration, *Geophysics* 73(2): R23–R35.

Symes, W. W. (2008b). Migration velocity analysis and waveform inversion, *Geophysical Prospecting* 56: 765–790.

Tang, Y. (2009). Target-oriented wave-equation least-squares migration/inversion with phase-encoded Hessian, *Geophysics* 74(6): WCA95–WCA107.

Tarantola, A. (1987). *Inverse problem theory: methods for data fitting and model parameter estimation*, Elsevier, Netherlands.

Trad, D. (2003). Interpolation and multiple attenuation with migration operators, *Geophysics* 68: 2043–2054.

Trad, D., Ulrych, T. J. & Sacchi, M. D. (2002). Accurate interpolation with high-resolution time-variant Radon transforms, *Geophysics* 67: 644–656.

Tygel, M., Schleicher, J. & Hubral, P. (1996). A unified approach to 3-D seismic reflection imaging, Part II: theory, *Geophysics* 61(3): 759–775.

Valenciano, A., Biondi, B. & Clapp, R. (2009). Imaging by target-oriented wave-equation inversion, *Geophysics* 74(6): WCA109–WCA120.

Valenciano, A., Biondi, B. & Guitton, A. (2006). Target-oriented wave-equation inversion, *Geophysics* 71(4): A35–38.

van Dedem, E. J. & Verschuur, D. J. (2005). 3D surface-related multiple prediction: A sparse inversion approach, *Geophysics* 70(3): V31–V43.

van Groenestijn, G. J. A. & Verschuur, D. J. (2009). Estimating primaries by sparse inversion and application to near-offset data reconstruction, *Geophysics* 74(3): A23–A28.

Virieux, J. & Operto, S. (2009). An overview of full-waveform inversion in exploration geophysics, *Geophysics* 74(6): WCC1–WCC26.

Wang, Y. & Yang, C. (2010). Accelerating migration deconvolution using a nonmonotone gradient method, *Geophysics* 75(4): S131–S137.

Yu, J., Hu, J., Schuster, G. & Estill, R. (2006). Prestack migration deconvolution, *Geophysics* 71(2): S53–S62.

Zhang, Y., Xu, S., Bleistein, N. & Zhang, G. (2007). True-amplitude, angle-domain, common-image gathers from one-way wave-equation migrations, *Geophysics* 72(1): S49–S58.

8

Using a Poroelastic Theory to Reconstruct Subsurface Properties: Numerical Investigation

Louis De Barros[1], Bastien Dupuy[2], Gareth S. O'Brien[1], Jean Virieux[2]
and Stéphane Garambois[2]

[1] *School of Geological Sciences, University College Dublin*
[2] *ISTerre, CNRS - Université J. Fourier, Grenoble*
[1]*Ireland*
[2]*France*

1. Introduction

The quantitative imaging of the Earth subsurface is a major challenge in geophysics. In oil and gas exploration and production, aquifer management and other applications such as the underground storage of CO_2, seismic imaging techniques are implemented to provide as much information as possible on fluid-filled reservoir rocks. Biot theory (Biot, 1956) and its extensions provide a convenient framework to connect the various parameters characterizing a porous medium to the wave properties, namely, their amplitudes, velocities and frequency contents. The poroelastic model involves more parameters than the elastodynamic theory, but on the other hand, the wave attenuation and dispersion characteristics at the macroscopic scale are determined by the intrinsic properties of the medium without having to resort to empirical relationships. Attenuation mechanisms at microscopic and mesoscopic scales, which are not considered in the original Biot theory, can be introduced into alternative poroelastic theories (see e.g. Pride et al., 2004).

The inverse problem, that is, the retrieval of poroelastic parameters from the seismic waveforms, is much more challenging. Porosity, permeability and fluid saturation are the most important parameters for reservoir engineers. The estimation of poroelastic properties of reservoir rocks from seismic waves is however still in its infancy. The classical way of estimating these is to first solve the elastic problem and then interpret the velocities in terms of poroelastic parameters by using deterministic or stochastic rock physics modelling. However, unlike Full Waveform Inversion (FWI), these methods do not make full use of the seismograms.

In the poroelastic case, eight model parameters are required to describe the medium, compared with only one or two in the acoustic case, and three in the elastic case if wave attenuation is not considered. The advantages of using a poroelastic theory in FWI are (1) to directly relate seismic wave characteristics to porous media properties; (2) to use information that cannot be described by viscoelasticity or elasticity with the Gassmann (1951) formulae and (3) to open the possibility to use fluid displacement and force to determine permeability and fluid properties.

As an example of geological target, figure 1 presents the data recorded by a seismic survey on a seashore in the South of France. As water is pumped inland, saltwater is intruding into the coastal aquifer. This can affect the ground water and lead to severe problems with water supplies in the area. The monitoring of this phenomenon requires knowledge of the soil characteristics, including the permeability and porosity, and the properties of the fluid. A simple elastic approach cannot fully solve this problem, however the poroelastic theory may offer an alternative solution. In this example, the medium comprises alternating layers of sand, silt and clay with varying levels of compaction and a wide range of porosity and permeability. This layering produces strong reflected waves as shown in figure 1. The aim of the paper is to investigate how a poroelastic theory can be used to monitor water flow, and identify preferential pathways, using reflected waves. However, at this early stage, this chapter will only focus on numerical tests.

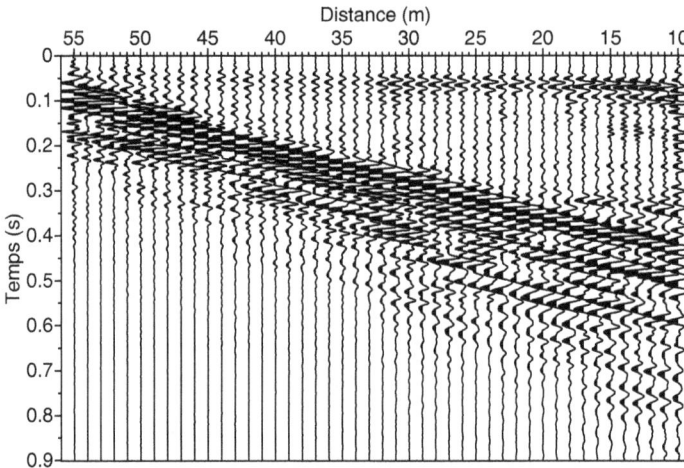

Fig. 1. Example of data for which a poroelastic-based interpretation may be useful. Data are recorded on a seashore in the South of France. Source is a hammer shot, and the 24 receivers are equally spaced between 10 and 55 m from the source. The medium is very soft and water saturated, with a direct P wave velocity of c. 1600m/s and a S wave velocity lower than 200 m/s.

We will investigate the gain and the feasibility of using a poroelastic approach, rather than the classical elastic one, in full waveform methods. The forward modelling is solved using different algorithms: a reflectivity approach, a 3D finite difference scheme and a 2D discontinuous Galerkin method. The comparison of synthetic data computed in the elastic and poroelastic cases shows that poroelastic modelling leads to some typical patterns that cannot be explained by elastic theory. This proves that the use of poroelastic theories may bring more insight to the model reconstruction, particularly, in relation to the fluid properties. Moreover, mesoscopic attenuation can be introduced in the poroelastic laws for double porosity medium, adding extra changes in the waveforms. This demonstrates the utility of using such theories to correctly reproduce measured seismic data.

Analytical formulae are then derived to compute the first-order effects produced by plane inhomogeneities on the point source seismic response of a fluid-filled stratified porous medium. The derivation is achieved by a perturbation analysis of the poroelastic wave equations in the plane-wave domain using the Born approximation. The sensitivity of the wavefields to the different model parameters can be investigated: the porosity, consolidation parameter, solid density, and mineral shear modulus emerge as the most sensitive parameters in the forward and inverse modelling problems. However, the amplitude-versus-angle response of a thin layer shows strong coupling effects between several model parameters.

The inverse problem is then tackled using a generalized least-squares, quasi-Newton approach to determine the parameters of the porous medium. Simple models consisting of plane-layered, fluid-saturated and poro-elastic media are considered to demonstrate the concept and evaluate the performance of such a full waveform inversion scheme. Numerical experiments show that, when applied to synthetic data, the inversion procedure can accurately reconstruct the vertical distribution of a single model parameter, if all other parameters are perfectly known. However, the coupling between some of the model parameters does not permit the reconstruction of several model parameters at the same time. To get around this problem, we consider composite parameters defined from the original model properties and from *a priori* information, such as the fluid saturation rate or the lithology, to reduce the number of unknowns. We then apply this inversion algorithm to time-lapse surveys carried out for fluid substitution problems, such as CO_2 injection, since in this case only a few parameters may vary as a function of time. A two-step differential inversion approach allows the reconstruction of the fluid saturation in reservoir layers, even though the medium properties are mainly unknown.

2. Wave propagation in stratified porous media

The governing equations for the poroelastodynamic theory were first derived by Biot (1956), and are thus often referred to as "Biot's theory". The main hypothesis behind these equations is that the seismic wavelengths are longer than the pore size; the medium can then be described by homogeneised laws. Poroelastic theories have since been derived and improved by many authors (e.g. Auriault et al., 1985; Geertsma & Smith, 1961; Johnson et al., 1987).

2.1 Governing equations

Assuming a $e^{-i\omega t}$ dependence, Pride et al. (1992) rewrote Biot's (1956) equations of poro-elasticity in the frequency domain in the form

$$[(K_U + G/3) \nabla\nabla + (G\nabla^2 + \omega^2\rho) \, I \,] \, . \, \mathbf{u} + [\, C\nabla\nabla + \omega^2\rho_f I \,] \, . \, \mathbf{w} = 0 \qquad (1)$$

$$[\, C\nabla\nabla + \omega^2\rho_f I \,] \, . \, \mathbf{u} + [\, M\nabla\nabla + \omega^2\tilde{\rho}I \,] \, . \, \mathbf{w} = 0 \, ,$$

where $\mathbf{u}$ and $\mathbf{w}$ respectively denote the average solid displacement and the relative fluid-to-solid displacement, ω is the angular frequency, I the identity tensor, $\nabla\nabla$ the gradient of the divergence operator and ∇^2 the Laplacian operator. The other quantities appearing in equations (1) are properties of the medium.

The bulk density of the porous medium ρ is related to the fluid density ρ_f, solid density ρ_s and porosity ϕ:

$$\rho = (1 - \phi)\rho_s + \phi\rho_f \, . \qquad (2)$$

K_U is the undrained bulk modulus and G is the shear modulus. M (fluid storage coefficient) and C (C-modulus) are mechanical parameters. In the quasi-static limit, at low frequencies, these parameters are real, frequency-independent and can be expressed in terms of the drained bulk modulus K_D, porosity ϕ, mineral bulk modulus K_s and fluid bulk modulus K_f (Gassmann, 1951):

$$K_U = \frac{\phi K_D + \left[1 - (1 + \phi) \dfrac{K_D}{K_s} \right] K_f}{\phi(1 + \Delta)},$$

$$C = \frac{\left[1 - \dfrac{K_D}{K_s} \right] K_f}{\phi(1 + \Delta)}, \quad M = \frac{K_f}{\phi(1 + \Delta)} \tag{3}$$

$$\text{with} \quad \Delta = \frac{1 - \phi}{\phi} \frac{K_f}{K_s} \left[1 - \frac{K_D}{(1 - \phi)K_s} \right].$$

It is also possible to link the frame properties K_D and G to the porosity and constitutive mineral properties (Korringa et al., 1979; Pride, 2005):

$$K_D = K_s \frac{1 - \phi}{1 + c_s \phi} \quad \text{and} \quad G = G_s \frac{1 - \phi}{1 + 3c_s \phi/2}, \tag{4}$$

where G_s is the shear modulus of the grains. The consolidation parameter c_s appearing in these expressions is not necessarily the same for K_D and G (Korringa et al., 1979). However, to minimize the number of model parameters, and following the recommendation of Pride (2005), we consider only one consolidation parameter to describe the frame properties. c_s typically varies between 2 to 20 in a consolidated medium, but can be much greater than 20 in a soft soil.

Finally, the wave attenuation is explained by a generalized Darcy's law which uses a complex, frequency-dependent dynamic permeability $k(\omega)$ defined via the relationship (Johnson et al., 1994):

$$\tilde{\rho} = i \frac{\eta}{\omega\, k(\omega)} \quad \text{with} \quad k(\omega) = k_0 / \left[\sqrt{1 - i \frac{4}{n_J} \frac{\omega}{\omega_c}} - i \frac{\omega}{\omega_c} \right]. \tag{5}$$

In equation (5), η is the viscosity of the fluid and k_0 the hydraulic permeability. Parameter n_J is considered constant and equal to 8 to simplify the equations.

The relaxation frequency $\omega_c = \eta/(\rho_f F k_0)$, with F the electrical formation factor, separates the low frequency regime where viscous losses are dominant from the high frequency regime where inertial effects prevail. We refer the reader to the work of Pride (2005) for more information on the parameters used in this study.

The solution of equation (1) leads to classical fast P- and S-waves, and to an additional slow P-wave (often called Biot wave). The fast P-wave has fluid and solid motion in phase, while the Biot wave has out-of-phase motions. At low frequency, the Biot wave has a diffusive pattern and can be seen as a fluid pressure diffusion wave. At high frequency, the inertial effects are predominant. This wave becomes propagative and can be seen in data, giving an experimental justification to the dynamic poroelasticity theory Plona (1980).

2.2 Mesoscopic attenuation and more complex theories

Although the slow P-wave does not appear on the seismograms at low frequency, it plays an important role in the attenuation process, as it produces loss of energy by wave-induced fluid-flow. However, the attenuation as described in the Biot theory is not strong enough to model the attenuation in geological media, especially at low (i.e. seismic) frequencies.

Attenuation processes can actually be separated into 3 different spatial scales, namely microscopic, mesoscopic and macroscopic (Pride et al., 2004). Within this classification, the Biot mechanism of attenuation takes place at macroscopic scale (on the order of the seismic wavelength). The microscopic attenuation is due to mechanisms that occur at the grain size, such as the squirt flow mechanism (Mavko & Jizba, 1991). This mechanism leads to wave attenuation mainly at high frequencies. The attenuation mechanism that prevails at low frequency comes from the mesoscopic scale (Pride et al., 2004), and it is due to fluid flow that occurs at boundaries between any medium heterogeneities whose sizes are between the grain sizes and the seismic wavelengths. This is particularly true for layered media (Gurevich et al., 1997; Pride et al., 2002) or when the medium contains 1) inclusions of different materials such as composite medium or double porosity medium (Berryman & Wang, 2000; Pride et al., 2004; Santos et al., 2006), or 2) different fluids (Santos et al., 1990) or patches of different saturation (Johnson, 2001).

Double porosity medium (DP in the text) refers to a porous medium which contains inclusions with different porosity and permeability (Pride et al., 2004; Santos et al., 2006). Assuming that the most compressible phase (patches, phase 2) is embedded into the least compressible one (host rock, phase1), the fluid flow inside the phase 2 can be eliminated from the homogenised equations. This assumption allowed Pride & Berryman (2003a, ,2003b) to write the DP equations under the form of the classical Biot theory (eq. 1). This involves the use of complex frequency dependent moduli K_U, C and M. Particularly, these parameters are functions of the respective volume of each phase and of the size of the patches (a denotes here the average radius of the inclusions in the host rock). Finally, as the patches are assumed to be spherical, the shear modulus of the medium is still real and not frequency dependent, and can be approximated by the geometrical mean of the modulus inside each phase. For the derivation and the detailed expressions of the parameters, please refer to the work of Pride et al. (2004) and Pride (2005).

To show how the mesoscopic attenuation due to DP media impacts the seismic properties, we look for the changes in the P-wave velocity and attenuation. The medium is composed of little patches of high permeable and high porosity in a less permeable host rock. Following the work of Liu et al. (2009), we consider a sandstone with 3% sand inclusions. The complex moduli K_U, C and M are computed using the DP effective theory of Pride et al. (2004), leading to the P-wave velocity and attenuation with respect to the frequency. The results are compared to the seismic properties of each single phase and to the response using an average single porosity medium, where moduli are computed by geometrical averages. Figure 2 shows the P-wave velocity and attenuation (via the inverse of the quality factor) for the double porosity medium (for the inclusion radius a equal to 1, 5 and 10 cm), for the average single porosity medium and for both single phase media.

The P-wave velocity is much more dispersive for the double porosity medium than for the equivalent single porosity medium. At high frequency, it is much higher in the double porosity medium than in the equivalent single porosity medium. It shows two main changes:

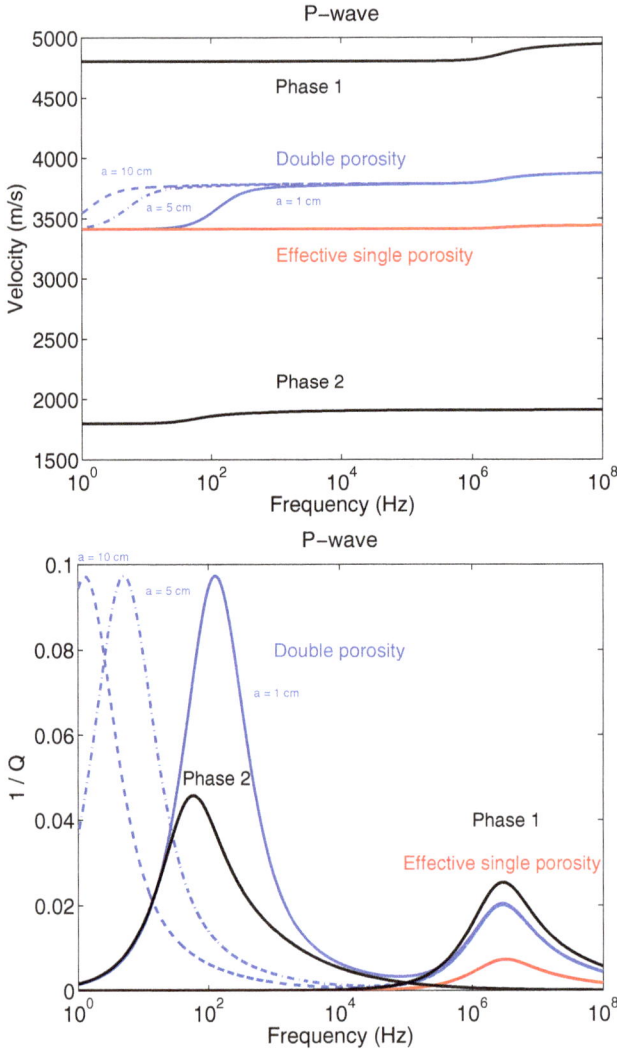

Fig. 2. P-wave velocities (top panel) and attenuations (bottom panel) with respect to the frequency (between 1 and 10^8 Hz). Blue lines are for the double porosity theory for inclusion sizes $a = 1$ cm (continuous lines), $a = 5$ cm (dotted-dashed lines) and $a = 10$ cm (dashed lines). Poroelastic responses of the individual phases are the black lines. The results for the average single porosity theory, computed using weighted average of each phase properties, are given by the red lines.

1) at the relaxation frequency of the host rock medium, which are at the transition between the diffusive and the propagative regime of the Biot wave, and 2) at low frequencies, around the relaxation frequency of the patches, with a strong dependance on a. It is worth noting that the size of inclusions, a, has an strong effect on the low frequency behaviour. We observe similar

patterns for the P-wave attenuation (inverse of the quality factor). In the double porosity medium, attenuation shows two main peaks, associated with the two phases, while the equivalent single poroelastic medium produces only one single peak at high frequency. At low frequency (seismic frequencies), the attenuation is very high for the double porosity medium, leading to quality factors that are in good agreement with quality factors in geological materials. This pattern strongly depends on the size of the inclusions. On the other hand, attenuation produced by single phase medium is too low to be realistic. This means that the fluid flow at the boundaries between heterogeneities plays a fundamental role in the attenuation process, that cannot be neglected. These P-wave characteristics will have a strong influence on the seismic waveforms (see section 3.2).

Using a poroelastic theory is much more complex than an equivalent visco-elastic theory. However, modelling seismic waves with poroelastic theories take into account the attenuation induced by fluid equilibration at layer interfaces or heterogeneity boundaries, whereas a viscoelastic approach neglects this attenuation process. As shown by Pride et al. (2004), this is the most important attenuation process at low frequency. As the shallow subsurface has strong lateral and vertical heterogeneities, one should solve the full poroelastic theory to deal with attenuation.

3. Numerical modelling of seismic waves in porous media

3.1 Forward modelling solution

The model properties $\mathbf{m}$, which are the material parameters introduced in the previous section, are nonlinearly related to the seismic data $\mathbf{d}$ via an operator f, i.e., $\mathbf{d} = f(\mathbf{m})$. The forward problem has been solved by many authors, using different methods. Analytical solutions have been derived for a homogeneous medium (Boutin et al., 1987; Philippacopoulos, 1997). The response of porous layered medium has been computed using reflectivity methods in the frequency-wavenumber domain, such as the Kennett (1983) approach (De Barros & Dietrich, 2008; Pride et al., 2002). This method was also used to solve the coupling between seismic and electromagnetic waves (Garambois & Dietrich, 2002; Haartsen & Pride, 1997). The poroelastic equations have been solved in 2D and 3D cases, mainly using finite difference schemes (Carcione, 1998; Dai et al., 1995; Masson & Pride, 2010; O'Brien, 2010) in the time-space domain. For discretisation issues, the equations (1) should be decomposed in propagative and diffusive parts, which have to be solved independently (Carcione, 1998). Other time domain numerical schemes have been used, such as finite elements (Morency & Tromp, 2008) or finite volume (de la Puente et al., 2008). Finally, Dupuy et al. (2011) solved this problem in the frequency domain using a discontinuous Galerkin approach. For a complete and precise review of the numerical modelling used to solve the poroelastic problem, we refer the reader to Carcione et al. (2010).

In this paper, we will use three different techniques to illustrate our points:

- a 3D Finite Difference scheme (FD; O'Brien, 2010): The solutions of Biot's equations are obtained by fourth-order in space and second-order in time staggered-grid and rotated-staggered-grid methods. Stability of the methods and accuracy of the solutions have been carefully checked in the low-frequency domain.

- a reflectivity approach (SKB; De Barros & Dietrich, 2008; De Barros et al., 2010): The 3D solution is obtained in the frequency-wavenumber domain for horizontally layered media by using the generalized reflection and transmission method of Kennett (1983). The synthetic seismograms are finally transformed into the time-distance domain by using the 3D axisymmetric discrete wavenumber integration technique of Bouchon (1981).

- a Discontinuous Galerkin Method (DGM; Dupuy et al., 2011): For 2 dimension medium, the discrete linear system for the Biot theory has been deduced in the frequency domain for a discontinuous finite-element method, known as the nodal discontinuous Galerkin method. Solving this system in the frequency domain allows accurate modelling of the wave propagation for all frequencies.

The last two approaches are in the frequency domain, which has several advantages: 1) there is no need to decompose the problem into diffusive and propagative parts; 2) all frequencies, i.e., in the low- and high-frequency regimes, can be accuarately treated; 3) solving more complex theories, such as double porosity or poroviscoelastic theories is straighforward and does not require any modifications of the solver; and 4) frequency domain has been shown to be the most efficient way to solve the Full waveform inverse problem, as the solution has to be calculated only for a few frequencies (Pratt et al., 1998).

3.2 Seismic waveforms in poroelastic medium

As already stated, the main improvement of using poroelastic versus elastic theories is in the description of the attenuation from intrinsic medium parameters. Figure 3 gives an example of poroelastic and elastic data computed by equivalent finite difference codes (FD, O'Brien, 2010). The medium is a complex 200m thick reservoir embedded in a homogeneous half-space. The reservoir, modified from Manzocchi et al. (2008) is composed of 7 different facies, with different mineral and frame properties (see Fig. 3, top). In order to mimic a time lapse survey for CO_2 geological storage in a saline aquifer, a baseline is first computed for fully brine saturated medium. CO_2 is then injected in the center of the reservoir and spread out according to the permeability, leading to areas containing gas of roughly 500 m and 2000 m diameter.

Figure 3 presents the differential data (data with CO_2 in the reservoir minus baseline) for both gas extensions. The same models are run using equivalent elastic properties (computed through the Gassmann formulation). As the velocities are equal, arrival times of the elastic and poroelastic waves are the same. However, changes appear in amplitude, mainly in the multiple reflected waves and coda. The amplitude differences are up to 40% of the data. It stresses the importance of using attenuation in the forward modelling and inversion processes. Even if the Biot theory cannot explain the full range of attenuation, it leads to significant changes in the seismic waveforms.

Poroelastic attenuation is due to the fluid movement. In the poroelastic theories, evidence for this lies in the relative fluid-to-solid motion and in the existence of the slow P-wave. Figure 4 reproduces the experiment performed by Plona (1980). The models are made using the Discontinuous Galerking method (DGM) of Dupuy et al. (2011) and are checked against the reflectivity approach (SKB) used by De Barros & Dietrich (2008). The SKB synthetic data have been corrected from the 3D effects, using an infinite line of sources, in order to be directly comparable to the DGM solutions. A P-wave is generated with an explosive source (central frequency of 200 Hz) in an quasi-elastic layer (the Biot wave is entirely diffusive in this layer) and is transmitted and converted into S-wave and Biot wave at the interface with a porous layer. Receivers are set into the second layer and record the three waves.

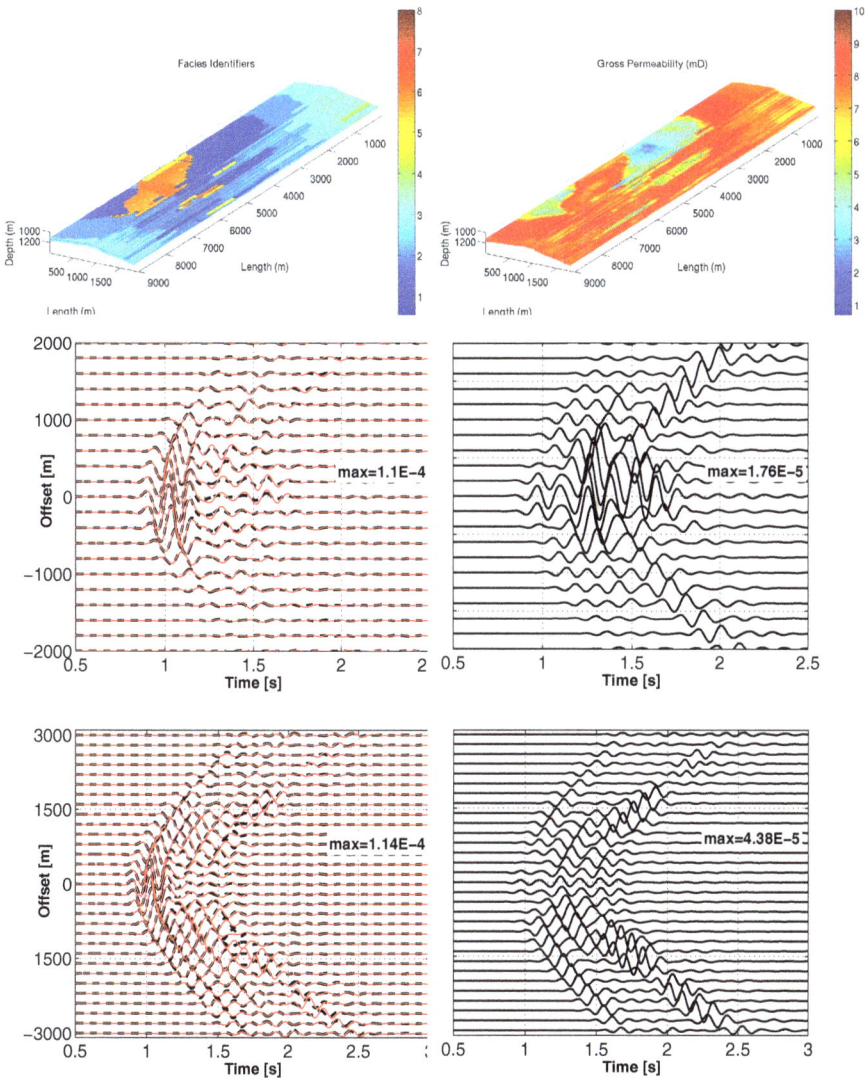

Fig. 3. Example of poroelastic and elastic numerical modelling to mimic differential data for CO_2 storage. Top) Resevoir models (left: Facies, right: Permeability) used in the modelling and modified from Manzocchi et al. (2008). Elastic properties are computed using the Gassmann relationships from the porous parameters. Middle) Left: Differential seismograms (model with CO_2 minus baseline) for elastic (red) and poroelastic data (black), and right: differences between the elastic and poroelastic differential data. Bottom) As the middle panels for a larger extension of CO_2. Receivers and source are located on the free surface. The explosive source has a 8Hz Ricker wavelet signature.

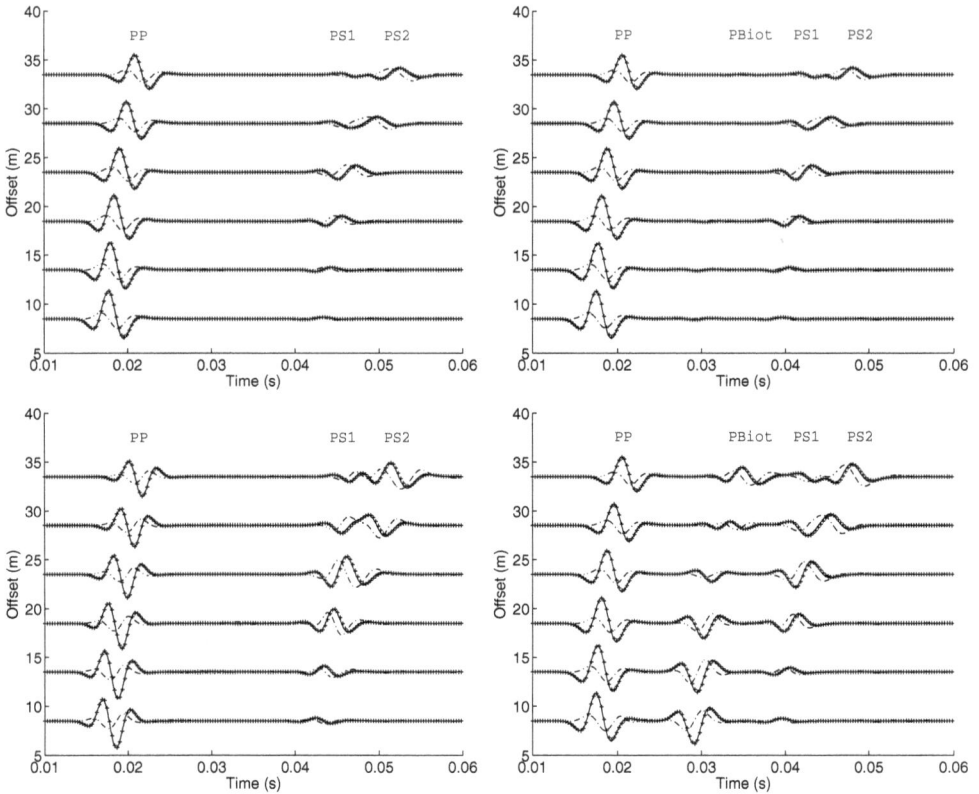

Fig. 4. Flat interface case: Left) Seismograms of (top) vertical solid u_z, and (bottom) relative fluid/solid displacement components w_z in the low-frequency regime; Right) As Left, but in the high-frequency domain. The SKB solution is indicated by a continuous line and the DGM by crosses; dashed-dotted lines indicate the differences between the two solutions (multiplied by a factor of 5). PP, PBiot and PS2 stand for the transmitted P and the converted Biot and S waves, respectively. PS1 stands for the conical wave associated with the direct P wave in the first layer. The explosive source is in a quasi-elastic layer, while receivers stand in a porous half space.

Two cases are studied: 1) a low frequency case, where the source frequency is smaller than the cut-off frequency (f_c = 6400 Hz). The Biot wave, in this case, is not propagative and cannot be seen; 2) a high frequency case, the source frequency is higher than the cut-off frequency (f_c = 0.64 Hz). This is obtained by decreasing the value of the fluid viscosity by 10000. The Biot wave becomes propagative, and can be observed, mainly in the fluid displacement data. In the seismograms of figure 4, the transmitted P-wave (PP), the conical wave associated with the P direct wave in the first layer (PS1), and the transmitted S-wave (PS2) are identified. These two S-waves (PS1 and PS2) will be uncoupled at further offsets. Finally, the very small differences between both modelling methods prove their accuracy.

A similar case is studied in a double porosity medium. Taking the same source/receiver layout with an explosive source in the first layer comprising a quasi-elastic sandstone and a line of receivers in the second layer comprising the double porosity medium described in the part 2.2 (sandstone with 3% of 1 cm spherical sand inclusions), we compute the seismograms and compare the double porosity results with the effective single porosity results. The seismograms of solid and relative fluid/solid displacements are given in figure 5. The influence of the double porosity homogeneization (via complex frequency dependent mechanic moduli) is clearly visible on the transmitted P-waves. Particularly, the waveforms are strongly distorted as the attenuation and dispersion are higher (see figure 2). As we are in the low frequency domain, the Biot waves are not visible in the seismograms, but they are responsible for the loss of seismic energy. As predicted by the theory, the converted S-waves are not impacted by the double porosity approach.

In these examples, we have demonstrated the importance of taking into account complex poroelastic theories in order to understand and reproduce real seismic signals whose waveforms are strongly impacted by the presence of fluid and medium heterogeneities.

4. Sensitivity analysis

In the next sections, we use the reflectivity algorithm SKB (De Barros & Dietrich, 2008) and focus on backscattered energy, i.e., we consider reflected seismic waves as in a seismic reflection experiment. We further assume that, whenever they exist, waves generated in the near surface (direct and head waves, surface and guided waves) are filtered out of the seismograms prior to the analysis. The assumption of plane-layered media is admittedly too simple to correctly describe the structural features of geological media, but it is nevertheless useful to explore the feasibility of an inversion process accounting for the rheology of porous media.

The sensitivity of the seismic waveforms to the model parameters is investigated for layered medium by computing the first-order derivatives of the seismic displacements with respect to the relevant poroelastic parameters. These operators, which are often referred to as the Fréchet derivatives, are expressed via semi-analytical formulae by using the Born approximation (De Barros & Dietrich, 2008). They can be readily and efficiently evaluated numerically because they are only functions of the Green's functions of the unperturbed medium. In each layer, we consider the eight following quantities as model parameters: 1) the porosity ϕ, 2) the mineral bulk modulus K_s, 3) the mineral density ρ_s, 4) the mineral shear modulus G_s, 5) the consolidation parameter c_s, 6) the fluid bulk modulus K_f, 7) the fluid density ρ_f and 8) the permeability k_0. This parameter set allows us to distinguish the parameters characterizing the solid phase from those describing the fluid phase. The fluid viscosity η is one of the input parameter but it is not considered in the inversion tests as its sensitivity is similar to the permeability.

Figure 6 (left) presents the sensitivity of the 8 parameters, for P and S waves. The sensitivity of the reflected wavefield varies drastically among the different parameters. We note that the reflected waves are especially sensitive to the mineral density ρ_s, porosity ϕ, shear modulus G_s and consolidation parameter cs. If we have some knowledge about the mineral properties (i.e., G_s, K_s and ρ_s are fixed), the porosity ϕ and consolidation factor cs are the most sensitive parameters and therefore the key parameters to consider in an inversion procedure. On the other hand, the viscosity η and permeability k_0 have only a weak influence (10^4 times smaller

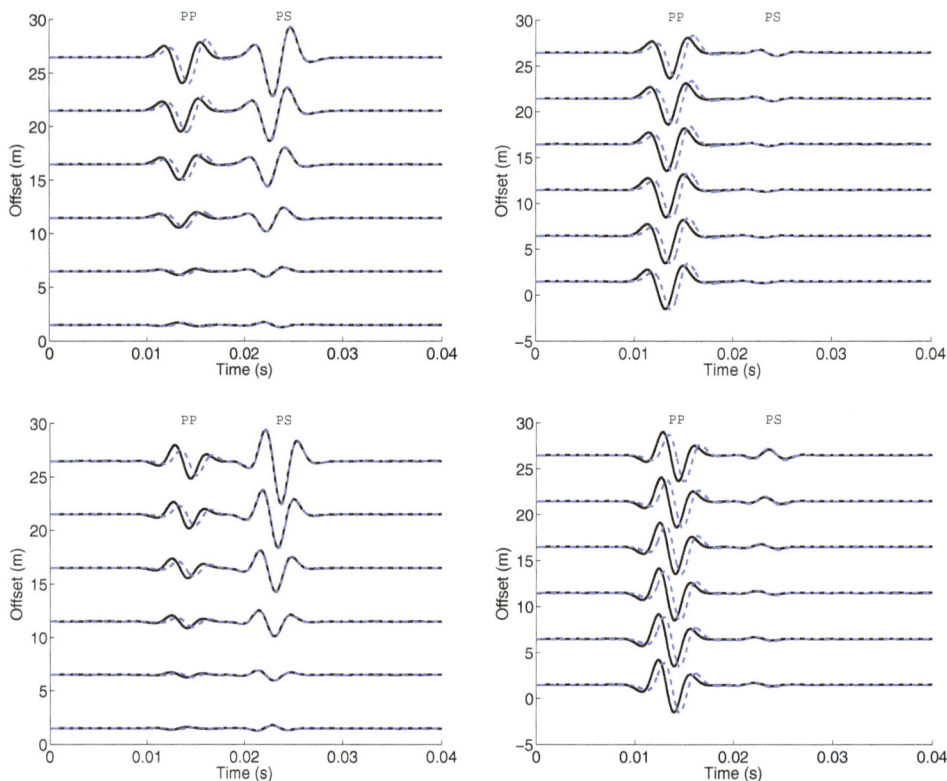

Fig. 5. Double porosity case: seismograms of top) solid displacement u_x (left) and u_z (right) and bottom) relative fluid/solid displacements w_x (left) and w_z (rigth). The setup is the same as for figure 4, the explosive source (Ricker with a 200Hz peak frequency) is in a quasi-elastic layer, while receivers stand in a double porosity half space. The double porosity solution is indicated by a continuous black line and the effective single porosity by a dashed blue line. PP and PS stand for the converted P and S waves, respectively.

than the porosity) on the wave amplitudes. The inversion for the parameters with such a low influence on the seismic waves will therefore be very delicate if other parameters are imperfectly known.

Figure 6 (right) shows the sensitivity of the P waves to the the fluid modulus K_f and the mineral solid modulus K_s. We note that the fluid modulus K_f has a stronger influence than the solid modulus K_s if the medium is poorly consolidated. The inverse is true for a consolidated medium. Similar patterns can be observed with the porosity: the higher the porosity, the stronger the influence of the fluid on the seismic waves. This means that it will be easier to determine fluid properties for an unconsolidated medium. For example, fluid substitution due to CO_2 injection leads to clear bright spot in Sleipner area (Norway, Arts et al., 2004), where the medium is poorly consolidated. The same set-up in stiff rocks does not produce such clear images, like in Weyburn field (Canada, White, 2009)

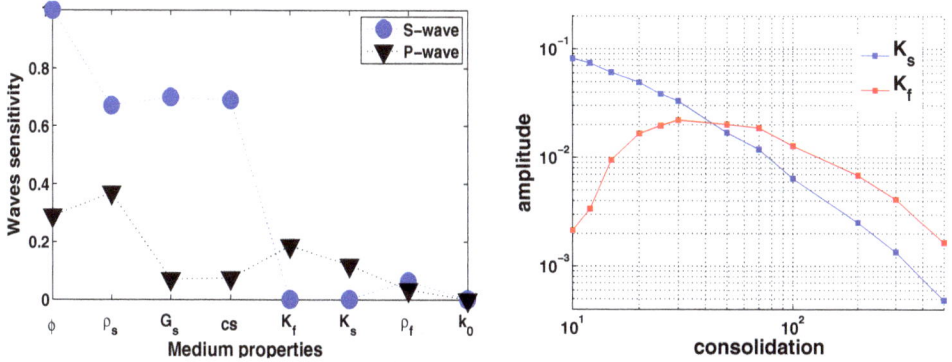

Fig. 6. Left) Sensitivity of the reflected waves to the porous parameters for the P (downward triangles) and S (circles) waves, i.e., maximum of the energy reflected back by a perturbation of the porous parameters. Right) Sensitivity of the reflected waves to the fluid K_f and the mineral bulk modulus K_s as a function of the consolidation parameter cs. Note that higher consolidation parameter cs corresponds to softer materials.

To evaluate the coupling between parameters, we look at the Amplitude Versus Angle (AVA) curves in figure 7 for the PP (left) and SS (right) reflected waves due to a small and localized perturbation of a model parameter. We note that for some parameters, the model perturbations lead to similar modifications of the seismic response. For example, perturbations in densities and permeability show identical AVA responses. The same is true for the bulk moduli. This strong coupling between parameters will prevent simultaneous reconstruction of these parameters in an inversion process.

Morency et al. (2009) also investigated the sensitivity of the seismic waves in porous media. They determined finite-frequency kernels based upon adjoint methods and investigated different parameter sets, in order to find the set that leads to the minimal coupling between parameters. They concluded that decomposing the input parameters into seismic velocities is the most stable approach in an inversion code.

5. Full waveform inversion

Full waveform inversion has shown to be an efficient and accurate tool to study the subsurface in the acoustic and elastic wave theory (Brossier et al., 2009). Historically, most of the FWI methods (Lailly, 1983; Tarantola, 1984) have been implemented under the acoustic approximation, for 2D model reconstruction (e.g. Gauthier et al., 1986; Pratt et al., 1998) or 3D structures (for instance, Ben-Hadj-Ali et al., 2008; Sirgue et al., 2008). Applications to real data are even more recent (Hicks & Pratt, 2001; Operto et al., 2006; Pratt & Shipp, 1999). The elastic case is more challenging, as the coupling between P and S waves leads to ill-conditioned problems. Since the early works of Mora (1987) and Kormendi & Dietrich (1991), the elastic problem has been addressed several times over the last years with methodological developments (Brossier et al., 2009; Choi et al., 2008; Gélis et al., 2007).

Using a poroelastic theory makes the problem even more difficult, especially because it adds much more unknowns. To the best of our knowledge, the first attempt to solve this

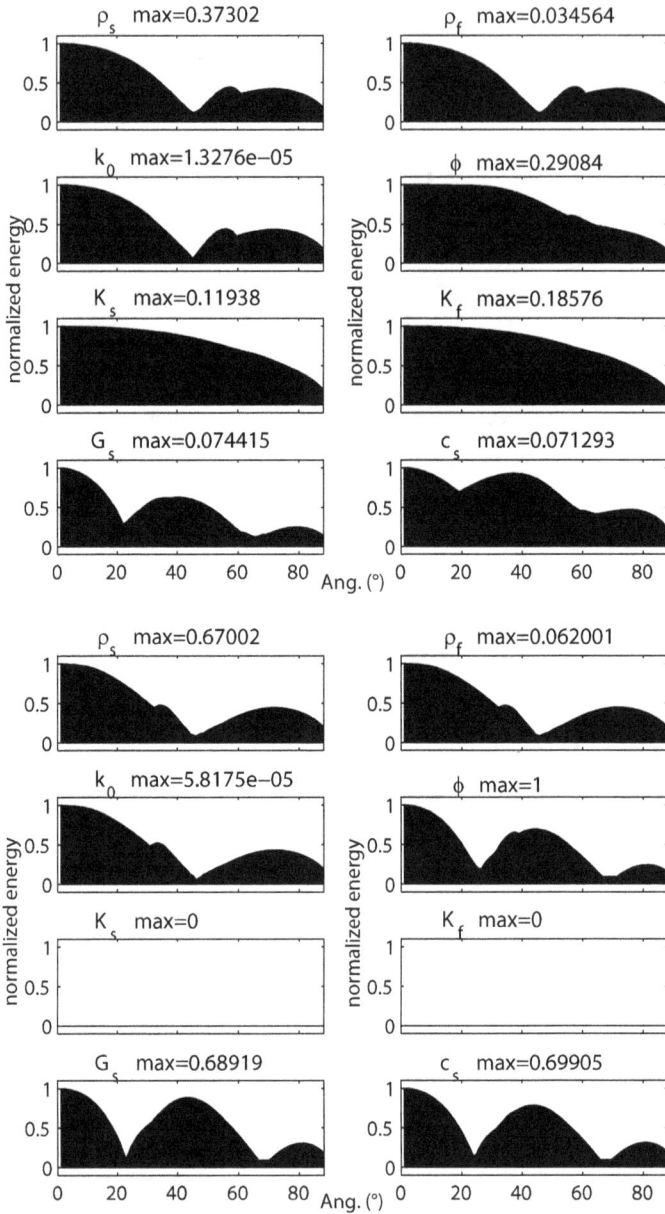

Fig. 7. Energy of plane waves reflected from perturbations in ρ_s, ρ_f, k_0, ϕ, K_s, K_f, G_s, and cs, as a function of incidence angle. The eight upper panels and eight lower panels correspond to PP and SS reflections, respectively. The curves are normalized with respect to the maximum value indicated above each panel.

problem was made by De Barros & Dietrich (2008) and De Barros et al. (2010) for stratified media and by Morency et al. (2009) and Morency et al. (2011) in 3-dimensional media. In the following sections, we will describe the main results obtained by De Barros & Dietrich (2008) and De Barros et al. (2010).

5.1 Inversion algorithm

Our method to determine the intrinsic properties of porous media is based on a full waveform iterative inversion procedure. It is carried out with a gradient technique to infer an optimum model which minimizes a misfit function. The latter is defined by a sample-to-sample comparison of the observed data d_{obs} with a synthetic wavefield $d = f(m)$ in the time-space domain, and by an equivalent term describing the deviations of the current model m with respect to an *a priori* model m_0, i.e.,

$$S(m) = \frac{1}{2} ||d - d_{obs}||_D + ||m - m_0||_M, \tag{6}$$

where the L2-norms $|| \cdot ||_D$ and $|| \cdot ||_M$ are defined in terms of a data covariance matrix C_D and an *a priori* model covariance matrix C_M Tarantola (1987). The model m contains the description of one or several parameters in layers whose thicknesses are defined by the peak content of the data (Kormendi & Dietrich, 1991). The model is updated using a quasi-Newton algorithm Tarantola (1987), which involves the Fréchet derivatives obtained earlier. As this problem is strongly non-linear, several iterations are necessary to converge toward an optimum model m, i.e, a model whose response d satisfactorily fits the observed data d_{obs}.

5.2 Numerical results

In order to determine the accuracy of the inversion procedure for the different model parameters considered, we first invert for a single parameter, in this case the mineral density ρ_s, and keep the others constant. The true model to reconstruct and the initial model used to initialize the iterative inversion procedure (which is also the *a priori* model) are displayed in figure 8. The other parameters are assumed to be perfectly known. Their vertical distributions consist of four 250 m thick homogeneous layers. Parameters ϕ, c_s and k_0 decrease with depth while parameters ρ_f, K_s, K_f and G_s are kept strictly constant.

Vertical-component seismic data (labelled DATA, fig 9) are then computed from the true model for an array of 50 receivers spaced 20 metres apart at offsets ranging from 20 to 1000 metres from the source. The latter is a vertical point force whose signature is a perfectly known Ricker wavelet with a central frequency of 25 Hz. Source and receivers are located at the free surface. As mentioned previously, direct and surface waves are not included in our computations to avoid complications associated with their contributions. Figure 9 also shows the seismogram (labelled INIT) at the beginning of the inversion, i.e., the seismogram computed from the starting model. Figure 8 shows that the true model, which consists in 10 metre thick layers from the surface to 1000 metre depth, is very accurately reconstructed by inversion. As there are no major reflectors in the deeper part of the model, very little energy is reflected toward the surface, which leads to some minor reconstruction problems at depth. In figure 9, we note that the final synthetic seismograms (SYNT) almost perfectly fit the input data (DATA) as shown by the data residuals (RES) which are very small.

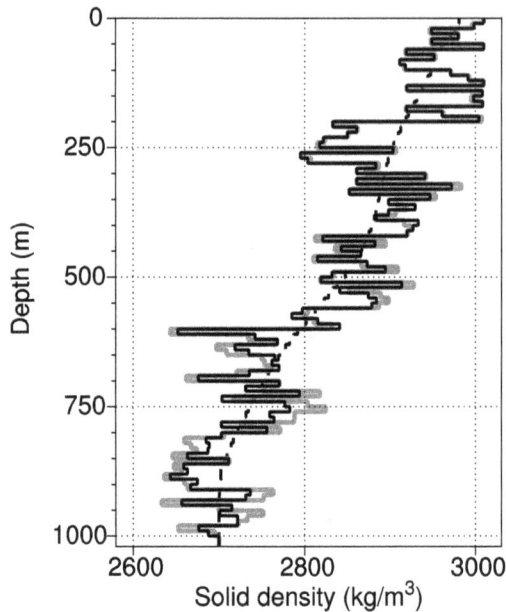

Fig. 8. Models corresponding to the inversion for the mineral density ρ_s: initial model, which is also the *a priori* model (dashed line), true model (thick grey line), and reconstructed model (black line). The corresponding seismograms are shown in figure 9.

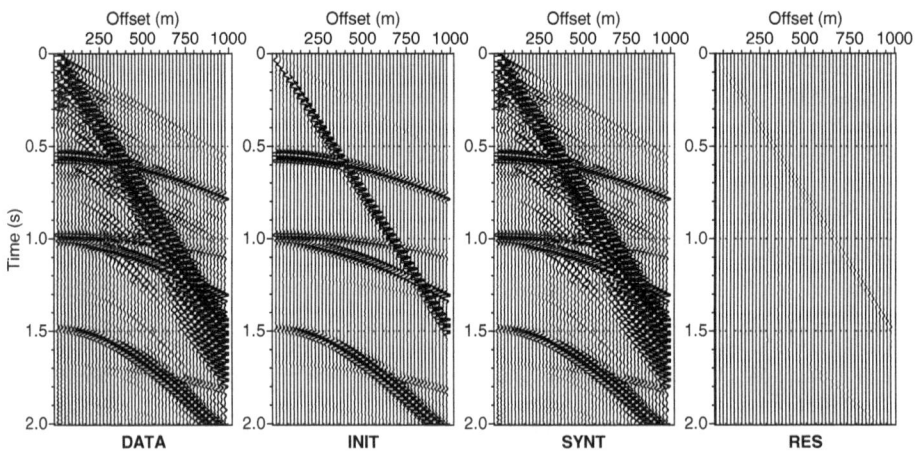

Fig. 9. Seismograms corresponding to the inversion for the mineral density ρ_s: synthetic data used as input (DATA), seismograms associated with the initial model (INIT), seismograms obtained at the last iteration (SYNT), and data residuals (RES) computed from the difference between the DATA and SYNT sections for the models depicted in figure 8. For convenience, all sections are displayed with the same scale, but the most energetic signals are clipped.

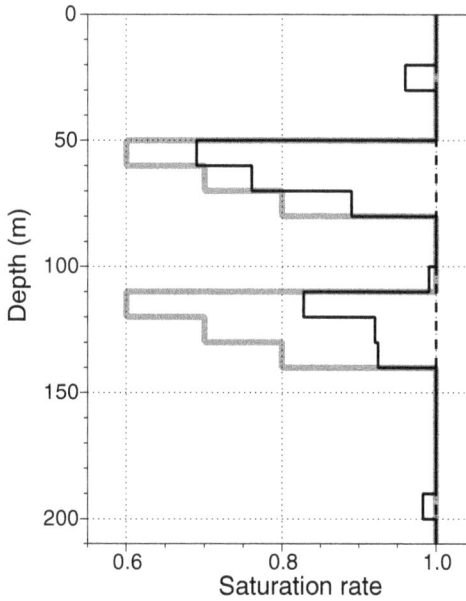

Fig. 10. Models corresponding to the differential inversion for the water saturation S_r : Initial model (dashed line), true model (thick grey line), and reconstructed model (black line).

The inversions carried out for the ϕ, ρ_f, K_s, K_f, G_s and c_s parameters (not shown) exhibit the same level of accuracy. However, as predicted by De Barros & Dietrich (2008) and Morency et al. (2009) with two different approaches, the weak sensitivity of the reflected waves to the permeability does not allow us to reconstruct the variations of this parameter. Being related to seismic wave attenuation and fluid flow, permeability appears as not only the most difficult parameter to estimate but also the one which would have the greatest benefits to the characterization of porous formations, notably in the oil industry (Pride et al., 2003). One possibility to estimate it is to measure the fluid motion, or, by reciprocity, to use fluid pressure sources.

As observed in the sensitivity study, parameters are strongly coupled. Multiparameter inversion is thus an ill-posed problem, which is, in most of the cases, not reliable, as errors on one parameter will map into the reconstruction of the other parameters. The use of analytical expressions for the sensitivity kernels allows an easy rearrangement of the parameter set, in order to invert for the most pertinent parameters. Using some *a priori* information, it is then possible to efficiently decrease the number of unknown parameters. For example, there is no reason to invert for both fluid parameters ρ_f and K_f, if we know that pores are filled by either gas or water. It is much more efficient to invert only for the saturation rate.

5.3 Differential inversion

To reduce the ambiguities of multiparameter inversion, a differential inversion has been considered and implemented (De Barros et al., 2010). Instead of dealing with the full complexity of the medium, we concentrate on small changes in the subsurface properties such as those occurring over time in underground fluid-filled reservoirs. This approach may be particularly useful for time-lapse studies to follow the extension of fluid plumes or to assess the fluid saturation as a function of time.

For example, the monitoring of CO_2 underground storage sites mainly aims at mapping the CO_2 extension. Time lapse studies performed over the Sleipner CO_2 injection site in the North Sea (see e.g. Arts et al., 2004) highlight the variations of fluid content as seen in the seismic data after imaging and inversion. In this fluid substitution issue, the parameter of interest is the carbon dioxide/saline water relative saturation. A differential inversion process will allow us to free ourselves from the unknown model parameters. This approach is valid for any type of fluid substitution monitoring problem, such as water-table variation, gas and oil extraction or hydrothermal activity.

The first step in this approach is to perform a base or reference survey to estimate the solid properties before the fluid substitution occurs. When performing a multiparameter inversion, the model properties are poorly reconstructed in general. However, the seismic data are reasonably well recovered. Thus, in spite of its defects, the reconstructed model respects the wave kinematics of the input data. In other words, the inverted model provides a description of the solid earth properties which can be used as a starting model for subsequent inversions. The latter would be used to estimate the fluid variations within the subsurface from a series of monitor surveys (second step).

To test this concept, we perform an inversion for two strongly coupled parameters, namely the porosity and the consolidation parameter. The resulting models are then used as starting models. We perturb the fluid properties of the true model to simulate a fluid variation over time. Two 30-metre thick layers located between 50 and 80 metres depth and between 110 and 140 metres depth are water depleted due to gas injection. The water saturation then varies between 60 and 80% in these two layers (figure 10). Our goal is to estimate the fluid properties by inverting the seismic data for the water saturation.

The model obtained is displayed in figure 10. We see that the location and extension of the gas-filled layers are correctly estimated. The magnitude of the water saturation curve, which defines the amount of gas as a function of depth, is somewhat underestimated in the top gas layer but is nevertheless reasonably well estimated. In the bottom gas layer, the inversion procedure only provides a qualitative estimate of the water saturation. These computations show that the differential inversion approach is capable of estimating, with reasonably good quality, the variations of fluid content in the subsurface without actually knowing all properties of the medium.

6. Conclusion

Using a poroelastic theory is much more complex than an elastic or a visco-elastic theory. However, poroelastic theories are an attempt to quantitatively describe the attenuation processes from the physical properties of the geological material. Furthermore, at seismic frequencies, attenuation is dominated by the mesoscopic scale mechanism, involving fluid flow at the boundaries of any heterogeneities. Poroelastic theories intrinsically take into account this loss of energy, while the equivalent visco-elastic approach neglects it. As the near surface media are stronly heterogeneous, with strong lateral and vertical contrasts, and different fluids involved, one has to deal with full poroelastic theories to accurately consider attenuation and fluid-solid motions.

The sensitivity to the different parameters varies hugely among parameters, and parameters are strongly coupled. Using a poroelastic theory to reconstruct model properties is in its nature an ill-conditioned problem. It shows however very promising possibility for differential inversion, and for certain issues where the problems can be reduced to the determination of only few parameters.

Poroelastic theories are, of course, not perfect yet, as they fail to give an universal law to explain seismic wave attenuation and propagation. They are however the direction to go if one wants to use full waveform inversion to make quantitative imagery of the rock physics and subsurface fluids. In particular, the permeability is a key parameter for exploration; the reconstruction of such a parameter from seismic waves will necessitate the use of complex poroelastic theories. Development in this direction still has to be continued. In particular, for imagery problems, data have to be improved to get around the problems of coupling between parameters. This can be done by using the information carried by fluid motions, which can give new insights into the permeability and the fluid properties, or by exploring in deeper details the coupling between seismic and electromagnetic waves.

7. Acknowledgements

L. De Barros and G. S. O'Brien were partly funded by the department of Communications, Energy and Natural Resources (Ireland) under the National Geosciences programme 2007-2013. B. Dupuy, S. Garambois and J. Virieux are supported by the National Research Agency (ANR) "Captage et Stockage de CO_2" program (ANR-07-PCO2-002). The numerical computations were performed by using the computational facilities from SFI/HEA ICHEC (Ireland), the HPC national computer centers (France, CINES and IDRIS under the allocation 2010-046091 GENCI) and the HPC center of the Grenoble observatory (OSUG).

8. References

Arts, R., Eiken, O., Chadwick, A., Zweigel, P., des Meer, L. V. & Zinszner, B. (2004). Monitoring of CO_2 injected at Sleipner using time-laps seismic data., *Energy* 29: 1323–1392.

Auriault, J.-L., Borne, L. & Chambon, R. (1985). Dynamics of porous saturated media, checking of the generalized law of Darcy, *J. Acoust. Soc. Am.* 77(5): 1641–1650.

Ben-Hadj-Ali, H., Operto, S. & Virieux, J. (2008). Velocity model-building by 3-D frequency-domain, full waveform inversion of wide aperture seismic data, *Geophysics* 73(5): 101–117.

Berryman, J. & Wang, H. (2000). Elastic wave propagation and attenuation in a double-porosity dual permeability medium, *Int. J. Rock Mech.* 37: 63–78.

Biot, M. (1956). Theory of propagation of elastic waves in a fluid-saturated porous solid. I. Low-frequency range, II. Higher frequency range, *J. Acoust. Soc. Am.* 28: 168–191.

Bouchon, M. (1981). A simple method to calculate Green's functions for elastic layered media, *Bull. Seism. Soc. Am.* 71(4): 959–971.

Boutin, C., Bonnet, G. & Bard, P. (1987). Green functions and associated sources in infinite and stratified poroelastic media, *Geophys. J. Roy. Astr. Soc.* pp. 521–550.

Brossier, R., Operto, S. & Virieux, J. (2009). Seismic imaging of complex onshore structures by 2D elastic frequency-domain full-waveform inversion, *Geophysics* 74(6): 63–76.

Carcione, J. (1998). Viscoelastic effective rheologies for modelling wave propagation in porous media, *Geophysical Prospecting* 46: 249–270.

Carcione, J. M., Morency, C. & Santos, J. E. (2010). Computational poroelasticity - a review, *Geophysics* 75: A229–A243.

Choi, Y., Min, D.-J. & Shin, C. (2008). Two-dimensional waveform inversion of multi-component data in acoustic-elastic coupled media, *Geophysical Prospecting* 56(19): 863–881.

Dai, N., Vafidis, A. & Kanasewich, E. (1995). Wave propagation in heterogeneous porous media: A velocity-stress, finite-difference method, *Geophysics* 60(2): 327–340.

De Barros, L. & Dietrich, M. (2008). Perturbations of the seismic reflectivity of a fluid-saturated depth-dependent poroelastic medium, *J. Acoust. Soc. Am.* 123(3): 1409–1420.

De Barros, L., Dietrich, M. & Valette, B. (2010). Full waveform inversion of seismic waves reflected in a stratified porous medium, *Geophy. J. Int.* 182(3): 1543–1556.

de la Puente, J., Dumbser, M., Käser, M. & Igel, H. (2008). Discontinuous galerkin methods for wave propagation in poroelastic media., *Geophysics* 73(5): 77–97.

Dupuy, B., De Barros, L., Garambois, S. & Virieux, J. (2011). Wave propagation in heterogeneous porous media formulated in the frequency-space domain using a discontinuous galerkin method, *Geophysics* 76: N13–N21.

Garambois, S. & Dietrich, M. (2002). Full waveform numerical simulations of seismoelectromagnetic wave conversions in fluid-saturated stratified porous media, *J. Geophys. Res.* 107(B7): 2148–2165.

Gassmann, F. (1951). Über die elastizität poröser medien, *Vierteljahrsschrift der Naturforschenden Gesellschaft in Zürich* 96: 1–23.

Gauthier, O., Virieux, J. & Tarantola, A. (1986). Two dimensionnal nonlinear inversion of seismic waveforms: numerical results, *Geophysics* 51: 1387–1403.

Geertsma, J. & Smith, D. (1961). Some aspects of elastic wave propagation in fluid-saturated porous solid, *Soc. Pet. Eng. J.* 26: 235–248.

Gélis, C., Virieux, J. & Grandjean, G. (2007). Two-dimensional elastic full waveform inversion using Born and Rytov formulations in the frequency domain, *Geophy. J. Int.* 168(2): 605–633.

Gurevich, B., Zyrianov, V. & Lopatnikov, S. (1997). Seismic attenuation in finely layered porous rocks: Effects of fluid flow and scattering, *Geophysics* 62(1): 319–324.

Haartsen, M. & Pride, S. (1997). Electroseismic waves from point sources in layered media, *J. Geophys. Res.* 102(B11): 745–769.

Hicks, G. & Pratt, R. (2001). Reflection waveform inversion using local descent methods: Estimating attenuation and velocity over a gas-sand deposit, *Geophysics* 66: 598–612.

Johnson, D. (2001). Theory of frequency dependent acoustics in patchy-saturated porous media, *J. Acoust. Soc. Am.* 110(2): 682–694.

Johnson, D., Koplik, J. & Dashen, R. (1987). Theory of dynamic permeability and tortuosity in fluid-saturated porous media, *Journal of Fluid Mechanics.* 176: 379–402.

Johnson, D., Plona, T. & Kojima, H. (1994). Probing porous media with first and second sound. I. dynamic permeability., *Journal of Applied Physics* 76(1): 104–125.

Kennett, B. (1983). *Seismic Wave Propagation in Stratified Media*, 342 p, Cambridge University Press, Cambridge.

Kormendi, F. & Dietrich, M. (1991). Nonlinear waveform inversion of plane-wave seismograms in stratified elastic media, *Geophysics* 56(5): 664–674.

Korringa, J., Brown, R., Thompson, D. & Runge, R. (1979). Self-consistent imbedding and the ellipsoidal model for porous rocks, *J. Geophys. Res.* 84: 5591–5598.

Lailly, P. (1983). The seismic inverse problem as a sequence of before stack migrations, *in* J. B. Bednar, R. Redner, E. Robinson & A. Weglein (eds), *Conference on Inverse Scattering: Theory and Application*, Soc. Industr. Appl. Math.

Liu, X., Greenhalgh, S. & Zhou, B. (2009). Transient solution for poro-viscoacoustic wave propagation in double porosity media and its limitations, *Geophy. J. Int.* 178: 375–393.

Manzocchi, T., Carter, J. N., Skorstad, A., Fjellvoll, B., Stephen, K. D., Howell, J., Matthews, J. D., Walsh, J. J., Nepveu, M., Bos, C., Cole, J., Egberts, P., Flint, S., Hern, C., Holden, L., Hovland, H., Jackson, H., Kolbjornsen, O., MacDonald, A., Nell, P., Onyeagoro, K., Strand, J., Syversveen, A. R., Tchistiakov, A., Yang, C., Yielding, G. & Zimmerman, R. (2008). Sensitivity of the impact of geological uncertainty on production from faulted and unfaulted shallow marine oil reservoirs - objectives and methods, *Petroleum Geoscience* 14: 3–15.

Masson, Y. & Pride, S. (2010). Finite-difference modeling of Biot's poroelastic equations across all frequencies, *Geophysics* 75(2): 33–41.

Mavko, G. & Jizba, D. (1991). Estimating grain-scale fluid effects on velocity dispersion in rocks, *Geophysics* 56: 1940–1949.

Mora, P. (1987). Nonlinear two-dimensional elastic inversion of multioffset seismic data, *Geophysics* 52: 1211–1228.

Morency, C., Luo, Y. & Tromp, J. (2009). Finite-frequency kernels for wave propagation in porous media based upon adjoint methods, *Geophy. J. Int.* 179(2): 1148–1168.

Morency, C., Luo, Y. & Tromp, J. (2011). Acoustic, elastic and poroelastic simulations of co2 sequestration crosswell monitoring based on spectral-element and adjoint methods, *Geophy. J. Int.* 185(2): 955–966.

Morency, C. & Tromp, J. (2008). Spectral-element simulations of wave propagation in porous media, *Geophy. J. Int.* 175(1): 301–345.

O'Brien, G. (2010). 3D rotated and standard staggered finite-difference solutions to Biot's poroelastic wave equations: Stability condition and dispersion analysis, *Geophysics* 75(4): T111–T119.

Operto, S., Virieux, J., Dessa, J. X. & Pascal, G. (2006). Crustal-scale seismic imaging from multifold ocean bottom seismometer data by frequency- domain full-waveform tomography: application to the eastern nankai trough, *J. Geophys. Res.* 111: B09306.

Philippacopoulos, A. (1997). Buried point source in a poroelastic half-space, *J. Engineering Mechanics* 123(8): 860–869.

Plona, T. (1980). Observation of a second bulk compressional wave in a porous medium at ultrasonic frequencies, *Appl. Phys. Lett.* 36(4): 259–261.

Pratt, R., Shin, C. & Hicks, G. (1998). Gauss-Newton and full Newton methods in frequency-space seismic waveform inversion, *Geophy. J. Int.* 133(22): 341–362.

Pratt, R. & Shipp, R. (1999). Seismic waveform inversion in the frequency domain, part II: Fault delineation in sediments using crosshole data, *Geophysics* 64(3): 902–914.

Pride, S. (2005). Relationships between seismic and hydrological properties, in *Hydrogeophysics*, Water Science and Technology Library, Springer, chapter 8, pp. 253–284.

Pride, S. & Berryman, J. (2003a). Linear dynamics of double-porosity dual-permeability materials, i. governing equations and acoustic attenuation, *Physical Review E* 68: 1–211.

Pride, S. & Berryman, J. (2003b). Linear dynamics of double-porosity dual-permeability materials, ii. fluid transport equations, *Physical Review E* 68: 1–211.

Pride, S., Berryman, J. & Harris, J. (2004). Seismic attenuation due to wave-induced flow, *J. Geophys. Res.* 109: B01201.

Pride, S., Gangi, A. & Morgan, F. (1992). Deriving the equations of motion for porous isotropic media, *J. Acoust. Soc. Am.* 92(6): 3278–3290.

Pride, S., Harris, J., Johnson, D., Mateeva, A., Nihei, K., Nowack, R., Rector, J., Spetzler, H., Wu, R., Yamamoto, T., Berryman, J. & Fehler, M. (2003). Permeability dependence of seismic amplitudes, *The Leading Edge* 22: 518–525.

Pride, S., Tromeur, E. & Berryman, J. (2002). Biot slow-wave effects in stratified rock, *Geophysics* 67: 201–211.

Santos, J. E., Douglas, J., Corbero, J. & Lovera, O. M. (1990). A model for wave propagation in a porous medium saturated by a two-phase fluid, *J. Acoust. Soc. Am.* 87(4): 1439–1448.

Santos, J. E., Ravazzoli, C. L. & Geiser, J. (2006). On the static and dynamic behavior of fluid saturated composite porous solids: A homogenization approach, *International Journal of Solids and Structures* 43(5): 1224–1238.

Sirgue, L., Etgen, J. & Albertin, U. (2008). 3D frequency domain waveform inversion using time domain finite difference methods, *Extended Abstracts, 70th EAGE Conference & Exhibition, Rome*, number F022.

Tarantola, A. (1984). The seismic reflection inverse problem, in *Inverse Problems of Acoustic and Elastic Waves*, SIAM, Philadelphia, pp. 104–181.

Tarantola, A. (1987). *Inverse Problem Theory: Methods for data fitting and model parameter estimation*, Elsevier, Amsterdam.

White, D. (2009). Monitoring CO2 storage during EOR at the Weyburn-Midale Field, *The Leading Edge* 28(7): 838–842.

9

Wave Propagation from a Line Source Embedded in a Fault Zone

Yoshio Murai
Hokkaido University
Japan

1. Introduction

The major crustal faults are not a single fault but form fault zones. The fault zone consists of several fault segments (e.g., Tchalenko, 1970; Tchalenko & Berberian, 1975). Each fault segment contains many cracks on a small scale. It is revealed from analyses of shear-wave splitting and P-wave polarization anomalies that parallel cracks are densely distributed in a fault zone (Leary et al., 1987; Li et al., 1987). Televiewer observations in boreholes also reveal the presence of a distribution of parallel cracks within a fault zone (Malin et al., 1988; Leary et al., 1987). Moreover it is revealed from seismic observations that a fault zone is characterized as a lower velocity zone than the surrounding intact rocks (Mooney & Ginzburg, 1986) and low-Q area (Kurita, 1975; Li et al., 1994). When an earthquake occurs in the fault zone, the following seismic waveforms are observed in the fault zone: the P and S headwaves refracted along the cross-fault material contrast (Ben-Zion & Malin, 1991; Hough et al., 1994) and seismic waves trapped in a low-velocity zone (e.g., Li & Leary, 1990; Li et al., 1994). It is important to determine the fault zone structure for the purposes of earthquake prediction and strong motion prediction. It is necessary to achieve it that the observational data should be simulated by means of theoretical studies.

In this study, we compute synthetic seismograms of the displacement field radiated from a seismic source embedded in a fault zone in order to simulate fault zone trapped waves. We assume a low-velocity zone and/or a zonal distribution of parallel cracks as a fault zone and investigate SH wave propagation in a 2-D elastic medium. We use the method introduced by Murai & Yamashita (1998) for the zonal distribution of parallel cracks. This method of analysis has advantages that multiple elastic wave scattering due to a large number of densely distributed cracks is easily treated and that the velocity contrast can be easily introduced. Finally, we try to simulate the fault zone trapped waves observed by Li et al. (1994) and estimate the crack size and the density of crack distribution.

2. Models of a fault zone

We assume following five models as a fault zone as shown in Fig. 1.

1. A zone of densely distributed parallel cracks (Fig. 1a). All the cracks are assumed to have the same length $2a$ and the same strike direction, which coincides with the X-axis. All the crack surfaces are assumed to be stress-free. Cracks are distributed periodically

with the spacings of d_X and d_Y in the X and Y directions, respectively. The number density of cracks ν is given by $\nu = 1/d_X d_Y$. The cracks are distributed along the line $Y=(j-1)d_Y$ ($j=1, \cdots, N$), where N is the number of crack arrays in the Y direction. We determine the X-coordinate of the centre of cracks in each array $p_j + l d_X$ ($0 \le p_j < d_X$, $j=1, \cdots, N$, $l=0, \pm 1, \pm 2, \cdots$) by generating N uniform random numbers between 0 and 1, which are multiplied by d_X.

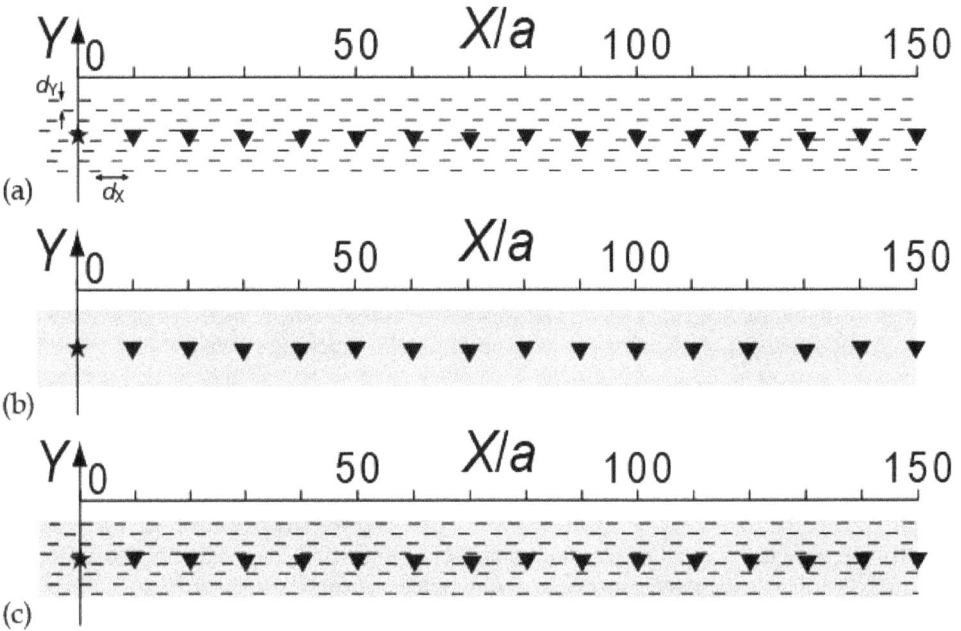

Fig. 1. Five models of a fault zone. A star denotes an isotropic line source located at the centre of the fault zone at $X=0$. The row of triangles represents observation stations. The spacing in the Y direction between the stations and the centre of the fault zone is $0.425a$, where a is half the crack length for fault zone models with distributed cracks and normalization length for those without cracks. (a) An example of Model (1). Cracks are distributed periodically with the density of $\nu a^2 = 0.1$. The crack spacings are $d_X = 5.88a$ and $d_Y = 1.7a$ in the X and Y directions, respectively. (b) Examples of fault zone Models (2), (3) and (4). The grey-shaded zone is an anisotropic zone, a low-velocity zone and an anisotropic low-velocity zone in Models (2), (3) and (4), respectively. The width of the grey-shaded zone is $h=13.6a$. The elastic constant of the anisotropic zone is $c_{2323}=0.711\mu$ for Model (2), where μ is the rigidity. The shear wave velocity β and density ρ of the low-velocity zone for Model (3) are $\beta/\beta_0 = 0.8$ and $\rho/\rho_0 = 0.93$, where β_0 and ρ_0 are shear wave velocity and density of the surrounding rocks, respectively. The elastic constant, density and rigidity of the anisotropic low-velocity zone for Model (4) are $c_{2323}=0.711\mu$, $\rho/\rho_0=0.93$ and $\mu/\mu_0=0.6$, respectively, where μ_0 is the rigidity of the surrounding rocks. (c) An example of Model (5). The grey-shaded zone is a low-velocity zone. The width, shear wave velocity and density of the low-velocity zone are $h=13.6a$, $\beta/\beta_0=0.8$ and $\rho/\rho_0=0.93$, respectively. The same crack distribution is assumed as Model (1) in (a)

2. A single anisotropic zone whose elastic constant is equivalent to that of the crack distribution Model (1) at the long-wavelength limit (Fig. 1b). The elastic constant c_{2323} is derived for the case of normal incidence to the crack surfaces by Murai (2007) as

$$\frac{c_{2323}}{\mu} = \frac{(2 - \pi v a^2)^2}{4},\tag{1}$$

where μ is the rigidity and the coordinate system (X, Y) is redefined as (x_1, x_3) for notation of the anisotropic media. The width of the cracked zone h is defined as $h=Nd_Y$.

3. A low-velocity zone with β/β_0 and ρ/ρ_0, where β is the shear wave velocity and ρ is the density, and the subscript 0 denotes the surrounding rocks (Fig. 1b). The rigidity μ is obtained by $\mu=\rho\beta^2$.
4. An anisotropic low-velocity zone (Fig. 1b). We assume the same elastic constant of c_{2323}/μ as Model (2) and the same β/β_0, ρ/ρ_0 and μ/μ_0 as Model (3).
5. A low-velocity zone with densely distributed parallel cracks (Fig. 1c). We assume the same crack distribution as Model (1) and the same β/β_0, ρ/ρ_0 and μ/μ_0 as Model (3).

3. Formulation

The seismic source displacement field is represented as a superposition of homogeneous and inhomogeneous plane waves propagating at discrete angles. This discretization results from a periodicity assumption in the description of the source (Bouchon & Aki, 1977).

The harmonic waves radiated from a line source in an infinite homogeneous medium can be represented as a continuous superposition of homogeneous and inhomogeneous plane waves. Therefore, the displacement u_s in wavenumber domain from the seismic source located at the origin of the coordinate system (x, y) can be written in the form,

$$u_s(x,y) = \int_{-\infty}^{\infty} f(s,y)e^{isx}ds,\tag{2}$$

where i is the imaginary unit and s is the x-component of the wavenumber. The time factor $\exp(-i\omega t)$ is omitted for brevity, where $\omega=k\beta$, k is the wavenumber. When such sources distribute along the x-axis at equal interval Δx_s, eq.(2) is reduced to

$$u_{ss}(x,y) = \frac{2\pi}{\Delta x_s} \sum_{l=-\infty}^{\infty} f(s_l,y)e^{is_l x},\tag{3}$$

according to Bouchon & Aki (1977), where u_{ss} is the displacement from periodically distributed sources and $s_l=2\pi l/\Delta x_s$. If the series converges, eq.(3) can be approximated by the finite sum equation

$$u_{ss}(x,y) = \frac{2\pi}{\Delta x_s} \sum_{l=-N_d}^{N_d} f(s_l,y)e^{is_l x}.\tag{4}$$

Thus the seismic source displacement field is represented as a superposition of the discrete plane waves.

Now we give the wavenumber a small imaginary part to remove the singularities of $f(k, y)$ as

$$k = k_R + ik_I. \tag{5}$$

The resulting attenuation is used to minimize the influence of the neighboring fictitious sources. The effect of the imaginary part of the wavenumber can be removed from the final time domain solution. When the solution in wavenumber domain by using the complex wavenumber is denoted by $U(k)$, the solution in time domain $u(t)$ is obtained through the relation

$$u(t) = \frac{\beta}{2\pi} e^{\beta k_I t} \int_{-\infty}^{\infty} U(k) e^{-i\beta k_R t} dk. \tag{6}$$

We consider an isotropic line source. The displacement field radiated from a source is written as

$$u_s(x, y) = \frac{i}{4} H_0^{(1)}(kR), \tag{7}$$

where $R^2 = x^2 + y^2$, and $H_0^{(1)}(\cdots)$ is the Hankel function. We employ the relation by Morse & Feshbach (1953)

$$H_0^{(1)}(kR) = \frac{1}{\pi} \int_{-\infty}^{\infty} \frac{\exp[i\{\sqrt{k^2 - s^2}\, |y| + sx\}]}{\sqrt{k^2 - s^2}} ds. \tag{8}$$

Then eq.(7) can be written as

$$u_s(x, y) = \frac{i}{4\pi} \int_{-\infty}^{\infty} \frac{\exp[i\{\sqrt{k^2 - s^2}\, |y|\}]}{\sqrt{k^2 - s^2}} e^{isx} ds. \tag{9}$$

Therefore, $f(k, y)$ in eq.(2) can be determined to be

$$f(s, y) = \frac{i}{4\pi} \frac{\exp[i\{\sqrt{k^2 - s^2}\, |y|\}]}{\sqrt{k^2 - s^2}}. \tag{10}$$

For the anisotropic media of Models (2) and (4), eq.(10) is modified as

$$f(s, y) = \frac{i}{4\pi} \sqrt{\frac{\mu}{c_{2323}}} \frac{\exp\left[i\sqrt{\frac{\mu}{c_{2323}}} \sqrt{k^2 - s^2}\, |y|\right]}{\sqrt{k^2 - s^2}}. \tag{11}$$

The velocity contrast can be introduced easily because we have only to treat plane waves. The wave field in a fault zone can be calculated by the wave propagator method (Kennett, 1983) or the reflection and transmission operator method (Kennett, 1984) by use of the discretization results. Moreover we can calculate the displacement field radiated from a seismic source embedded in a fault zone for Models (1) and (5) on the basis of this expansion because the seismic wave propagation in a zone of densely distributed parallel cracks for incident plane waves can be calculated by the method introduced by Murai & Yamashita (1998).

4. Synthetic seismograms

We consider the fault zone Model (1) as shown in Fig. 1(a). We assume 8 crack arrays which are $1.7a$ apart each other in the Y direction. A seismic source is located at $(0, 5.95a)$ and is assumed to be isotropic. Observation stations are located along the line $Y=6.375a$. Thus both the source and the stations are located near the centre of the fault zone. The synthetic seismograms for 15 stations in the range $10a \leq X \leq 150a$ are shown in Fig. 2. The seismograms in the time domain are obtained by the Fourier transform of the wavenumber domain solutions for 134 wavenumbers in the range from $ka=0.025$ to $ka=3.35$. We use the Ricker wavelet as the source time function; the characteristic nondimensional wavenumber of the wavelet, $k_c a$, is assumed to be 1.0. Fig. 2 shows the wave trains scattered by cracks following the direct wave at only the stations neighboring the source. Moreover the wave trains contain the dominant high wavenumber components. Thus we cannot simulate the relatively long-period fault zone trapped waves for the events with various focal distances. Because the cracked zone of Model (1) is equivalent to a single anisotropic zone at the low wavenumber limit, we consider the fault zone Model (2) as shown in Fig. 1(b). The elastic constant c_{2323} is obtained as $c_{2323}=0.711\mu$ for $va^2=0.1$ by eq.(1). The width of the cracked zone h is defined as $h=Nd_Y$. The synthetic seismograms for 15 stations in the range $10a \leq X \leq 150a$ are shown in Fig. 3. We can see neither scattered waves nor fault zone trapped waves.

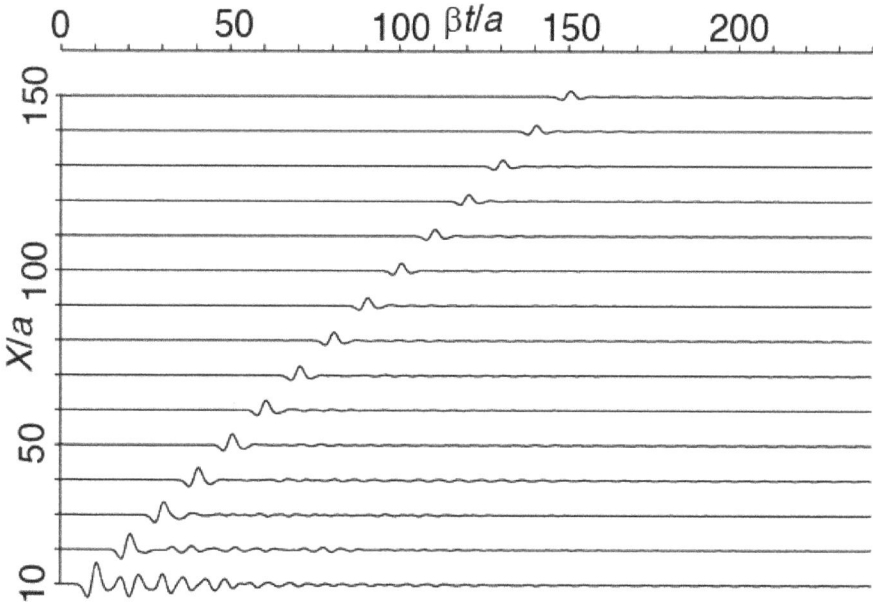

Fig. 2. The synthetic seismograms calculated for the fault zone Model (1) of crack distribution shown in Fig. 1(a). The characteristic nondimensional wavenumber of Ricker wavelet, $k_c a$, is 1.0

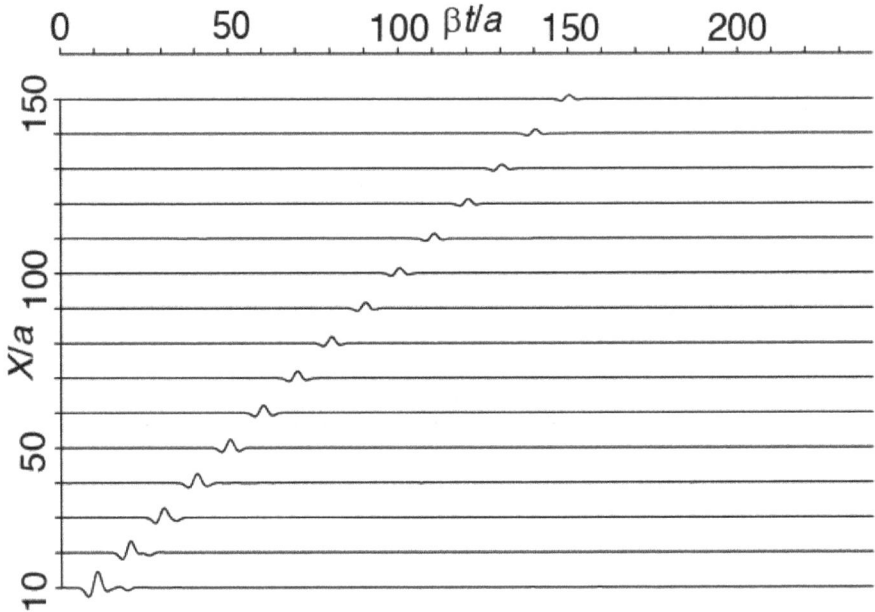

Fig. 3. The synthetic seismograms calculated for the fault zone Model (2) of the single anisotropic zone shown in Fig. 1(b). $k_c a$ is assumed to be 1.0

The results of Models (1) and (2) suggest that a low-velocity fault zone is necessary to excite trapped waves. Actually a fault zone is characterized as a low-velocity zone as stated in section 1. It is certainly the case that the phase velocity decreases remarkably in the fault zone for the waves propagating normal to the cracks, but the velocity is almost the same as the shear wave velocity of the matrix for the waves propagating parallel to the cracks in the assumed Models (1) and (2) because we consider only SH waves. The low-velocity zone is, however, considered to be attributed to fault gouge (Mooney & Ginzburg, 1986), which might include not only the parallel cracks but also randomly oriented microcracks; a velocity reduction is observed for the waves propagating to any direction.

We now consider the low-velocity fault zone Model (3) as shown in Fig. 1(b). The width of the low-velocity fault zone is the same as that of the anisotropic fault zone Model (2). The shear wave velocity and density of the low-velocity zone are assumed to be β/β_0=0.8 and ρ/ρ_0=0.93, respectively, which correspond to the rigidity of μ/μ_0=0.6. The synthetic seismograms are shown in Fig. 4. This figure shows the relatively long-period wave trains with relatively large amplitude closely following the direct waves with small amplitude at all the stations. These long-period wave trains are understood as trapped waves in the low-velocity zone because they are observed only in the low-velocity fault zone. Thus we can simulate the relatively long-period fault zone trapped waves. Therefore, an actual fault zone is considered to be low-velocity. In addition, Fig. 4 shows the headwaves refracted along the cross-fault material contrast, which are observed in actual fault zones as stated in section 1 (Ben-Zion & Malin, 1991; Hough et al., 1994). We cannot see scattered waves because there is no crack in the fault zone.

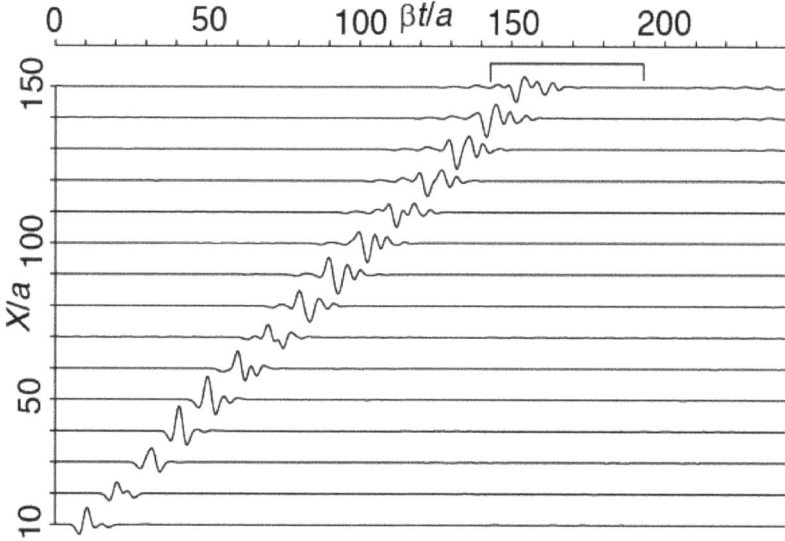

Fig. 4. The synthetic seismograms calculated for the fault zone Model (3) of the single low-velocity zone shown in Fig. 1(b). $k_c a$ is assumed to be 1.0. The bracket denotes a spectral time window of $50\beta/a$ including the direct wave and the trapped wave trains for the seismogram of the station at $X=150a$. The amplitude spectra are shown in Fig. 7(a)

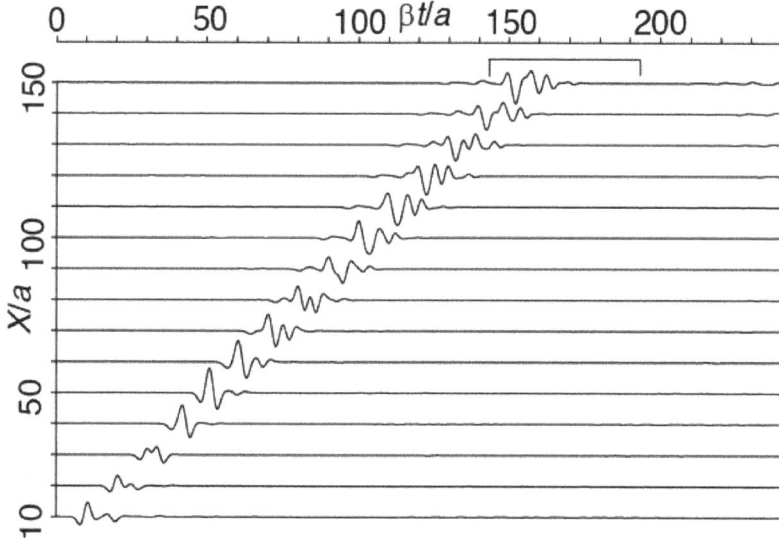

Fig. 5. The synthetic seismograms calculated for the fault zone Model (4) of the single anisotropic low-velocity zone shown in Fig. 1(b). $k_c a$ is assumed to be 1.0. The bracket denotes a spectral time window of $50\beta/a$ including the direct wave and the trapped wave trains for the seismogram of the station at $X=150a$. The amplitude spectra are shown in Fig. 7(b)

Next, we consider the anisotropic low-velocity fault zone Model (4) as shown in Fig. 1(b) in order to consider the effect of parallel crack distribution. The width of the anisotropic low-velocity fault zone is the same as those of Models (2) and (3). The elastic constant c_{2323}, density and rigidity of the anisotropic low-velocity zone are assumed to be $c_{2323}=0.711\mu$, $\rho/\rho_0=0.93$ and $\mu/\mu_0=0.6$, respectively as the same for Models (2) and (3), which correspond to the crack density of $\nu a^2=0.1$ and shear wave velocity of $\beta/\beta_0=0.8$. The synthetic seismograms are shown in Fig. 5. This figure shows the same characteristics as Fig. 4: we can observe the refracted headwaves, the direct waves with relatively small amplitude and the trapped wave trains and cannot see scattered waves. This is because the crack length is assumed to be much smaller than the incident wavelength in the anisotropic zone.

Finally, we consider a low-velocity zone with densely distributed parallel cracks (Fig. 1c) in order to consider the effect of a wavenumber dependence of the crack interactions in a fault zone. The width of the low-velocity fault zone is the same as those of Models (2), (3) and (4). We assume the same crack distribution as Model (1) and the same shear wave velocity and density of the low-velocity zone as Model (3): $\nu a^2=0.1$, $\beta/\beta_0=0.8$ and $\rho/\rho_0=0.93$ are assumed.

The synthetic seismograms are shown in Fig. 6. This figure shows the same characteristics as Figs. 4 and 5 of the wave propagation in a low-velocity fault zone. In addition, we can observe the wave trains scattered by cracks following the fault zone trapped waves.

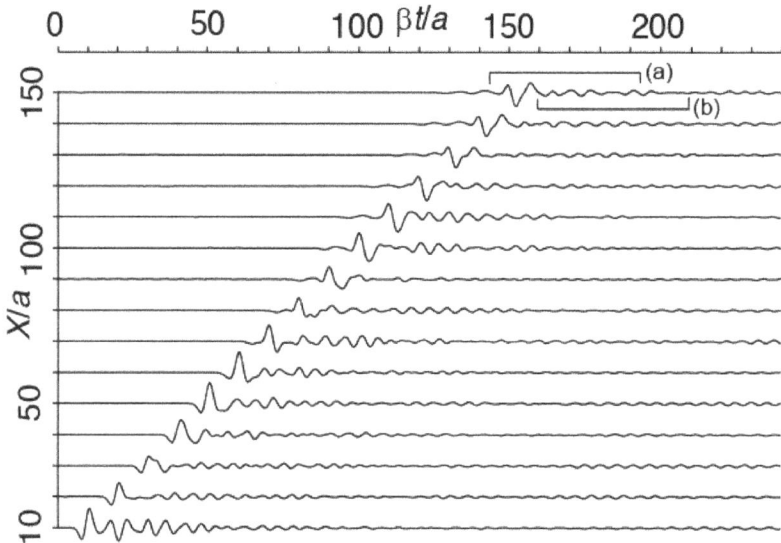

Fig. 6. The synthetic seismograms calculated for the fault zone Model (5) of the low-velocity zone with densely distributed parallel cracks shown in Fig. 1(c). We assume the same crack distribution as for Fig. 2. $k_c a$ is assumed to be 1.0. The brackets denote spectral time windows of $50\beta/a$ for the seismogram of the station at $X=150a$. The time window (a) is including the direct wave, the wave trains trapped in the low-velocity zone and the waves scattered by cracks whereas the time window (b) is including only the scattered wave trains. The amplitude spectra for the time windows (a) and (b) are shown in Figs. 7(c) and (d), respectively

5. Spectral analyses

In this section, we make a short analysis on the amplitude spectra for the fault zone trapped waves calculated in the previous section. Figs. 7(a) and (b) show the amplitude spectra calculated for the fault zone Models (3) and (4) illustrated in Fig. 1(b), respectively, and Figs. 7 (c) and (d) show those for Model (5) illustrated in Fig. 1(c). The amplitude spectra U are calculated by the following procedure. First, we compute the synthetic seismograms with the Ricker wavelet as the source time function; we assume the 4 Ricker wavelets whose $k_c a$ are 0.25, 0.5, 1.0 and 2.0. Next, we calculate the amplitude spectrum for each Ricker wavelet source time function in a time window of $50\beta/a$ using a cosine type window with $5\beta/a$ edge length. The spectral windows are shown in Figs. 4, 5 and 6 for the seismograms of the station at $X=150a$ with $k_c a=1.0$ as an example. Finally, the amplitude spectrum for each Ricker wavelet is normalized by that of each source time function to eliminate the contribution of source spectra. The amplitude spectra $U(k)$ are calculated for the wavenumber range $2k_c/3 \leq k \leq 4k_c/3$ from each Ricker wavelet with the characteristic wavenumber of k_c. Because 4 Ricker wavelets with $k_c a=0.25$, 0.5, 1.0 and 2.0 are assumed, the amplitude spectra are discontinuous at $ka=1/3$, $2/3$ and $4/3$.

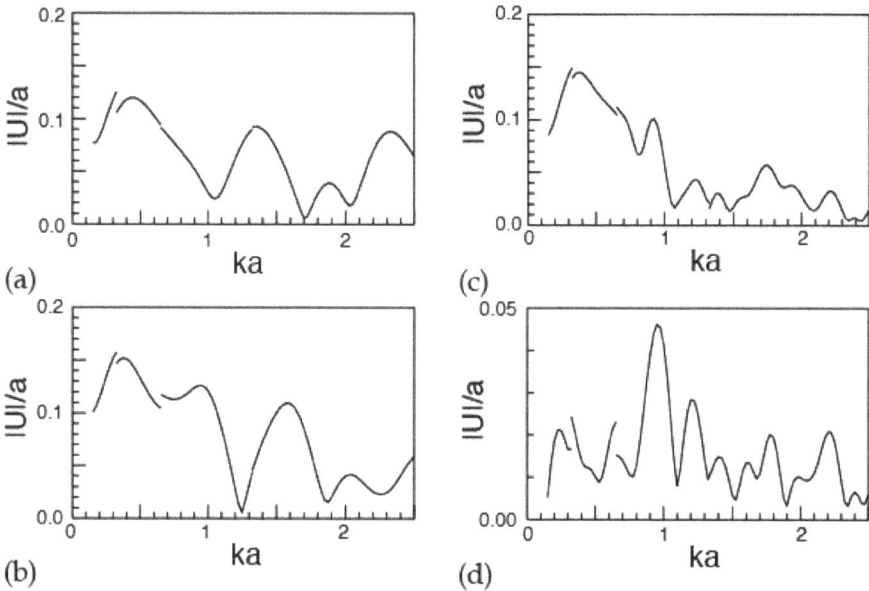

Fig. 7. The normalized amplitude spectra $|U(ka)|/a$ calculated for the seismograms of the observation station at $X=150a$. $|U(ka)|/a$ are calculated for the wavenumber ranges of $0.150 \leq ka \leq 0.325$, $0.325 \leq ka \leq 0.650$, $0.650 \leq ka \leq 1.325$ and $1.325 \leq ka \leq 2.650$ from Ricker wavelet source time functions with $k_c a=0.25$, 0.5, 1.0 and 2.0, respectively. (a) The amplitude spectra calculated for the fault zone Models (3) illustrated in Fig. 1(b). (b) The amplitude spectra calculated for the fault zone Models (4) illustrated in Fig. 1(b). (c) The amplitude spectra calculated for the fault zone Models (5) illustrated in Fig. 1(c). The amplitude spectra in (a), (b) and (c) correspond to the direct wave and the trapped wave trains of the seismogram in Figs. 4, 5 and 6, respectively. (d) The same as in (c) except for corresponding to the wave trains scattered by cracks

Fig. 7(a) shows the normalized amplitude spectra for the low-velocity fault zone Model (3). This figure corresponds to the seismogram of the station at $X=150a$ in Fig. 4. The amplitude spectra show the prominent peak at $ka=0.45$, which is understood to be formed by the waves trapped in the low-velocity zone. Fig. 7(b) shows the normalized amplitude spectra for the anisotropic low-velocity fault zone Model (4), which correspond to the seismogram of the station at $X=150a$ in Fig. 5. This figure also shows the prominent peaks at $ka=0.375$ and 0.95 formed by the fault zone trapped waves. The spectral peak split into two peaks because the interference of resonated waves in the layer occurs at the wavenumbers different from that for the isotropic medium due to the wave speed depending on the propagation direction. Fig. 7(c) shows the normalized amplitude spectra for the fault zone Model (5) of the low-velocity zone with densely distributed parallel cracks, which correspond to the seismogram in the time window (a) of the station at $X=150a$ in Fig. 6. The amplitude spectra show the prominent peaks at $ka=0.325$ in relatively low wavenumber range and at around $ka=1.0$ in relatively high wavenumber range. The low-wavenumber spectral peak is considered to be formed by the waves trapped in the low-velocity zone because it is seen in Figs. 7(a) and (b). The peak amplitude in the low wavenumber range is higher in Fig. 7(c) than that in Fig. 7(a). Moreover the existence of the cracks lowers the peak wavenumber at which the amplitude spectra take the peak value in the low wavenumber range. These phenomena occur because the crack distribution lowers the overall rigidity and velocity in the fault zone. The high-wavenumber spectral peak is caused only for the fault zone Model (5). Now we calculate the amplitude spectra for the wave trains scattered by cracks in order to investigate the high-wavenumber peak. The spectral time window (b) is denoted by the bracket in Fig. 6 for $k_c a=1.0$ as an example. Fig. 7(d) shows the normalized amplitude spectra for the scattered waves. This figure shows the most prominent peak at around $ka=1.0$ in relatively high

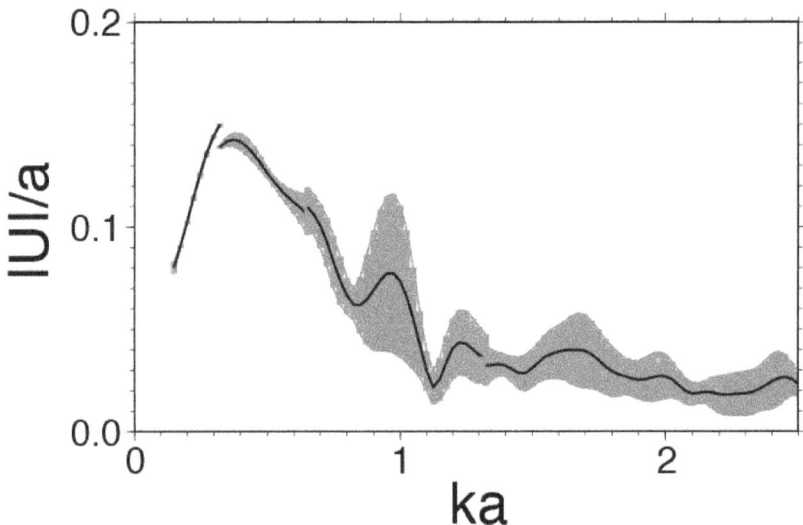

Fig. 8. The normalized amplitude spectra calculated from the direct wave and the trapped and scattered wave trains of the synthetic seismograms for the fault zone Model (5) illustrated in Fig. 1(c). Solid curve and grey-shaded range represent the mean values and the standard deviations of 201 stations in the range $100a \leq X \leq 150a$ along the line $Y=6.375a$

wavenumber range, therefore it is understood to be formed by scattered waves. This means that we can estimate the crack length in a fault zone from the peak frequency in the high frequency range if the spectral peak caused by the waves scattered by cracks is observable.

We calculate synthetic seismograms and amplitude spectra from the direct wave and the trapped wave trains of 201 stations in the range $100a{\leq}X{\leq}150a$ along the line $Y=6.375a$ of the fault zone Models (5) illustrated in Fig. 1(c) in order to investigate the spatial variation of the spectral peaks in relatively low and high wavenumber ranges. Fig. 8 shows the mean values and the standard deviations of the normalized amplitude spectra for the 201 stations. The amplitude of the low-wavenumber spectral peak is not attenuated among these observation stations because the long-period wave trains are trapped and propagating without geometrical spreading in the low-velocity zone. On the other hand, the amplitudes of the high-wavenumber spectral peak fluctuate greatly among stations although it is seen at most of the stations. The mean values of the amplitude spectra show the low-wavenumber peak is caused at $ka=0.375$ and the amplitude is $|U|/a=0.143$, and the high-wavenumber peak is at $ka=0.975$ with the amplitude of $|U|/a=0.077$.

6. Wave propagation in a fault zone containing densely distributed parallel cracks

In the previous section, the amplitude spectra show the prominent peaks in relatively low and high wavenumber ranges for the fault zone Model (5) of the low-velocity zone with densely distributed parallel cracks when both the source and stations are located near the centre of the fault zone. In this section, we investigate the amplitude spectra for the fault zone Model (5) with crack distributions different from that illustrated in Fig. 1(c). We assume the same width, shear wave velocity and density of the low-velocity zone as shown in Fig. 1(c). First, we consider 10 crack distributions with the same crack spacings of d_X and d_Y, which correspond to the same crack density of $va^2=0.1$ as shown in Fig. 1(c). Each of the above 10 crack distributions is determined by generating an independent sequence of random numbers in order to determine the X-coordinates of the centre of cracks. For each model, the mean values of the amplitude spectra are obtained by averaging over 201 stations by the same procedure as stated in the previous section. Each of 10 curves in Fig. 9(a) shows the mean values of the normalized amplitude spectra from the 201 stations for each of 10 models. The spectral peaks in relatively low and high wavenumber ranges are detected from each of the 10 curves in Fig. 9(a). The mean values and the standard deviations of both the wavenumbers and amplitudes of the spectral peaks for all the 10 models are computed. The low-wavenumber peak is caused at $ka=0.38{\pm}0.01$ and the amplitude is $|U|/a=0.142{\pm}0.001$, and the high-wavenumber peak is at $ka=0.94{\pm}0.16$ with the amplitude of $|U|/a=0.077{\pm}0.016$.

Next, we assume the crack densities of $va^2=0.075$ and 0.05 lower than above for the fault zone Model (5) although the same width, shear wave velocity and density of the low-velocity zone are assumed. We assume 2 sets of d_X and d_Y for each crack density: $d_X=5.88a$ and $d_Y=2.38a$ (Fig. 10a), and $d_X=7.84a$ and $d_Y=1.7a$ (Fig. 10b) for $va^2=0.075$, and $d_X=5.88a$ and $d_Y=3.97a$ (Fig. 11a), and $d_X=11.76a$ and $d_Y=1.7a$ (Fig. 11b) for $va^2=0.05$. We consider 10 crack distributions for each model with the different crack spacings, i.e., 20 distributions for each crack density in total.

Each of 20 curves in Fig. 9(b) shows the mean values of the normalized amplitude spectra from the 201 stations for each of 20 models for the crack density of va^2=0.075. The spectral peaks in relatively low and high wavenumber ranges are detected from each of the 20 curves in Fig. 9(b) and the mean values and the standard deviations of both the wavenumbers and amplitudes of the spectral peaks for all the 20 models are computed. The low-wavenumber peak is caused at ka=0.38±0.02 and the amplitude is $|U|/a$=0.137±0.001, and the high-wavenumber peak is at ka=0.98±0.14 with the amplitude of $|U|/a$=0.071±0.012. Each of 20 curves in Fig. 9(c) shows the mean values of the normalized amplitude spectra from the 201 stations for each of 20 models for the crack density of va^2=0.05. The low-wavenumber peak is caused at ka=0.39±0.01 and the amplitude is $|U|/a$=0.131±0.002, and the high-wavenumber peak is at ka=0.92±0.13 with the amplitude of $|U|/a$=0.074±0.011.

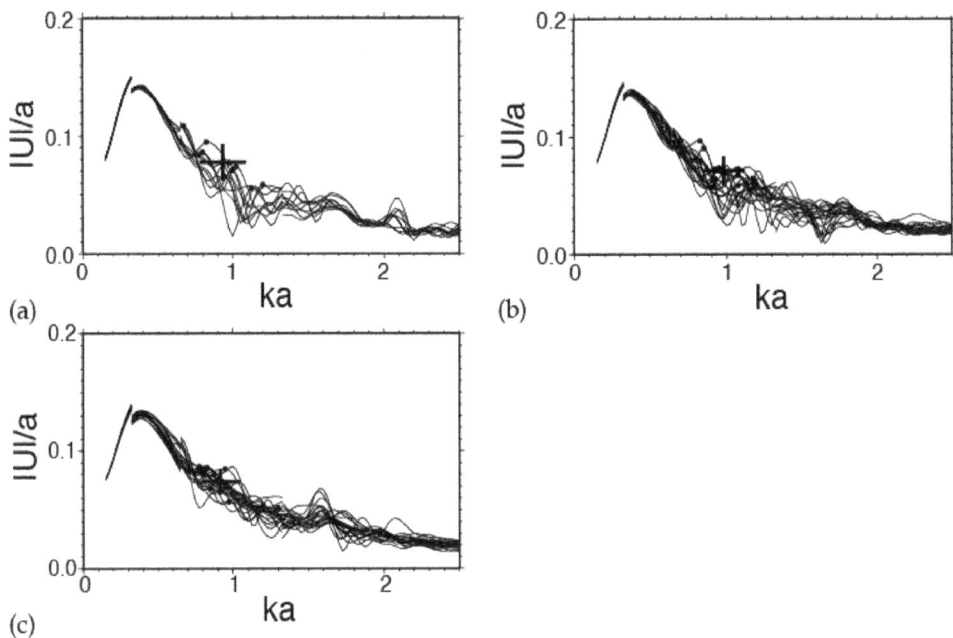

Fig. 9. The mean values of the normalized amplitude spectra obtained by averaging over 201 stations for each fault zone Model (5). The high-wavenumber spectral peak detected from each curve is denoted by a dot. The standard deviations of both the wavenumber and amplitude of the high-wavenumber spectral peaks for all the models are denoted by bars. The intersecting point of the bars represents the mean values. (a) Each of 10 curves shows the mean values of the amplitude spectra for each of 10 models with d_X=5.88a and d_Y=1.7a (va^2=0.1). (b) Each of 20 curves shows the mean values of the amplitude spectra for each of 20 models with va^2=0.075. Ten of 20 curves show the models with d_X=5.88a and d_Y=2.38a and other 10 of 20 show the models with d_X=7.84a and d_Y=1.7a. (c) Each of 20 curves shows the mean values of the amplitude spectra for each of 20 models with va^2=0.05. Ten of 20 curves show the models with d_X=5.88a and d_Y=3.97a and other 10 of 20 show the models with d_X=11.76a and d_Y=1.7a

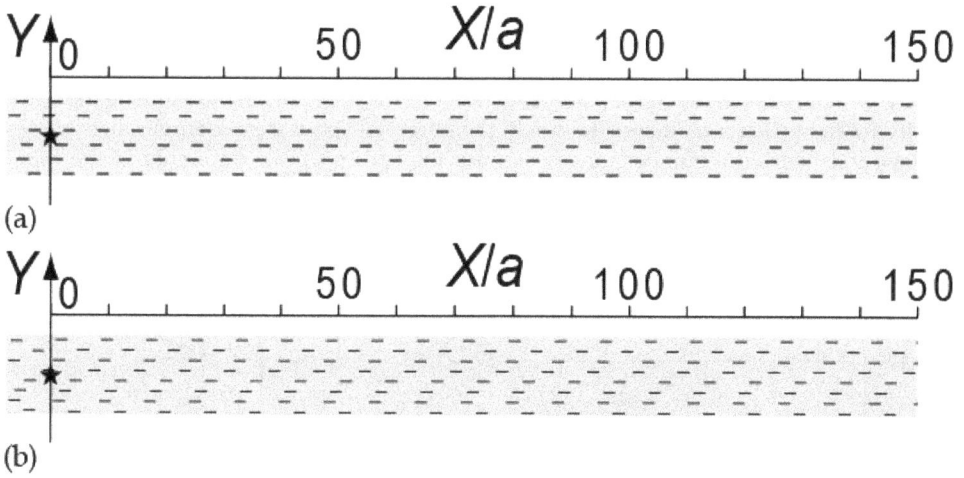

Fig. 10. The same as Fig. 1(c) except for $va^2=0.075$. (a) An example of fault zone Model (5) with $d_X=5.88a$ and $d_Y=2.38a$. (b) An example of fault zone Model (5) with $d_X=7.84a$ and $d_Y=1.7a$

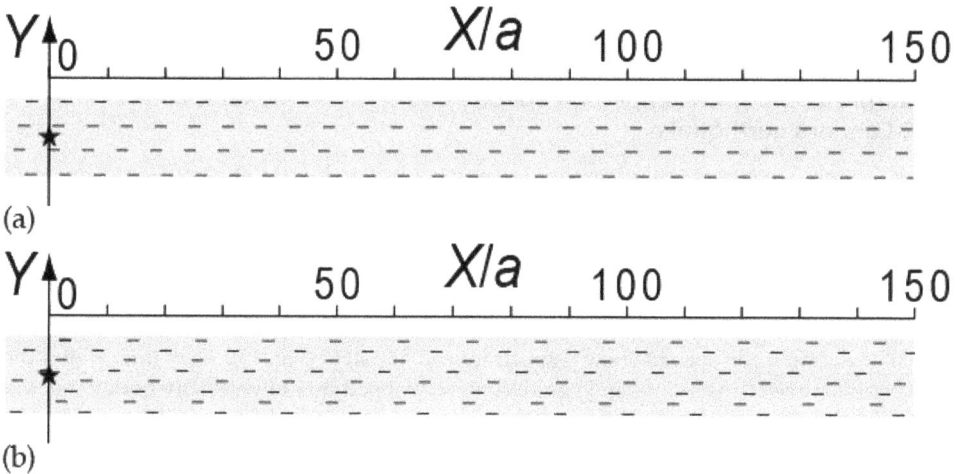

Fig. 11. The same as Fig. 10 except for $va^2=0.05$. (a) An example of fault zone Model (5) with $d_X=5.88a$ and $d_Y=3.97a$. (b) An example of fault zone Model (5) with $d_X=11.76a$ and $d_Y=1.7a$

We summarize the crack density dependence of the amplitudes of the spectral peaks in the relatively low and high wavenumber ranges (Fig. 9) in Figs. 12(a) and (b), respectively. Fig. 12(a) shows the larger amplitude for the higher crack density as for the spectral peaks at around $ka=0.4$ in the relatively low wavenumber range. On the other hand, the amplitudes of the spectral peaks at around $ka=1.0$ in the relatively high wavenumber range fluctuate greatly among models of crack distribution and show no clear dependency on the crack density (Fig. 12b). Thus the spectral peak amplitude in the low-wavenumber range becomes larger relative to that in the high-wavenumber range for higher crack density. The spectral peak amplitude in the low-wavenumber range does not depend only on the crack density but also depends on the shear wave velocity, density and the width of the low-velocity zone because it is formed by the waves trapped in the low-velocity zone. Therefore, the crack density cannot be estimated from it alone. However, it will be possible to estimate the crack density by modelling a fault zone to satisfy the observed spectral peak amplitudes in both the low and high wavenumber ranges because the spectral peak amplitude in the high-wavenumber range can be used as the reference to that in the low-wavenumber range.

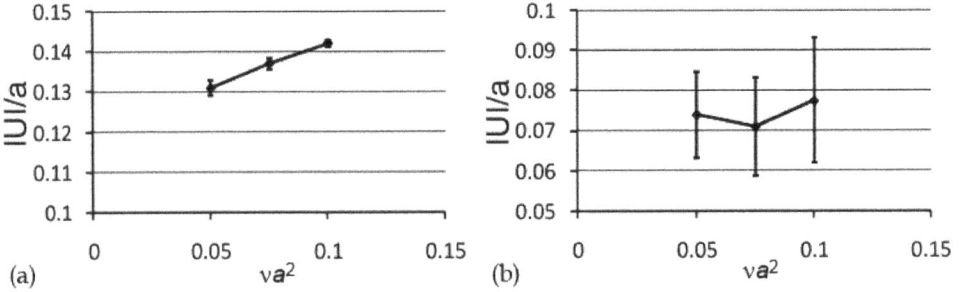

Fig. 12. Crack density dependence of the spectral peak amplitudes. (a) and (b) show the spectral peak amplitudes in the relatively low $(ka\sim0.4)$ and high $(ka\sim1.0)$ wavenumber ranges, respectively

7. Interpretation of the amplitude spectra observed in the fault zone of the 1992 Landers earthquake

In this section, we compare the amplitude spectra observed by Li et al. (1994) in the fault zone of the 1992 Landers earthquake with the synthesis calculated for the fault zone Model (5) of the low-velocity zone with densely distributed parallel cracks. Li et al. (1994) deployed a seismic array across the fault trace of the M7.4 Landers earthquake of June 28, 1992. They found the distinct wave train with a relatively long period following the direct S waves that shows up only when both the stations and the events are close to the fault trace. The coda-normalized amplitude spectra show a spectral peak at 3-4Hz (Fig. 13). They interpreted the long-period wave trains as a seismic guided wave trapped in a low-velocity fault zone and estimated a waveguide width of about 180m and a shear wave velocity of 2.0-2.2 km/s. The amplitude spectra show also a spectral peak of the high frequency at 8-15Hz (Fig. 13). The observed amplitude spectra can be simulated by modelling the fault zone as Model (5) of the low-velocity zone with densely distributed parallel cracks in this study (e.g., Fig. 7c). Although Li et al. (1994) did not infer the origin of the high frequency spectral components, each of the peaks in the low and high wavenumber range is interpreted from our simulation to be formed by the waves trapped in the low-velocity zone and the waves scattered by the cracks, respectively.

First, we try to estimate the dominant crack length in the fault zone. Because the amplitude spectra show the prominent peak at around $ka=1.0$ in relatively high wavenumber range, we can estimate the crack length by using the estimated shear wave velocity of 2.0 km/s and the observed frequency of 10Hz of the spectral peak at the high frequency by Li et al. (1994).

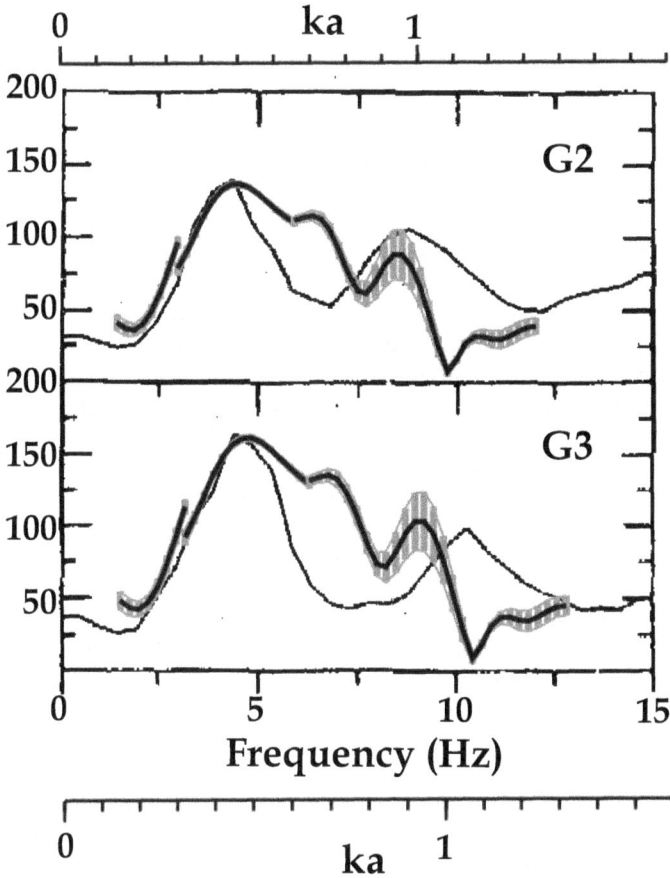

Fig. 13. The amplitude spectra observed in the fault zone of the 1992 Landers earthquake by Li et al. (1994) (thin curves) [reproduced by permission of American Geophysical Union.] and those calculated from the direct wave and the trapped and scattered wave trains of the synthetic seismograms for the fault zone Model (5) with β/β_0=0.7, ρ/ρ_0=0.895, h=8.4a, d_X=3.97a, d_Y=2.1a and va^2=0.12 (bold curves). Thin curves represent the coda-normalized amplitude spectra of horizontal components (parallel to the mainshock fault trace) of seismograms recorded at stations (G2 and G3) located close to the mainshock fault trace for an event occurred within the fault zone. Bold curves and grey-shaded ranges represent the mean values and the standard deviations of 201 stations in the range 100a≤X≤150a along the line Y=3.5a for an isotropic line source located at the centre of the fault zone (0, 3.15a), respectively. The low-velocity zone is bounded by lines of Y=7.35a and Y=-1.05a

The dominant crack length is estimated as

$$2a \sim \frac{2}{k} = \frac{2\beta}{\omega} = \frac{\beta}{\pi f} = \frac{2000(\mathrm{m/s})}{10(\mathrm{Hz})\pi} = 60\mathrm{m}, \tag{12}$$

where ω is the angular frequency and f is the frequency of the spectral peak at the high frequency. Next, we try to estimate the crack density in the fault zone by simulating the observed amplitude spectra. Because Li et al. (1994) estimated the shear wave velocities of the low-velocity fault zone and the surrounding rock as 2.0-2.2 km/s and 3.0 km/s, respectively, we assume the shear wave velocity and density of the low-velocity zone as $\beta/\beta_0=0.7$ and $\rho/\rho_0=0.895$ in our simulation. The peak wavenumber of the relatively low-wavenumber range becomes lower for a low-velocity zone with the larger width. The amplitude spectra also depend on the crack density. The spectral peak amplitudes in the low-wavenumber range become larger for the higher crack density and its dependency is heavier for a fault zone with the smaller width. On the other hand, the spectral peak amplitudes at around $ka=1.0$ show considerable variation among the spatial distributions of cracks and the observation stations even if the same crack density is assumed and do not obviously depend on the crack density. We find an example of the fault zone model which satisfies both of the spectral peak amplitudes in the low and high wavenumber ranges as the width of the low-velocity zone of $h=8.4a$ and the crack spacings in the X and Y directions of $d_X=3.97a$ and $d_Y=2.1a$, respectively (Fig. 13). Thus the width of the low-velocity zone is estimated as 252m by eq.(12), which is a little larger than 180m estimated by Li et al. (1994), and the crack density is $va^2=0.12$, which represents dense distribution of parallel cracks in the fault zone.

8. Conclusion

We compute the synthetic seismograms of the displacement field radiated from a seismic source embedded in a fault zone. We assume following five models as a fault zone and investigate SH wave propagation in a 2-D elastic medium.

1. A zone of densely distributed parallel cracks.
2. An anisotropic zone whose elastic constants are equivalent to those of the crack distribution model (1) at the long-wavelength limit.
3. A low-velocity zone.
4. An anisotropic low-velocity zone.
5. A low-velocity zone with densely distributed parallel cracks.

For Models (1) and (2), we cannot simulate the fault zone trapped waves. We therefore have to consider a low-velocity fault zone to excite trapped waves. For Models (3), (4) and (5), the seismograms show fault zone trapped waves and headwave refracted along the cross-fault material contrast. For Model (5), the seismograms show the waves scattered by cracks in addition to the fault zone trapped waves. Next, we investigate the amplitude spectra. We calculate the amplitude spectrum for each Ricker wavelet source time function in a time window including the direct wave and the trapped and scattered wave trains. For Models (3), (4) and (5), the amplitude spectra show the prominent peak in relatively low-wavenumber range corresponding to the fault zone trapped waves. For Model (4), the low-wavenumber spectral peak splits into two peaks because the interference of resonated waves in the layer occur at the wavenumbers different from that for the isotropic medium due to the wave speed depending on the propagation direction. For Model (5), the amplitude spectra show the prominent peak at $ka\sim1.0$ in relatively high-wavenumber range corresponding to the scattered waves in addition to the low-wavenumber spectral peak corresponding to the fault zone trapped waves.

Finally, we investigate the amplitude spectra for Model (5) of the low-velocity zone with densely distributed parallel cracks. The amplitude spectra depend on the width and velocity of the low-velocity zone and the crack density. We compare the amplitude spectra observed by Li et al. (1994) in the fault zone of the 1992 Landers earthquake with the synthesis calculated for the fault zone Model (5). We find an example of the fault zone model which satisfies both of the spectral peak amplitudes in the low and high wavenumber ranges. Thus we can estimate the dominant crack length as about 60m and the crack density as $\nu a^2 = 0.12$, which represents dense distribution of parallel cracks in the fault zone. Such an estimated model might be ambiguous because the spectral peak amplitudes in the high-wavenumber range show considerable variation among the spatial distributions of cracks and the observation stations even if the same crack density is assumed. Therefore, a statistical analysis will be required for results calculated for many crack distributions and observation stations.

We assumed cracks distributed periodically with the same spacings in a fault zone but such distribution is not realistic. We have to consider randomly distributed cracks in a fault zone for example. If the crack length has some distribution, the broader spectral peak will be observed. The dominant crack length can, however, be estimated in this case as well. Actually the frequency band of the seismic observation is limited. The length of the cracks can not be estimated by the proposed method when the crack length is shorter than the observable seismic wavelengths. Therefore, the crack length estimated in this study should be regarded as the dominant length in the range of the observable wavelengths and might reflect the distribution of relatively long cracks. Because the distribution of the relatively long cracks is considered to be effective to the large earthquake occurrence, the estimation stated here is significant to the monitoring of the preparation process of large earthquakes. It is a future work that the present computations will include the effects of microcrack distribution as the macroscopic parameters such as Q value. Although we investigate SH wave propagation in a 2-D elastic medium, this is the first theoretical study of elastic wave propagation in a low-velocity zone with densely distributed cracks without the assumption of low wavenumber approximation. The results obtained here will be the basis to estimate crack distribution in a fault zone. Further study is required to extend the present computations to 3-D simulations.

9. Acknowledgment

I thank Prof. T. Yamashita for helpful comments. This study was partly supported by a Grant-in-Aid from the Ministry of Education, Science, Sports and Culture of Japan (project 19540434) and the Earthquake Research Institute (ERI), the University of Tokyo, cooperative research programs (2006-W-06, 2007-W-05, 2008-W-03, 2009-W-01, 2010-W-04 and 2011-W-03). For this study, I have used the computer systems of Earthquake Information Center of ERI.

10. References

Ben-Zion, Y. & Malin, P. (1991). San Andreas fault zone head waves near Parkfield, California. Science, Vol.251, No.5001, (March 1991), pp. 1592-1594, ISSN 0036-8075.
Bouchon, M. & Aki, K. (1977). Discrete wave-number representation of seismic-source wave fields. Bull. seism. Soc. Am., Vol.67, No.2, (April 1977), pp. 259-277, ISSN 0037-1106.

Hough, S. E., Ben-Zion, Y., & Leary, P. (1994). Fault-zone waves observed at the southern Joshua Tree earthquake rupture zone. Bull. seism. Soc. Am., Vol.84, No.3, (June 1994), pp. 761-767, ISSN 0037-1106.

Kennett, B. L. N. (1983). Seismic wave propagation in stratified media, Cambridge University Press, ISBN 0-521-23933-8, Cambridge.

Kennett, B. L. N. (1984). Reflection operator methods for elastic waves II - composite regions and source problems. Wave Motion, Vol.6, No.4, (July 1984), pp. 419-429, ISSN 0165-2125.

Kurita, T. (1975). Attenuation of shear waves along the San Andreas fault zone in central California. Bull. seism. Soc. Am., Vol.65, No.1, (February 1975), pp. 277-292, ISSN 0037-1106.

Leary, P. C., Li, Y.-G., & Aki, K. (1987). Observation and modelling of fault-zone fracture seismic anisotropy - I. P, SV and SH travel times. Geophys. J. R. astr. Soc., Vol.91, No.2, (November 1987), pp. 461-484, ISSN 0952-4592.

Li, Y.-G., Aki, K., Adams, D., & Hasemi, A. (1994). Seismic guided waves trapped in the fault zone of the Landers, California, earthquake of 1992. J. geophys. Res., Vol.99, No.B6, (June 1994), pp. 11705-11722, ISSN 0148-0227.

Li, Y.-G. & Leary, P. C. (1990). Fault zone trapped seismic waves. Bull. seism. Soc. Am., Vol.80, No.5, (October 1990), pp. 1245-1271, ISSN 0037-1106.

Li, Y.-G., Leary, P. C., & Aki, K. (1987). Observation and modelling of fault-zone fracture seismic anisotropy - II. P-wave polarization anomalies. Geophys. J. R. astr. Soc., Vol.91, No.2, (November 1987), pp. 485-492, ISSN 0952-4592.

Malin, P. E., Waller, J. A., Borcherdt, R. D., Cranswick, E., Jensen, E. G., & Van Schaack, J. (1988). Vertical seismic profiling of Oroville microearthquakes: velocity spectra and particle motion as a function of depth. Bull. seism. Soc. Am., Vol.78, No.2, (April 1988), pp. 401-420, ISSN 0037-1106.

Mooney, W. D. & Ginzburg, A. (1986). Seismic measurements of the internal properties of fault zones. Pageoph, Vol.124, No.1/2, (January 1986), pp. 141-157, ISSN 0033-4553.

Morse, P. M. & Feshbach, H. (1953). Methods of theoretical physics. McGraw-Hill, ISBN 007043316X, New York.

Murai, Y. (2007). Scattering attenuation, dispersion and reflection of SH waves in two-dimensional elastic media with densely distributed cracks. Geophys. J. Int., Vol.168, No.1, (January 2007), pp. 211-223, ISSN 0956-540X.

Murai, Y. & Yamashita, T. (1998). Multiple scattering of SH waves by imperfectly bonded interfaces with inhomogeneous strengths. Geophys. J. Int., Vol.134, No.3, (September 1998), pp. 677-688, ISSN 0956-540X.

Tchalenko, J. S. (1970). Similarities between shear zones of different magnitudes. Geol. Soc. Am. Bull., Vol.81, No.6, (June 1970), pp. 1625-1640, ISSN 0016-7606.

Tchalenko, J. S. & Berberian, M. (1975). Dasht-e Bayaz fault, Iran: earthquake and earlier related structures in bed rock. Geol. Soc. Am. Bull., Vol.86, No.5, (May 1975), pp. 703-709, ISSN 0016-7606.

10

Effects of Random Heterogeneity on Seismic Reflection Images

Jun Matsushima
The University of Tokyo
Japan

1. Introduction

The reflection seismic method, which is a technique to map geologic structure and stratigraphic features, has been adopted in a variety of applications such as oil and gas explorations, fundamental geological studies, engineering and hydrological studies. Furthermore, recently time-lapse seismic monitoring is considered as a promising technology to monitor changes in dynamic physical properties as a function of time by analyzing differences between seismic data sets from different epochs (e.g., Lumley, 2001). On the other hand, its widespread use has often revealed a weakness in seismic reflection methods when applied to complex structures. It is widely believed that highly dense spatial sampling increases the quality of final seismic reflection sections. However, the quality of final seismic sections obtained in real fields is often very poor for a variety of reasons such as ambient noise, heterogeneities in the rocks, surface waves, reverberations of direct waves within the near-surface, and seismic scattering, even if highly dense spatial sampling is adopted. In most of reflection seismic explorations, people implicitly assume that the subsurface target heterogeneities are sufficiently large and strong that other background heterogeneities only cause small fluctuations to the signals from the target heterogeneity. In this case, a clear distinction can be made between target structures and the small-scale background heterogeneities. However, if the small-scale heterogeneities are significantly strong and are of comparable size to the seismic wavelength, complicated waveforms often appear. This complication causes much difficulty when investigating subsurface structures by seismic reflection. In deep crustal studies (Brown et al., 1983) or geothermal studies (Matsushima et al., 2003), seismic data often have a poor signal-to-noise ratio. Complicated seismic waves are due to seismic wave scattering generated from the small-scale heterogeneities, which degrades seismic reflection data, resulting in attenuation and travel time fluctuations of reflected waves, and the masking of reflected waves by multiple scattering events. In this case, the conventional single-scattering assumption of migration may not be applicable; in other words, multiple scattering caused by strong heterogeneities may disturb the energy distribution in observed seismic traces (Emmerich et al., 1993).

The understanding of seismic wave propagation in random heterogeneous media has been well advanced by many authors on the basis of theoretical studies (Sato and Fehler, 1997), numerical studies (Frankel and Clayton, 1986; Hoshiba, 2000), and experimental studies (Nishizawa et al., 1997; Sivaji et al., 2001; Matsushima et al., 2011). Since scattered waves seem incoherent and the small-scale heterogeneity is presumed to be randomly distributed, the statistical properties of seismic wave fluctuation relate to the statistical properties of this small-scale heterogeneity. Seismologists conclude that coda waves are one of the most convincing pieces of evidence for the presence of random heterogeneities in the Earth's interior. Seismic evidence suggests random heterogeneity on a scale ranging from tens of meters to tens of kilometers. In addition, geologic studies of exposed deep crustal rocks indicate petrologic variations in the lithosphere on a scale of meters to kilometers (Karson et al., 1984; Holliger and Levander, 1992). Well-logging data suggest that small-scale heterogeneities have a continuous spectrum (Shiomi et al., 1997).

From the viewpoint of seismic data processing, many authors have pointed out the disadvantages of the conventional CMP method proposed by Mayne (1962) when applied to complex structures. Based on a layered media assumption, the CMP stacking method does not provide adequate resolution for non-layered media. Since the 1970s, several prestack migration methods have been studied as improvements on CMP stacking. Sattlegger and Stiller (1974) described a method of prestack migration and demonstrated its advantages over poststack migration in complex areas. Prestack migration is divided into two types of techniques: prestack time migration (PSTM) and prestack depth migration (PSDM). PSTM is acceptable for imaging mild lateral velocity variations, while PSDM is required for imaging strong lateral velocity variations such as salt diapirism or overthrusting. A better image is obtained by PSDM when an accurate estimate of the velocity model exists; however, the advantage of PSTM is that it is robust and much faster than PSDM. From the viewpoint of the S/N ratio, Matsushima et al. (2003) discussed the advantages of prestack migration over synthetic data containing random noise.

Wave phenomena in heterogeneous media are important for seismic data processing but have not been well recognized and investigated in the field of seismic exploration. There are only several studies which have taken into account the effect of scattering in the seismic reflection data processing. Numerical studies by Gibson and Levander (1988) indicate that different types of scattered noise can have different effects on the appearance of the final processed section. Gibson and Levander (1990) showed the apparent layering in CMP sections of heterogeneous targets. Emmerich et al. (1993) also concluded that the highly detailed interpretation, which is popular in crustal reflection seismology, is less reliable than believed, as far as the internal structure of scattering zones and scatterer orientations are concerned. Sick et al. (2003) proposed a method that compensates for the scattering attenuation effects from random isomorphic heterogeneities to obtain a more reliable estimation of reflection coefficients for AVO/AVA analysis. It is important to understand how scattered waves caused by random heterogeneities affect data processing in seismic reflection studies and how these effects are compensated for. From the viewpoint of spatial sampling in time-lapse seismic survey, Matsushima and Nishizawa (2010a) reveal the effects of scattered waves on subsurface monitoring by using a numerical simulation of the seismic wave field and comparing the different responses of the final section by applying two different types of data processing: conventional CMP stacking and poststack migration. Matsushima and Nishizawa (2010a) demonstrate the existence of a small but significant difference by differentiating two sections with different spatial sampling. This small difference is attributed to the truncation artifact which is due to geometrical limitation and that cannot be practically prevented during data acquisition. Furthermore, Matsushima and Nishizawa (2010b) indicate that this small difference is also attributed to normal moveout

(NMO)-stretch effect which cannot be practically prevented during data acquisition and processing.

A primary concern of this article is to study effects of random heterogeneity on seismic reflection images. We investigate the effect of spatial sampling on the images of seismic reflection, by comparing two set of images: one reproduced from simulated seismic data having a superimposed random noise in time series, and the other generated from numerically simulated wave fields in a same medium but containing random heterogeneity. We also investigate the relationship between the spatial sampling interval and the characteristic size of heterogeneities and also investigate from the viewpoint of spatial sampling how noise-like scattered wave fields that are produced from random isotropic heterogeneity influence the seismic section. We consider the adoption of highly dense spatial sampling with intervals smaller than the Nyquist interval to improve the final quality of a section. In this paper, three types of data processing, conventional CMP stacking, poststack migration and prestack migration are compared to examine different responses to the migration effect of different spatial sampling intervals. We generate 2-D finite-difference synthetic seismic data as input to this study. Our numerical models have a horizontal layered structure, upon which randomly distributed heterogeneities are imposed.

2. Spatial sampling interval in seismic reflection

According to the Nyquist sampling theorem, sampling at two points per wavelength is the minimum requirement for sampling seismic data over the time and space domains; that is, the sampling interval in each domain must be equal to or above twice the highest frequency/wavenumber of the continuous seismic signal being discretized. The phenomenon that occurs as a result of undersampling is known as aliasing. Aliasing occurs when recorded seismic data violate the criterion expressed in equation (1).

$$\Delta x \le \Delta x_N = \frac{v_{\min}}{2 f_{\max} \cdot \sin \theta}, \tag{1}$$

where Δx is the spatial sampling interval which should be equal to or smaller than the spatial Nyquist sampling intervals Δx_N, $v_{\min}$ is the minimum velocity, $f_{\max}$ is the maximum frequency, and θ is the dip angle of the incident plane-wave direction.

On the other hand, in the case of zero-offset, the spatial sample interval should be equal to or smaller than a quarter-wavelength (Grasmueck et al., 2005). Aliasing occurs when recorded seismic data violate the criterion expressed in equation (2).

$$\Delta x \le \Delta x_N = \frac{v_{\min}}{4 f_{\max} \cdot \sin \theta}. \tag{2}$$

In the presence of structural dips or significant lateral velocity variations, adequate sampling becomes important for both vertical and lateral resolution. For the case of the maximum dip (θ=90), the spatial Nyquist sampling interval becomes a quarter-wavelength. Thus, quarter-wavelength spatial sampling is a minimum requirement for adequate recording. Vermeer (1990) defined the term "full-resolution recording" for unaliased shooting and recording of the seismic wave field at the basic signal-sampling interval. In practice, however, seismic data are often irregularly and/or sparsely sampled in the space domain because of limitations such as those resulting from difficult topography or a lack of resources. In many cases, proper sampling is outright impossible. In order to avoid aliasing, standard seismic imaging methods discard some of the high frequency components of recorded signals. Valuable image resolution will be lost through processing seismic data (Biondi, 2001). Once seismic data are recorded, it is difficult to suppress aliasing artifacts without resurveying at a finer spatial sampling (Spitz, 1991).

In the case of migration processing, there are three types of aliasing (Biondi, 2001), associated with data, operator, and image spacing. Data space aliasing is the aliasing described above. Operator aliasing, which is common in Kirchhoff migration algorithms, occurs when the migration operator summation trajectory is too steep for a given input seismic trace spacing and frequency content. Kirchhoff migration approximates an integral with a summation and is subject to migration operator aliasing when trace spacings do not support the dip of the migration operator. In contrast, migration algorithms such as the f-k method or finite-difference methods only require that the input data volume be sampled well enough to avoid aliasing of the input volume (Abma et al., 1999). Adequate solution for operator aliasing is to control the frequency content (e.g., low-pass filtering at steep dips). The anti-aliasing constraints to avoid operator aliasing can be easily derived from the Nyquist sampling theorem. The resulting anti-aliasing constraints are (Biondi, 1998):

$$f \leq \frac{1}{2\Delta x_{data} \cdot p_{op}}, \tag{3}$$

where Δx_{data} is the sampling rate of the data x-axis and p_{op} is the operator dip.

Image space aliasing occurs when the spatial sampling of the image is too coarse to adequately represent the steeply dipping reflectors that the imaging operator attempts to build during the imaging process. Image space aliasing can be avoided simply by narrowing the image interval. But for a given spatial sampling of the image, to avoid image space aliasing we need to control the frequency content of the image. Similarly to the case of operator aliasing, the anti-aliasing constraints to avoid image space aliasing can be easily derived from the Nyquist sampling theorem. The resulting anti-aliasing constraints are (Biondi, 1998):

$$f \leq \frac{1}{2\Delta x_{image} \cdot p_{ref}}, \tag{4}$$

where Δx_{image} is the image sampling rate of the x-axis and p_{ref} is the reflector dip.

From the viewpoint of the S/N ratio, dense spatial sampling increases the number of sources/receiver pairs (i.e., stacking fold), which raises the effect of signal enhancement, that is, increases the S/N ratio. The expected improvement in S/N is proportional to the square root of the stacking fold under the assumption that it is purely random noise which has a flat power spectrum. Thus, highly dense spatial sampling improves the S/N ratio of the section, even if the interval of spatial sampling becomes shorter than the Nyquist sampling interval.

3. Construction of synthetic data and seismic reflection imaging

We constructed two data sets. One is synthetic seismic data set generated from two-layer model where each layer has a constant velocity everywhere inside the layer. Random noise was added to the synthetic seismic data (random noise model=RN model). The other is synthetic seismic data set generated from two-dimensional random heterogeneous media where random velocity variation is superimposed on a layer above a reflector (random heterogeneous model=RH model). The second model will generate incoherent events by scattering of waves in the random heterogeneous media.

3.1 Random noise (RN) model

A numerical simulation model and source/receiver arrangements are shown in Figure 1a. A reflector is placed at a depth of 2000 m, separating two layers having a constant velocity of 3800 m/s and 4200 m/s, respectively. Three different source-receiver intervals 80, 20, and 5 m were employed; each requiring 26, 101, and 401 sources and receivers, respectively. The reflected waves generated by a flat reflector were obtained by using the 2-D finite difference method as described below. In order to remove direct wavelets, the wavefield without the reflector was subtracted from the total wavefiled of the reflector model. We then obtain the wavefield containing only reflected waves. Random noise is added to the data containing only signal components (reflections) so that the S/N ratio was 0.3. The S/N ratio is defined as the following equation (5):

$$S/N = \frac{S_{MAX}}{\sqrt{\dfrac{1}{N}\displaystyle\sum_{i=1}^{N} Noise(i)^2}}, \tag{5}$$

where S_{MAX} is the absolute value of the maximum amplitude of signal events in a stacked trace obtained from data consisting of only signal components, Noise (i) is the amplitude of the i-th sample in a stacked trace obtained from the random noise, and N is the total number of samples. The denominator of equation (5) equals the root-mean-square (rms) amplitude of the noise.

3.2 Layered model overlapped with random heterogeneity

Random heterogeneous media are generally described by fluctuations of wave velocity and density, superposed on a homogeneous background. Their properties are given by an autocorrelation function parameterized by the correlation lengths and the standard deviation of the fluctuation. Random media with spatial variations of seismic velocity were generated by the same method as described in Frankel and Clayton (1986). The outline of the scheme is as follows:

1. Assign a velocity value $v(x, z)$ to each grid point using a random number generator.
2. Fourier transform the velocity map into the wave number space.
3. Apply the desired filter in the wavenumber domain.
4. Inverse Fourier transform the filtered data back into the spatial domain.
5. Normalize the velocities by their standard deviation, centered on the mean velocity.

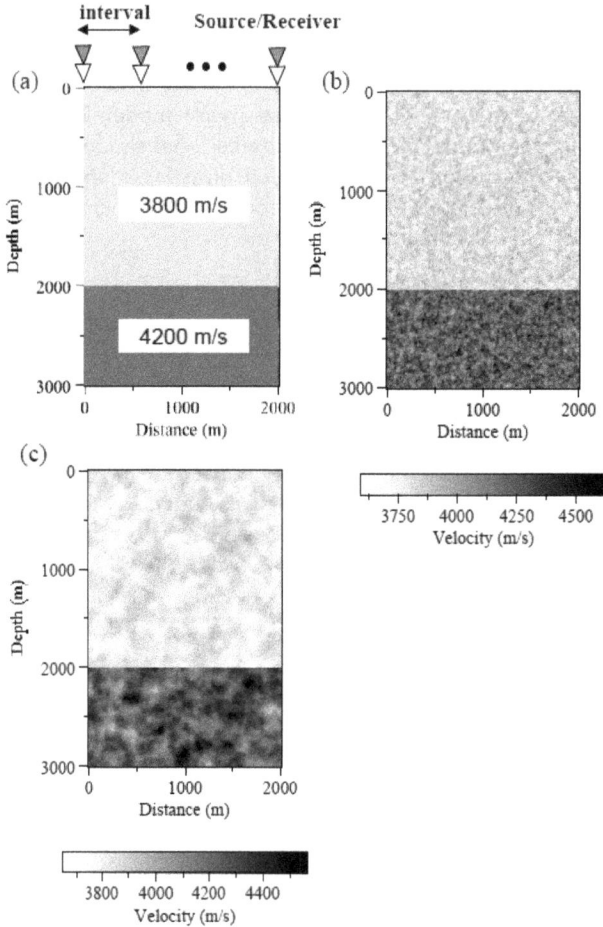

Fig. 1. (a) A single-interface model for numerical simulation examining specifications of data acquisition in reflection seismic surveys. A reflector is placed at a depth of 2000 m. (b) The first two-layered random media model for two-dimensional acoustic wave simulation using the finite-difference method. The average velocity of the upper layer is 3800 m/s with 3% standard deviation and correlation distance 10 m. (c) The second two-layered random media model with the same average velocity and standard deviation as for (b), except for a correlation distance of 50 m

In this paper, the applied filter (Fourier transform of autocorrelation function, which is equal to the power spectral density function) has a von Karman probability distribution described by equation (6):

$$P(k,a) = \frac{4\pi\beta a^2}{\left(1 + k^2 a^2\right)^{\beta+1}},$$

(6)

where k is the wavenumber, β is the Hurst number that controls the components of small scale random heterogeneities, and a is the correlation distance indicating the characteristic heterogeneity size. The wavenumber k we use here is defined by equation (7):

$$k = \frac{2\pi}{\lambda},$$
(7)

where λ is the wavelength. We use the above von Karman-type heterogeneous media with β =0.1. Saito et al. (2003) described that the value β =0.1 is nearly the same as the value for the power spectral density function of velocity fluctuation obtained from well-log data at depths shallower than 10 km (e.g., Shiomi et al., 1997; Goff and Holliger, 1999).

A homogeneous model and source/receiver arrangements are the same as the case of the RN model. To estimate the relationship between the spatial sampling interval and the characteristic size of heterogeneities, two types of random heterogeneities were generated and implemented in the layered model as shown in Figures 1b and 1c. The velocity perturbations shown in Figure 1b were normalized to have a standard deviation 3% of the 3800 m/s (upper) and 4200 m/s (lower) layers on average and a characteristic heterogeneity size of 10 m (a=10 m). Figure 1c is the same as Figure 1b except for characteristic heterogeneity sizes of 50 m (a=50 m).

The level of scattering phenomena is a function of the wavelength and the average scale of heterogeneities. If the wavelength of a seismic wave is much longer than the scale length of heterogeneity, the system is considered a homogeneous material. Although scattering phenomenon is important only at wavelengths comparable to the scale length of heterogeneity, small-scale heterogeneities influence the seismic waveform with respect to the size of heterogeneities. Wu and Aki (1988) categorized the scattering phenomena into several domains. When $ka < 0.01$ (Quasi-homogeneous regime), the heterogeneous medium behaves like an effective homogeneous medium where scattering effects may be neglected. When $0.01 < ka < 0.1$ (Rayleigh scattering regime), scattering effects may be characterized by Born approximation which is based on the single scattering assumption. When $0.1 < ka < 10$ (Mie scattering regime), the sizes of the heterogeneities are comparable to the wavelength. The scattering effects are most significant. When $ka > 10$ (Forward scattering regime), the heterogeneous medium may be treated as a piecewise homogeneous medium where ray theory may be applicable.

3.3 Wave field calculation

We employed a second-order finite difference scheme for the constant density two-dimensional acoustic wave equation described in the equation (8).

$$\frac{\partial^2 P}{\partial t^2} = V(x,z)\left(\frac{\partial^2 P}{\partial x^2} + \frac{\partial^2 P}{\partial z^2}\right),$$
(8)

where P is the pressure in a medium and $V(x,z)$ is the velocity as a function of x and z. The source wavelet was the Ricker wavelet with a dominant frequency of 20 Hz. The dominant frequency (20 Hz) and the average velocity (3800 m/s) yielded the dominant wavelength (190 m). A uniform grid was employed in the x-z plane. To minimize grid dispersion in finite difference modeling, the grid size was set to be about one eighteenth of the shortest wavelength, which was calculated from the minimum velocity of 3600 m/s, the maximum frequency of around 40 Hz (f_{max}=40), and a 5-m grid spacing. All edges of the finite-difference grid were set to be far from source/receiver locations so that unnecessary events would not disturb the synthetic data. Source/receivers were not located on the edge of the model, but within the model body. In this situation, scattered wave fields generated in the heterogeneous media above the source/receiver locations would be included in the synthetic data. However, this does not affect the conclusions of this article.

The reflected waves generated by a flat reflector were obtained by using the 2-D finite difference method. In order to remove direct wavelets, the wavefield without the reflector was subtracted from the total wavefiled of the reflector model. We then obtain the wavefield containing only reflected waves. Figure 2a shows an example of the shot gather of reflected wavefield. In the case of the RN model, band-limited random noise (5-50 Hz) was added to the synthetic data containing only signal components (reflections) so that the S/N ratio was 0.3 (Figure 2b). In Figure 2b, reflected waves can hardly be detectable due to masking effect by random noise.

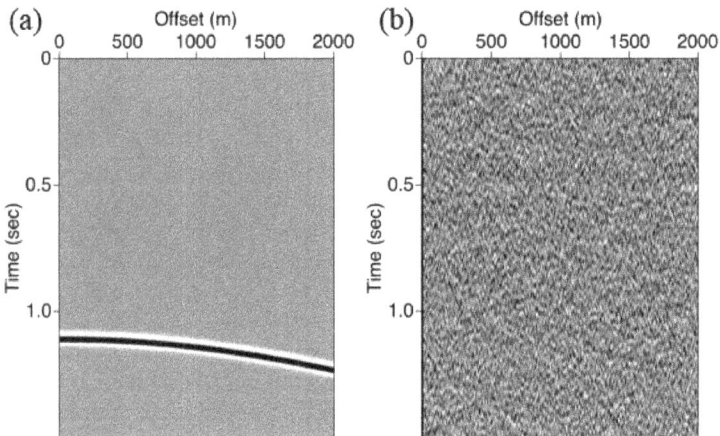

Fig. 2. (a) An example of common-shot gather of reflected wavefield calculated for the model shown in Figure 1a. (b) Common-shot gather containing time-series random noises in the traces shown in (a). The signal to noise ratio is 0.3

In the case of the RH model, on the other hand, to compare results between random media of different characteristic lengths, wavelengths have to be described with reference to the characteristic lengths of random media. The product of the wavenumber k and the characteristic length a is used as an index for describing effects of random heterogeneity on seismic waves. In the present cases, the ka values at the dominant wavelengths are about 0.33 (a=10 m) and 1.65 (a=50 m), respectively. According to the classification by Wu and Aki (1988), our heterogeneous models are categorized as "Mie scattering regime" where strong scattering may occur and full waveform modeling is required. In order to remove direct wavelets, the total wave field calculated with the model shown in Figures 1b and 1c was subtracted from the wave field in a model with a constant velocity of 3800 m/s to produce the wave field containing the reflected/scattered wave field. Figure 3a shows an example of the shot gather from the scattered wave field in the case of $a=10$ m for source-receiver intervals of 5 m. Similarly, Figure 3b shows an example of the shot gather from the scattered wave field in the case of $a=50$ m for source-receiver intervals of 5 m. Although we can clearly see the reflection event in each shot gather shown in Figure 3, the shot gathers are full of chaotic diffraction patterns originating from random heterogeneities.

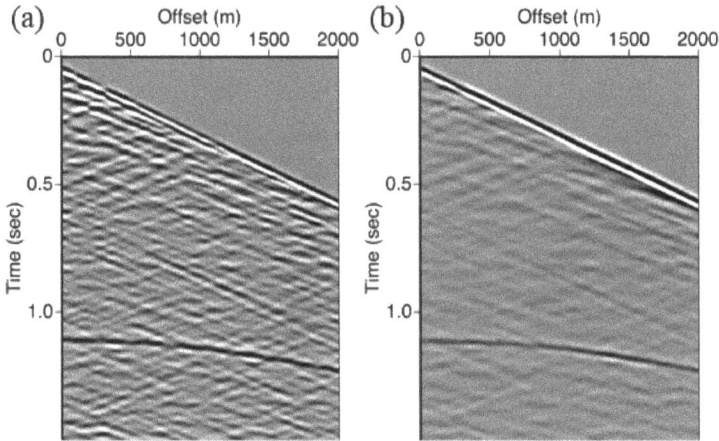

Fig. 3. Examples of common-shot gather of a scattered wave field calculated for the model with different spatial sampling intervals and characteristic heterogeneity sizes: spatial sampling intervals of 5 m for the case of for $a=10$ m (Fig. 1b) and spatial sampling intervals of 5 m for the case of for $a=50$ m (Fig. 1c)

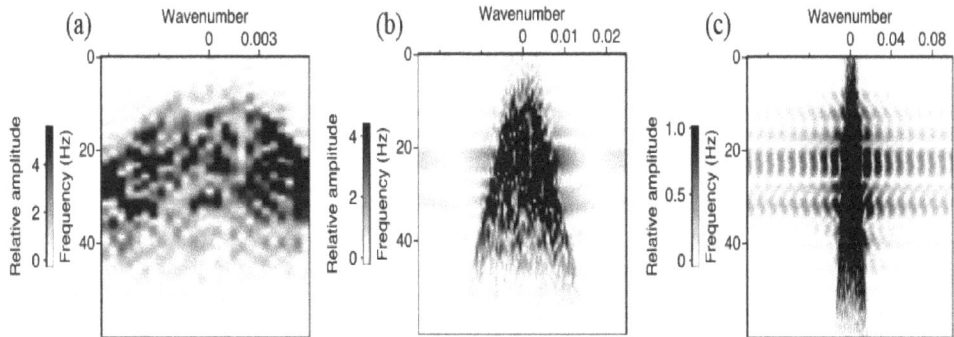

Fig. 4. Frequency-wavenumber (f-k) plots of the extracted shot gather of a scattered wave field with different spatial sampling intervals and characteristic heterogeneity sizes: spatial sampling intervals of (a) 80 m, (b) 20 m, and (c) 5 m for the case of for $a=10$ m

The frequency-wavenumber (f-k) diagram is helpful for visualizing the sampling of a continuous wave field (Vermeer, 1990). The time window (from 0.65 to 1.05 s) including only scattered wave fields was extracted from each shot gather to calculate an f-k plot. Figures 4a through 4c show f-k plots of the extracted shot gather from the scattered wave field in the case of a=10 m for source-receiver intervals of 80, 20, and 5 m, respectively. Similarly, Figures 5a through 5c show f-k plots of the extracted shot gather from the scattered wave field in the case of a=50 m for source-receiver intervals of 80, 20, and 5 m, respectively. According to the spatial Nyquist sampling criterion defined in equation (1), Δx_N becomes 45 m (f_{max} =40, v_{min} =3600, θ=90). Thus, spatial sampling less than 45 m is sufficient to prevent spatial aliasing of the scattered wave field. In the case of the 80 m spatial sampling interval of Figure 4a and 5a, the sector of strong amplitudes in the f-k plot would be severely truncated, causing wrap-around effects. On the other hand, in the case of zero-offset defined by equation (2), spatial sampling of less than 22.5 m is sufficient to prevent spatial aliasing.

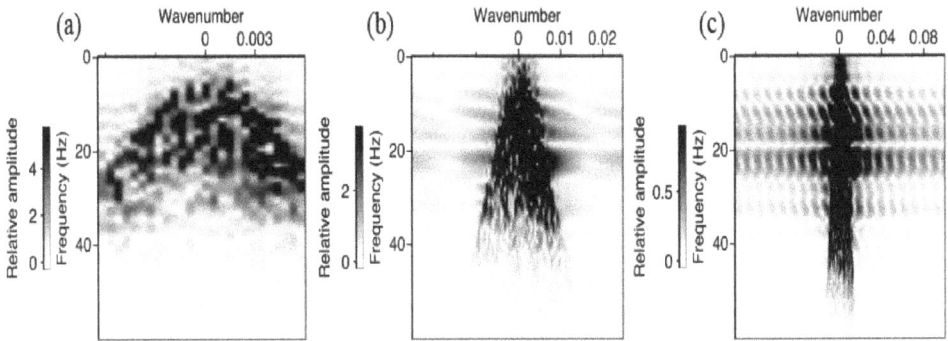

Fig. 5. Frequency-wavenumber (f-k) plots of the extracted shot gather of a scattered wave field with different spatial sampling intervals and characteristic heterogeneity sizes: spatial sampling intervals of (a) 80 m, (b) 20 m, and (c) 5 m for the case of for a=50 m

4. Results

Three types of data processing, conventional CMP stacking, poststack migration and prestack time migration (PSTM), were applied to the two types of model described above.

4.1 CMP stacked sections

Conventional CMP stacking was applied to both random noise (RN) model and random heterogeneous (RH) model. The shot gathers were sorted into CMP gathers and corrected for NMO using constant velocity of 3800 m/s, and finally stacked. Figures 6a through 6c show the CMP stacked sections for the RN model data shown in Figure 2b with different source/receiver intervals at 80, 20 and 5 m, respectively. We can see that the S/N ratio becomes larger with denser source/receiver arrangements. The difference of the S/N ratio among Figures 6a through 6c becomes larger with increasing the numbers of sources/receivers. The low quality of both sides of CMP sections is due to the low fold in the CMP gathers at the margins of target area.

Figures 7a through 7c show CMP stacked sections for the RH model in the case of a characteristic heterogeneity size of 10 m (a=10 m) with different source/receiver intervals at 80, 20, and 5 m, respectively. Similarly, Figure 8 is the same as Figure 7 except for characteristic heterogeneity sizes of 50 m (a=50 m). The CMP intervals of each model are 40,

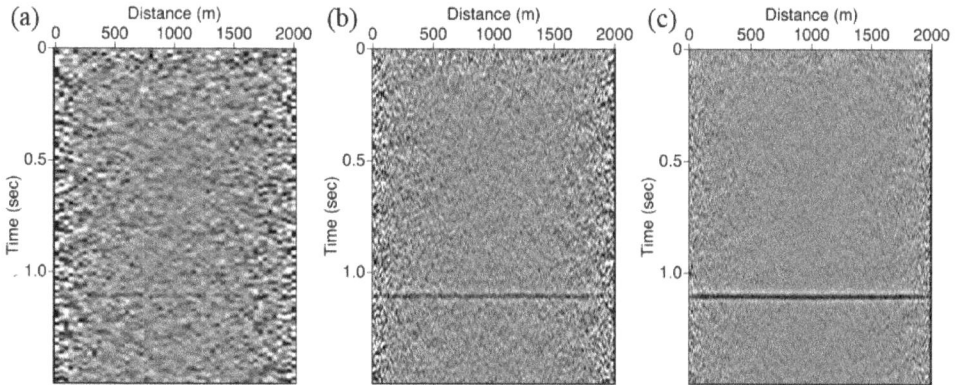

Fig. 6. CMP stacked sections for the synthetic time-series random noise data with an S/N ratio of 0.3 for different spatial sampling intervals: (a) 80 m, (b) 20 m, and (c) 5m

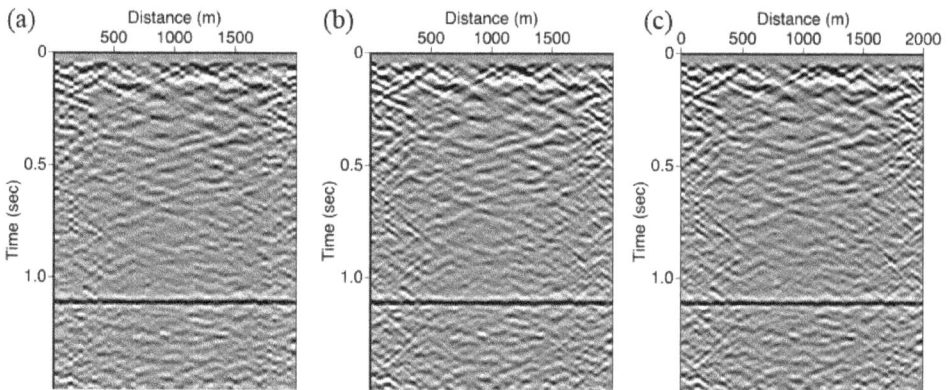

Fig. 7. CMP stacked sections with different spatial sampling intervals and characteristic heterogeneity sizes: spatial sampling intervals of (a) 80 m, (b) 20 m, and (c) 5 m for the case of for a=10 m

10, and 2.5 m, respectively. Although CMP stacking can act as a powerful mechanism for suppressing multiples and for the attenuation of many types of linear event noises such as airwaves and ground roll, we can see no significant differences among Figures 7a through 7c and among Figures 8a through 8c. However, a close examination of these sections reveals that image space aliasing occurs in the case of a CMP interval of more than 22.5 m (Figure 7a and 8a). Note that the effect of image space aliasing in the case of a=10 m is larger than the case of a=50 m. In each section of Figures 7 and 8, we can see a reflector at around 1.1 sec. and many discontinuously subhorizontal and dipping events that partly correlate with velocity heterogeneities of the model. Gibson and Levander (1988) mentioned that the limited bandwidth of the propagating seismic signal and spatial filtering attributable to CMP stacking cause these events, bearing no simple relation to the velocity anomalies of the model. While the reflector can be seen clearly from the chaotic background noise, we can see some arrival time fluctuations and amplitude variations in the observed reflector. These variations are attributed to the scattering effect of the heterogeneous media whose scale is smaller than the wavelength. In Figures 7 and 8, we can see no significant arrival time fluctuations but some amplitude variations in the observed reflector. These amplitude variations are attributed to the scattering attenuation (sometimes called apparent attenuation) in the heterogeneous media. When the heterogeneous scale is small, the amplitude is affected by the heterogeneity but the travel time is not strongly affected by the heterogeneity. In this situation, the assumptions of CMP stacking and simple hyperbolic reflection pattern can be fulfilled.

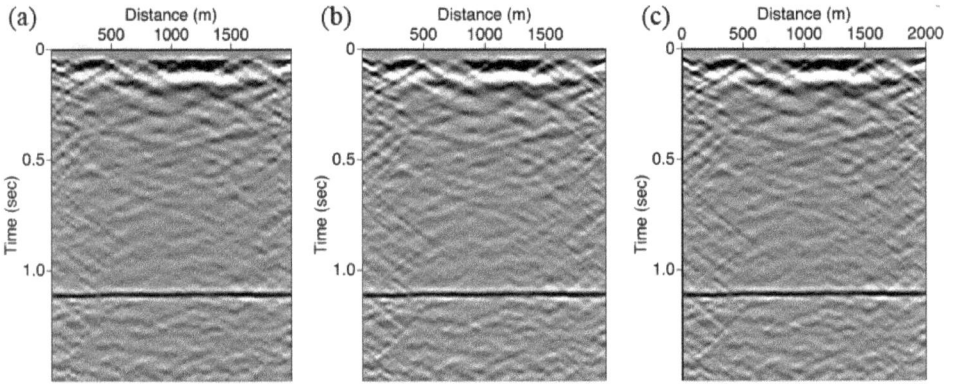

Fig. 8. CMP stacked sections with different spatial sampling intervals and characteristic heterogeneity sizes: spatial sampling intervals of (a) 80 m, (b) 20 m, and (c) 5 m for the case of for a=50 m

4.2 Poststack migrated sections

Figures 9a through 9c show poststack migrated sections using f-k migration (Stolt, 1978) for a RN model with different source/receiver intervals at 80, 20, and 5 m, respectively. The

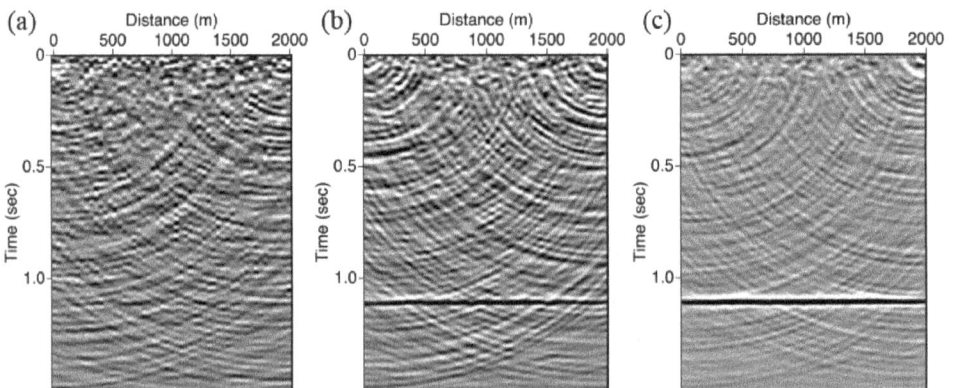

Fig. 9. Poststack migrated sections for the synthetic time-series random noise data with an S/N ratio of 0.3 for different spatial sampling intervals: (a) 80 m, (b) 20 m, and (c) 5m

trace intervals of each section shown in Figure 9 are 40, 10, and 2.5 m, respectively. Although the resulting migrated sections suffer from the inadequate cancellation of migration smiles, we can see that the S/N ratio becomes larger with denser source/receiver arrangements.

Figures 10a through 10c show poststack migrated sections using f-k migration (Stolt, 1978) with a random heterogeneous model for the case of a characteristic heterogeneity size of 10 m (a=10 m) with different source/receiver intervals at 80, 20, and 5 m, respectively. Similarly, Figure 11 is the same as Figure 10 except for characteristic heterogeneity sizes of 50 m (a=50 m). We can see that numerous small segments are still detectable even after the poststack migration and that the results of poststack migration for the different heterogeneous models differ with different source/receiver intervals. Although we can see

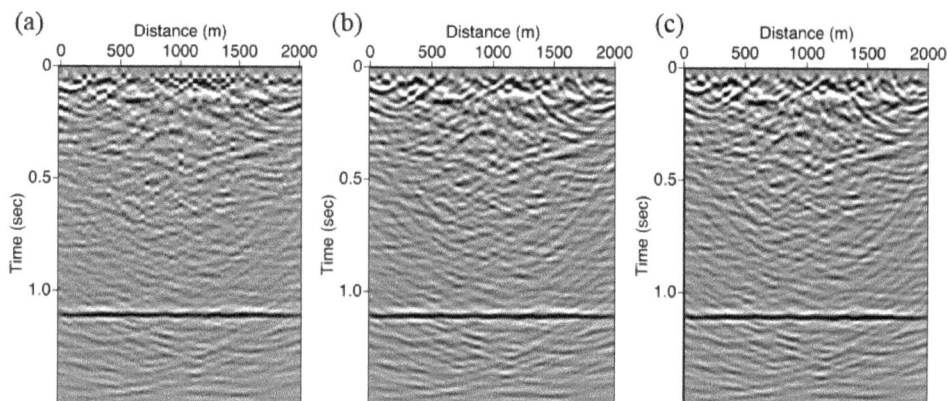

Fig. 10. Poststack migrated sections with different spatial sampling intervals and characteristic heterogeneity sizes: spatial sampling intervals of (a) 80 m, (b) 20 m, and (c) 5 m for the case of for a=10 m

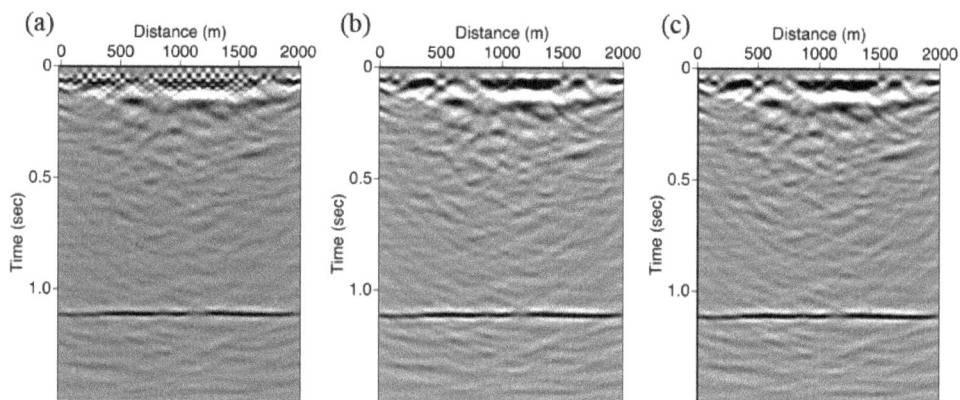

Fig. 11. Poststack migrated sections with different spatial sampling intervals and characteristic heterogeneity sizes: spatial sampling intervals of (a) 80 m, (b) 20 m, and (c) 5 m for the case of for a=50 m

no significant differences among Figures 10a through 10c and among Figures 11a through 11c, a close examination of these sections reveals that image space aliasing occurs in the case of a trace interval of more than 22.5 m (Figure 10a and 11a). Note that the effect of image space aliasing in the case of a=10 m is larger than the case of a=50 m. In general, migration can improve lateral resolution by correcting the lateral mispositioning of dipping reflectors or collapsing diffraction patterns caused by a point scatterer. However, the application of poststack migration here does not improve seismic images in heterogeneous media. It is thought that the reason is that multiple-scattering effects in small-scale heterogeneities do not satisfy the assumption of migration theory based on single scattering. Although migration techniques assume that the seismic data to be migrated consists only of primary reflections and diffractions, these wave fields are attenuated and distorted by heterogeneities and multiple scattered wave fields are generated, producing apparent discontinuities in reflectors or diffractors.

4.3 Prestack time migrated sections

In this paper, we obtained PSTM sections using a diffraction stacking method proposed by Matsushima et al. (2003). Figures 12a through 12c show PSTM sections for a RN model with different source/receiver intervals at 80, 20, and 5 m, respectively. We can see that the S/N ratio becomes larger with denser source/receiver arrangements.

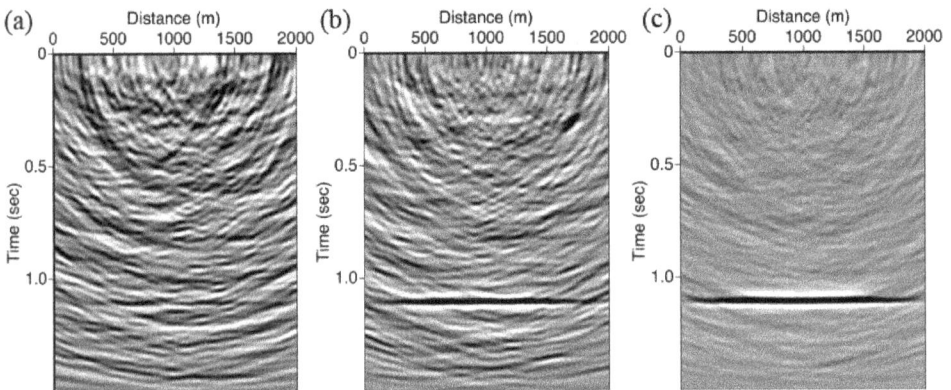

Fig. 12. PSTM sections for the synthetic time-series random noise data with an S/N ratio of 0.3 for different spatial sampling intervals: (a) 80 m, (b) 20 m, and (c) 5m

Figures 13a through 13c show the PSTM sections using a diffraction stacking method (Matsushima et al., 2003) for a random heterogeneous model with a characteristic heterogeneity size of 10 m (a=10 m), as shown in Figure 1b with different source/receiver intervals at 80, 20, and 5 m, respectively. Similarly, Figure 14 is the same as Figure 13 except for characteristic heterogeneity sizes of 50 m (a=50 m). Each PSTM section is full of migration smiles, producing the appearance that the section is heavily over-migrated, thus reducing the quality of the image. A possible explanation of this phenomenon is that the wave field is distorted by heterogeneities, which in turn produce apparent discontinuities in reflectors or diffractors. These discontinuities do not have associated diffraction hyperbolae, so that the migration, instead of collapsing the absent hyperbolae, propagates the noise represented by the discontinuity along wavefronts. As a result, the seismic section is full of migration smiles that are heavily over-migrated. Warner (1987) pointed out that deep continental data are often best migrated at velocities that are up to 50 % less than appropriate interval velocities from crustal refraction experiments or directly from stacking velocities. His explanation for this behavior is that near surface features distort and attenuate the seismic wave field and produce apparent discontinuities in deep reflections. During the process of migration, reflections are invented in order to cancel out the missing diffractions thereby producing a smiley section that appears over-migrated. Although PSTM is expected to provide more realistic images compared to conventional poststack migration (Gibson and Levander, 1988), we can see no significant differences among Figures 13a through 13c, and also among Figures 14a through 14c. Similar to the case of poststack migration, the reason is thought to be that multiple-scattering effects in small-scale heterogeneities do not satisfy the assumption of migration theory based on single scattering.

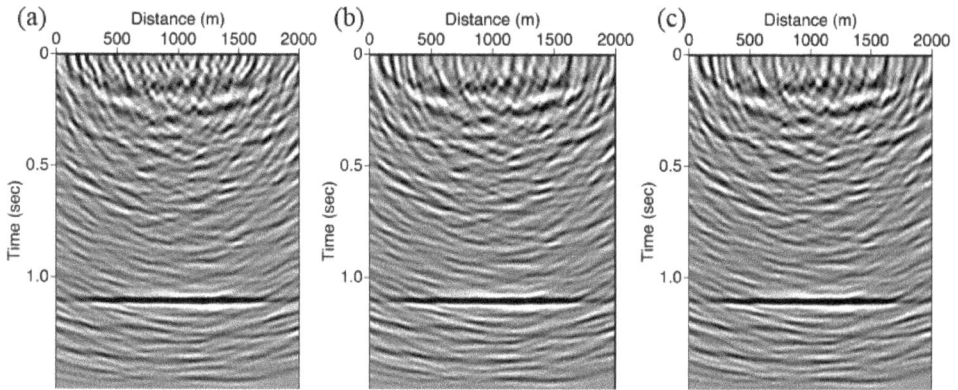

Fig. 13. PSTM sections with different spatial sampling intervals and characteristic heterogeneity sizes: spatial sampling intervals of (a) 80 m, (b) 20 m, and (c) 5 m for the case of for a=10 m

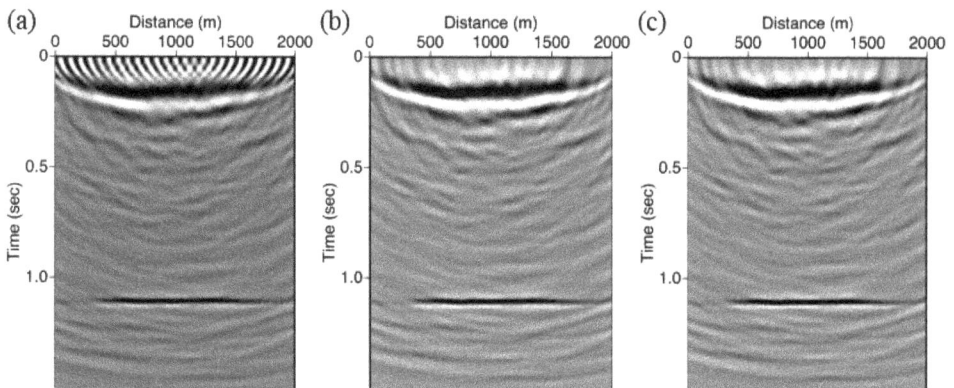

Fig. 14. PSTM sections with different spatial sampling intervals and characteristic heterogeneity sizes: spatial sampling intervals of (a) 80 m, (b) 20 m, and (c) 5 m for the case of for a=50 m

4.4 Comparison between the data processing variants

Figures 15a thorough 15c show the center trace of the corresponding section in the case of the RN model with three different spatial sampling intervals. We can see that there is little difference of the S/N ratio among the data processing variants when the spatial sampling

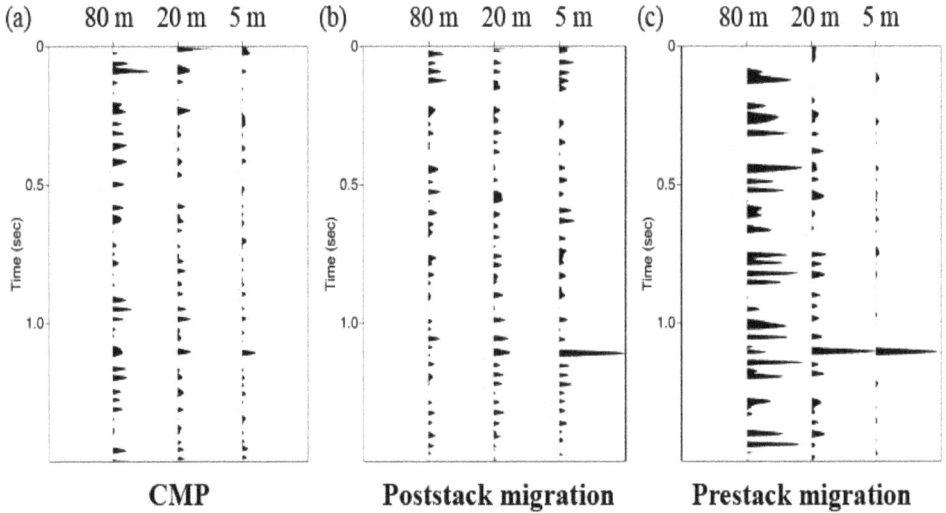

Fig. 15. Comparison of the center trace of the corresponding section in the case of the RN model with three different data processing and three different spatial sampling intervals

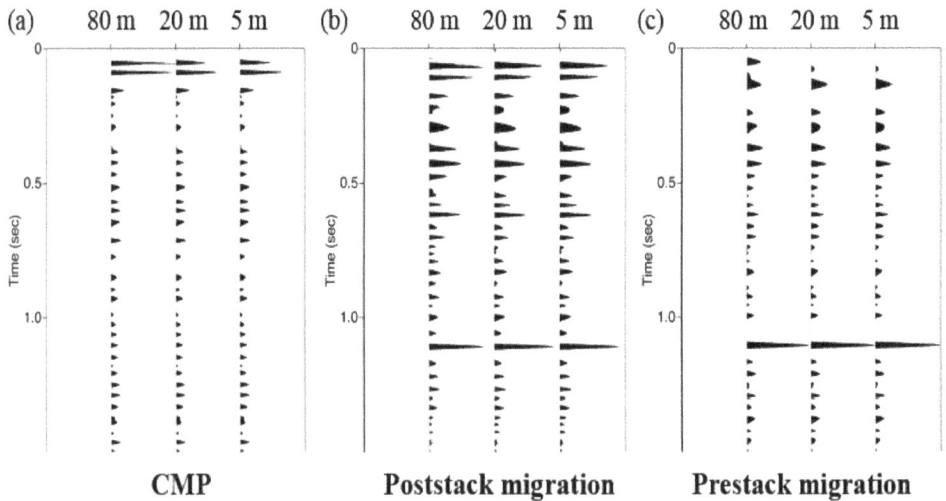

Fig. 16. Comparison of the center trace of the corresponding section in the case of the RH model (a=10 m) with three different data processing and three different spatial sampling intervals

interval is 80 m (i.e., the number of sources/receivers is small). However, the difference of the S/N ratio becomes larger with shortening the spatial sampling interval (i.e., increasing numbers of sources/receivers), and the PSTM does a much better job of imaging the reflector. Huygens' principle explains this mechanism as follows. A reflector is presumed to consist of Huygens' secondary sources, in which case imaging a reflector is considered to be equivalent to imaging each point scatterer separately and summing the imaged point scatterers at the end (Matsushima et al., 1998). A point scatterer can be delineated more appropriately by PSTM than by CMP stacking or poststack migration. In this case, an adequate zero-offset section cannot be obtained by CMP stacking without dip moveout (DMO) corrections.

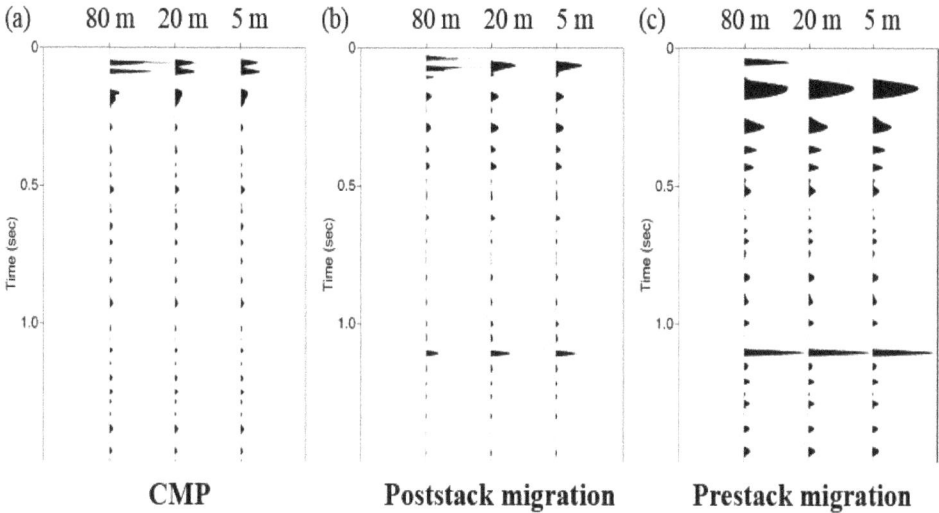

Fig. 17. Comparison of the center trace of the corresponding section in the case of the RH model (a=50 m) with three different data processing and three different spatial sampling intervals

Figures 16a thorough 16c, and Figures 17a thorough 17c show the center trace of the corresponding section with three different spatial sampling intervals in the case of the RH model (a=10) and RH model (a=50), respectively. We can see that there is little difference of the S/N ratio between different spatial sampling intervals except the shallow part of each section (less than 0.2 sec.) in each data processing. However, the difference of the S/N ratio among the data processing variants is obvious, that is, the PSTM does a much better job of imaging the reflector in the randomly heterogeneous media. The reason can be explained by the Huygens' principle as described above.

5. Discussion

It is important to discriminate between two different types of noise: a random noise in time series and a noise-like wave field produced from random heterogeneity. One may regard the scattered waves generated from heterogeneous media as a random noise in field seismic data. Some authors (e.g., Matsushima et al., 2003) have added random noise to their synthetic data for simulating field seismic data. However, the noise is a consequence of the wave phenomena in heterogeneous media, and is not same as the noise that randomly appears in the time-series (Levander and Gibson, 1991). Scales and Snieder (1998) concluded that the noise in a seismic wave is not merely a time-series which is independent from the original seismic wave but a signal-induced wave mostly consisting of scattered waves. This is important for seismic data processing but not well recognized in the field of seismic explorations. To generate the signal induced noise, the noise should be calculated from the interaction between the small-scale random heterogeneity and the original seismic wave. However, we should also note that the small-scale random heterogeneities are not known and should be estimated by other methods like numerical experiments.

We demonstrate that one can obtain better final section in terms of its S/N ratio as the intervals of spatial sampling becomes shorter (with increasing the numbers of sources/receivers) for the case of random noises model where the added random noise is a completely independent time-series against seismic traces. Thus, this type of random noise cancels each other by applying CMP stacking, poststack migration, and PSTM. On the other hand, scattering waves generated from random media is now recognized as a mutually dependent noise among the seismic traces, which indicates the interaction between the short-wavelength heterogeneity and the source and reflected wavelet. Although these scattered waves appear as random noises, they are thought to be an accumulation of many scattered waves which themselves partially coherent. Thus, this type of scattering noise should be categorized into coherent noise if we classify noise types. In general, coherent noise can not be reduced after processing the data, merely by increasing the source strength or shortening the sampling interval.

It is widely believed that highly dense spatial sampling increases the quality of final seismic sections. There are two aspects to the improvement of the quality. One is that a shorter spatial sampling interval can reduce the migration noise caused by spatial aliasing. The other is that the increase in the number of sources/receivers raises the effect of signal enhancement to increase the S/N (signal to noise) ratio.

In random heterogeneous media, three types of data processing, conventional CMP stacking, poststack migration, and PSTM, were applied and compared to examine different responses to different sampling intervals. Each data process without data space aliasing achieves very similar final sections for different sampling intervals. Safar (1985) studied the effects of spatial sampling on the lateral resolution of a surface seismic reflection survey when carrying out scatterer point imaging by applying migration, and found almost no effect of spatial sampling on lateral resolution. Safar (1985) also demonstrated the generation of migration noise caused by a large sampling interval. Migration noise is a consequence of spatial aliasing that is related to frequency, velocity, and dip of a seismic event. A shorter sampling interval cannot improve spatial resolution very much, even if there is no noise. The same conclusion was obtained by Vermeer (1999). The results we have obtained correlate well with those of these previous studies. Our numerical experiments indicate that the highly dense spatial sampling does not improve resolution of the section except the shallow part of the section when the subsurface structure contains random heterogeneity, even if the interval of spatial sampling becomes shorter than the Nyquist sampling interval. However, we found the existence of a significant difference among the data processing variants. We demonstrate that the prestack migration method has the advantage of imaging reflectors with higher S/N ratios than typically obtained with the conventional CMP stacking method with/without the poststack migration. We explained the possible mechanism by the Huygens' principle. A point scatterer can be delineated more appropriately by PSTM than by CMP stacking or poststack migration.

In our numerical experiments for RH models, two different heterogeneity sizes (a=10, 50 m) with three different spatial sampling (5, 20, 80 m) were applied. Our numerical experiments show that the effect of image space aliasing depends on the relationship between the heterogeneity size and the spatial sampling interval. Frequency components of scattering waves generated from random media depend on the heterogeneity size. When spatial sampling is too coarse, steeper-dip events are relatively aliased. To avoid spatial aliasing in heterogeneous media, it is important to know how dense the source/receiver arrangements should be in data acquisition. Narrower interval in spatial sampling can provide a clearer image of heterogeneous media. Qualitatively, spatial sampling should be smaller than the size of heterogeneities. Further consideration on quantifying the relationship between spatial sampling and the size of heterogeneities is needed. We also note that the small-scale random heterogeneities are not known and cannot be effectively estimated prior to data acquisition.

6. Conclusions

We have shown from the viewpoint of spatial sampling how the two different types noise, a random noise in time series and a noise-like wavefield produced from random isotropic heterogeneity, influence the final section. We use a 2-D finite difference method for numerically modeling acoustic wave propagation. In the presence of the time-series random noise, a final section can be obtained with a higher S/N ratio with shortening the interval of spatial sampling, that is, the increasing the numbers of sources/receivers improve the reflection image. On the other hand, in the case of random heterogeneous model, a final section is influenced by the interval of spatial sampling in different way as that of time-series random noise. Highly dense spatial sampling does not seem to improve the final quality of a section regardless of the relationship between the spatial sampling interval and the characteristic size of heterogeneities, even when the interval of spatial sampling is smaller than the Nyquist interval. We have pointed out the importance of discrimination between two different types of noise: a random noise in time series and a noise-like wave field produced from random heterogeneity. We have also demonstrated that the prestack migration method has the advantage of imaging reflectors with higher S/N ratios than typically obtained with the conventional CMP stacking method with/without the poststack migration in both RN and RH model, which can be explained by the Huygens' principle.

7. Acknowledgments

The author greatly acknowledges the thorough reviews and constructive comments of an anonymous reviewer, which helped increase the quality of the manuscript. This study was supported by a Grant-in-Aid for Scientific Research from the Ministry of Education, Culture, Sports, Science and Technology of Japan (Grant No. 21360445).

8. References

Abma, R., Sun, J., and Bernitsas, N., 1999, Antialiasing methods in Kirchhoff migration, Geophysics, 64, 1783-1792.

Biondi, B., 1998, Kirchhoff imaging beyond aliasing, Stanford Exploration Project, Report, 97, 13-35.

Biondi, B., 2001, Kirchhoff imaging beyond aliasing, Geophysics, 66, 654-666.

Brown, L., Serpa, L., Setzer, T., Oliver, J., Kaufman, S., Lillie, R., Steiner, D., and Steeples, D.W., 1983, "Intracrustal complexity in the United States midcontinent: Preliminary results from COCORP surveys in northeastern Kansas, Geology, 11, 25-30.

Emmerich, H., Zwielich, J., and Müller, G., 1993, Migration of synthetic seismograms for crustal structures with random heterogeneities, Geophysical Journal of International, 113, 225-238.

Frankel, A., and Clayton, R., 1986, Finite difference simulations of seismic scattering: Implications for the propagation of short-period seismic waves in the crust and models of crustal heterogeneity, Journal of Geophysical Research, 91, 6465-6489.

Gibson, B. S., and Levander, A. R., 1988, Modeling and processing of scattered waves in seismic reflection surveys, Geophysics, 53, 466-478.

Gibson, B. S., and Levander, A. R., 1990, Apparent layering in common-midpoint stacked images of two-dimensionally heterogeneous targets, Geophysics, 55, 1466-1477.

Goff, J. A., and Holliger, K., 1999, Nature and origin of upper crustal seismic velocity fluctuations and associated scaling properties: Combined stochastic analyses of KTB velocity and lithology logs, Journal of Geophysical Research, 104, 13169-13182.

Grasmueck, M., Weger, R., and Horstmeyer, H., 2005, Full-resolution 3D GPR imaging, Geophysics, 70, K12-K19.

Holliger, K., and Levander, A. R., 1992, A stochastic view of lower crustal fabric based on evidence from the Ivrea zone, Geophysical Research Letters, 19, 1153-1156.

Hoshiba, M., 2000, Large fluctuation of wave amplitude produced by small fluctuation of velocity structure, Phys. Earth Planet Inter., 120, 201-217.

Karson, J. A., Collins, J. A., and Casey, J. F., 1984, Geologic and seismic velocity structure of the crust/mantle transition in the Bay of Islands ophiolite complex, Journal of Geophysical Research, 89, 6126-6138.

Levander, A. R. and Gibson, B.S., 1991, Wide-angle seismic reflections from two-dimensional random target zones, Journal Geophysical Research, 96, 10,251-10,260.

Lumley, D., 2001, Time-lapse seismic reservoir monitoring, Geophysics, 66, 50–53.

Matsushima, J., Rokugawa, S., Yokota, T., Miyazaki, T., Kato, Y., 1998, On the relation between the stacking process and resolution of a stacked section in a crosswell seismic survey, Exploration Geophysics, 29, 499-505.

Matsushima, J., Okubo, Y., Rokugawa, S., Yokota, T., Tanaka, K., Tsuchiya, T., and Narita, N., 2003, Seismic reflector imaging by prestack time migration in the Kakkonda geothermal field, Japan, Geothermics, 32, 79-99.

Matsushima, J. and Nishizawa, O., 2010a, Effect of Spatial Sampling on Time-lapse Seismic Monitoring in Random Heterogeneous Media. In: Junzo Kasahara, Valeri Korneev and Michael Zhdanov, editors: Active Geophysical Monitoring, Vol 40, Handbook of Geophysical Exploration: Seismic Exploration, Klaus Helbig and Sven Treitel. The Netherlands, Elsevier, pp. 397-420.

Matsushima, J. and Nishizawa, O., 2010b, Difference image of seismic reflection sections with highly dense spatial sampling in random heterogeneous media, Journal of Seismic Exploration, Vol. 19, pp. 279-301.

Matsushima, J., Suzuki, M., Kato, Y., and Rokugawa, S., 2011, Estimation of ultrasonic scattering attenuation in partially frozen brines using magnetic resonance images, Geophysics, Vol. 76, pp. T13-T25.

Mayne, W.H., 1962. Common reflection point horizontal data stacking techniques, Geophysics, 27, 927–938.

Nishizawa, O., Satoh, T., Lei, X., and Kuwahara, Y., 1997, Laboratory studies of seismic wave propagation in inhomogeneous media using a laser Doppler vibrometer, Bulletin of the Seismological Society of America, 87, 809-823.

Safar, M. H., 1985, On the lateral resolution achieved by Kirchhoff migration, Geophysics, 50, 1091–1099.

Saito, T., Sato, H., Fehler, M., and Ohtake, M., 2003, Simulating the envelope of scalar waves in 2D random media having power-law spectra of velocity fluctuation, Bulletin of the Seismological Society of America, 93, 240-252.

Sato, H. and Fehler, M., 1997, Seismic wave propagation and scattering in the Heterogeneous Earth, Springer-Verlag, New York.

Sattlegger, J.W., Stiller, P.K., 1974, Section migration, before stack, after stack, or inbetween, Geophysical Prospecting, 22, 297–314.

Scales, J. A. and Snieder, R., 1998, What is noise?, Geophysics, 63, 1122–1124.

Shiomi, K., Sato, H. and Ohtake, M., 1997, Broad-band power-law spectra of well-log data in Japan, Geophysical Journal International, 130, 57-64.

Sick, C.M.A., Müller, T.M., Shapiro, S.A., Buske, S., 2003, Amplitude corrections for randomly distributed heterogeneities above a target reflector, Geophysics, 68, 1497-1502

Sivaji, C., Nishizawa, O., and Fukushima, Y., 2001, Relationship between fluctuations of arrival time and energy of seismic waves and scale length of heterogeneity: an inference from experimental study, Bulletin of the Seismological Society of America, 91, 292-303.

Spitz, S., 1991, Seismic trace interpolation in the F-X domain, Geophysics, 56, 785-794.

Stolt, R., 1978, Migration by Fourier transform, Geophysics, 43, 23-48.

Vermeer G. J. O., 1990, Seismic wavefield sampling, Geophysical Reference Series 4, SEG.

Vermeer, G.J.O., 1999, Factors affecting spatial resolution, Geophysics, 64, 942-953.

Warner, M., 1987, Migration – why doesn't it work for deep continental data ?, Geophysical Journal of the Royal Astronomical Society, 89, 21-26.

Wu, S. R., and Aki, K., 1988, Introduction: Seismic wave scattering in three-dimensionally heterogeneous earth, Pure and Applied Geophysics, 128, 1–6.

11

Seismic Modeling of Complex Geological Structures

Behzad Alaei
Rocksource ASA,
Norway

1. Introduction

Seismic forward modeling is seismic forward realization of a given geological model (Carcione et al., 2002; Fagin, 1991; Krebes, 2004; Sayers & Chopra, 2009). Two main stages of seismic modeling are geological model building, and numerical computation of seismic response for the model. It describes the forward process of propagating waves from sources to scatterers down in the subsurface and back to the receivers. The quality of the computed seismic response is partly related to the type of model that is built. Therefore the model building approaches become equally important as seismic forward realization methods. Models are considered to be representations of real objects (Ellison, 1993) and can be 1D, 2D, or 3D. 1D models are usually generated at well locations to predict the seismic response of the geological model and further to investigate the link between the geological beds at the well to the real reflection seismic data (seismic to well tie analysis).

The increasing amount of data which new technologies (such as advanced multi-component 3D seismic surveys) are able to provide, together with the development of more powerful and numerically efficient computing systems, have led to the rapid growth of subsurface modeling techniques (e.g. Alaei & Petersen, 2007; Mallet, 2008). Model building techniques developed significantly over the past decades. Khattri & Gir (1976) used a series of lithological elements through a cyclic succession (for example sand and shale) to create different 1D seismic models. The seismic response of such models have been predicted using ray theory approach. May & Hron (1978) carried out zero offset ray tracing for primary P waves to predict seismic response of a series of simple 2D geological models including stratigraphic wedge, unconformity, anticline, reef, normal fault, growth fault, thrust fault, salt dome flank, and overhang salt dome. The 2D models consisted of homogenous layers separated with curved interfaces.

Gjøystdal et al (1985) introduced solid modeling technique to build 3D models of complex geological structures. The model consisted of a series of columns or solids and the properties such as P and S wave velocities and density varied continuously within solids and discontinuously across model interfaces or boundaries. They have used the 3D models to run dynamic ray tracing. Open model building technique (Åstebøl, 1994 as cited in Vinje et al., 1999) unlike the solid modeling technique may contain holes and cracks in interfaces. A wide range of computer aided design (CAD) methods have been developed to build complex geological models. Mallet (2008) gave a thorough review of theses methods. Patel & McMechan (2002) provided an algorithm to create 2D geological models from controlled horizons and well log data.

The main goal of this chapter is to introduce seismic forward modeling as a powerful tool to investigate the seismic wave propagation in different geological settings with a special reference to complex geological structures. The source of complexities of seismic wave transmission and reflection in subsurface will be explained. Different model building approaches will be described with examples. Three different seismic forward realizations including asymptotic (ray tracing methods), integral, and direct (e.g. finite difference algorithms) methods will be presented.

2. Sources of complexities of seismic wave propagation

Most of the problems in seismic wave propagation of geological settings are due to the complexities in structure (structure dependent complexities) and rock types (structure independent or stratigraphic complexities). The term 'complex' is used for those geological settings which cannot be easily imaged (Fagin, 1991) due to special characteristics of structural or stratigraphic complexities. Examples of structural complexities are: steep dipping beds, faults with steep dips, complex faulted folding, folds with complex geometry, closely spaced folds and faults. Complex faulted/folded salt basins are good examples of complex geological settings. Near surface problems add more complexity to the seismic wave propagation in particular to the land seismic data with variable topography. Some of the near surface problems are: i) seismic data distortion due to near surface velocity variations, ii) topographic variations, iii) irregular data coverage caused by rugged topography, and iv) illumination problem caused by near surface complex velocity fields. In seismology, illumination is the amount of seismic wave energy that falls on a reflector and thus available to be reflected (Sheriff, 2004). The complication is caused by propagation of body waves through the complex near surface layers and source generated noise that are trapped in the near surface (Al-Ali & Verschuur, 2006).

2.1 Near surface problems

A near surface, low velocity layer (LVL) causes delay of seismic travel times. The term low velocity layer is often used for material above water table or to geologically unconsolidated deposits on harder consolidated rocks (Cox, 1999; Marsden, 1993). This seismic weathering layer despite its terminology is different from geologic weathered layer. The variability in both thickness and velocity of the near surface layers is the main source of problem. The LVL is usually above the water table and the pore spaces of rocks are filled with air rather than water which considerably lower the seismic velocity. Corrections must be applied to seismic travel time to compensate for the delay caused by the LVL. These corrections are part of the static corrections applied to seismic data and there are several methods available such as up-hole based statics, and first break statics. The main assumption behind the conventional static corrections is that raypath through a relatively simple near surface is almost vertical and therefore a vertical time shift can be used to reference the acquired data to a flat datum (Cox, 1999).

Topographic variation is one of the most complicating factors affecting reflection seismic imaging. Vertical near surface propagation assumption explained earlier may not be valid in case of rugged acquisition topography. The angular dependence of statics should be considered otherwise the diffractions would not be handled correctly in the subsequent imaging steps. Most of the conventional imaging methods require data has to be collected on a level or datum and with regular grid. In the case of land seismic, data are acquired along irregularly-sampled surface with varying topography. Redatuming with static shift can be used to remove the topographic variations. The objective is to determine the reflection events arrival times which, would have been observed if all recording were made on a flat datum. The limitation of conventional static corrections is known from before (e.g. Shtivelman & Canning, 1988) and alternative methods such as wave-equation datuming has been used instead (Al-Ali & Verschuur, 2006; Bevc, 1997; Reshef, 1991). Fig. 1 (Yang et al., 2009) shows a Prestack Depth Migrated (PSDM) seismic image from the Chinese Foothills using conventional static and wave equation based datuming. This example illustrates that it is necessary to compensate for the effect of complex propagation.

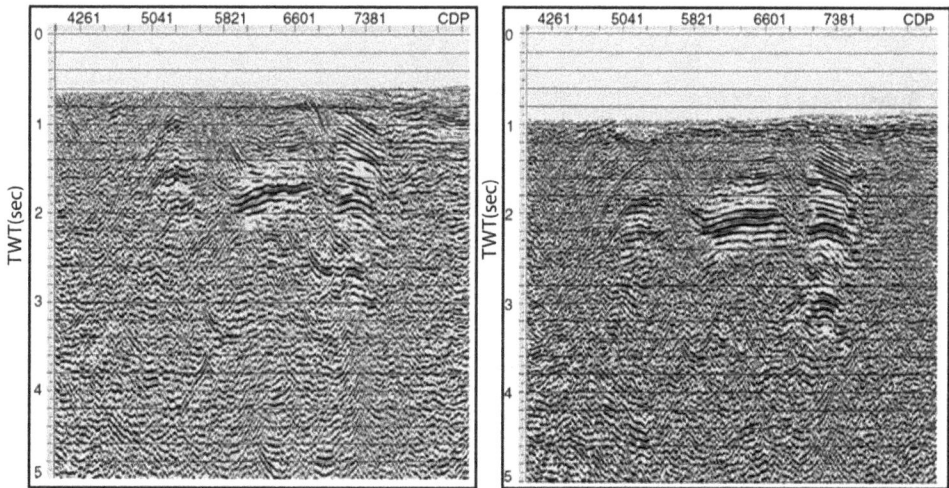

Fig. 1. PSDM seismic image from Chinese Foothills after static (left) and after wave equation datuming (right). The image qulaity improved using wave equation datuming in particluar in the deeper section (Yang et al., 2009)

The complexity of near surface can also cause poor illumination of deeper targets. Seismic wave propagation in near surface beds composed of incompetent rock types such as gypsum is complicated due to the sever heterogeneity of the rocks. Internal faulting and folding of such layers will add to the complexity of the wave propagation (Alaei & Pajchel, 2006). Fig. 2 shows a Prestack Time Migrated (PSTM) section from Zagros fold belt (Alaei, 2006). Incompetent material exposed at the surface of the line cause significant illumination problem for the deeper targets. Unusually high velocities at the near surface can also cause illumination problems. An example for that could be high velocity limestone near the sea floor in the Norwegian Barents Sea that act as a strong scatterer and complicate the wave propagation.

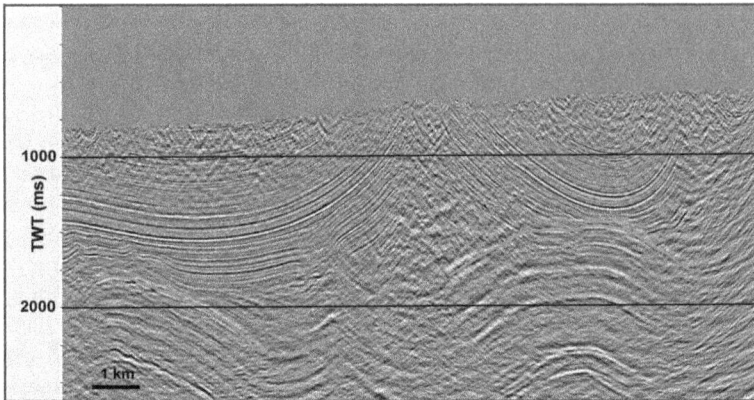

Fig. 2. PSTM seismic image from Zagros fold and thrust belt. The incompetent beds exposed at the surface (central part of the line) cause illumination problem for the deeper targets. The picked line (green) illustrates the boundary between competent and incompetent rocks. (Alaei, 2006)

2.2 Subsurface problems

Subsurface complexities also complicate the wave propagation and vary depending on the rock type and dominant structural patterns. Among the various geological settings, salt-related structures and structures of fold and thrust belts cause greater challenges for the propagation of seismic waves compared to other geologic settings. However, seismic modeling has been used to improve imaging in the salt basins and fold and thrust belts (Fagin, 1991).

Salt-related complexities: Complex structure and strong velocity contrast of salt with sediments around in salt-related geological settings is a great challenge for most of the seismic imaging algorithms (Albertin et al., 2001; Ray et al., 2004; Seitchick et al., 2009). Signal to noise ratio is usually low in the vicinity of salt bodies in particular below the salt. Examples of such settings are Gulf of Mexico, Nordkapp Basin in the Norwegian Barents Sea, and Santos Basin offshore Brazil. Seismic wave propagation through such large velocity contrast and structural complexity is associated with many wave phases including primary reflections, diffractions, and diffracted reflections. Seismic modeling has been extensively used to plan accurate seismic acquisition surveys over complex salt related structures (Gjøystdal et al., 2007) and improve seismic processing flows (Aminzadeh et al., 1997; Gjøystdal et al., 2007; Huang et al., 2010). Fig. 3 shows a seismic image from the Nordkapp Basin, Norwegian Barents Sea.

Fold and thrust belt complexities: Fold and thrust belts (such as Zagros fold belt, Canadian Rocky Mountain, and Andean fold belt) are dominated by a series of thrust faults and steeply dipping rock units. Fold geometry, internal structure complexity, highly dipping layers, and faulting associated with folding complicate the wave propagation (Alaei, 2005; Lines et al., 2000). Reflection seismic images from fold and thrust belts have frequently failed to give the correct picture of the subsurface structures when tested by drilled wells (Lingrey, 1991). Due to velocity and structural complexity, rays are bent and there are non-hyperbolic arrival times in addition to the hyperbolic arrival times. Fig. 4 (Alaei, 2006) shows an example of the ambiguity in seismic images from the structures in the Zagros fold and thrust belt. Parts of the stratal geometry is clear while the central part (indicated by yellow circle) shows a lack of reflection signal. Different seismic imaging algorithms, acquisition designs, and interpreted geologic models of fold and thrust belts can be tested using seismic forward modeling technique.

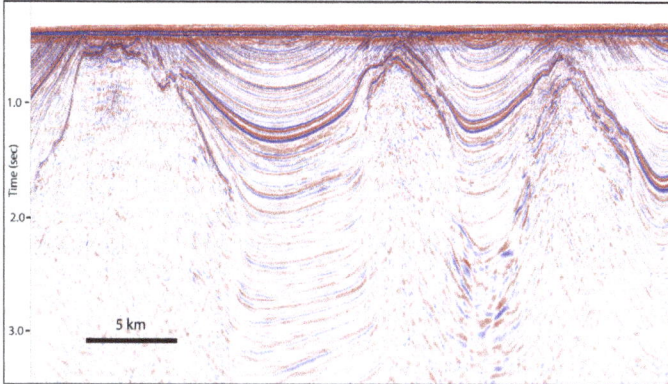

Fig. 3. Seismic image from Nordkapp Basin, Norwegian Barents Sea. The image is complex around and under the salt bodies

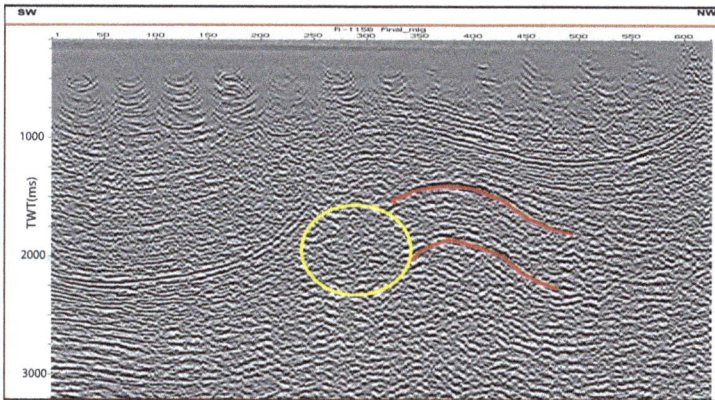

Fig. 4. 2D migrated seismic profile from the Zagros fold and thrust belt. The seismic image quality is good in one flank (indicated by red lines) and poor in other flank illustrated by yellow circle (Alaei, 2006)

Fault shadows: Seismic wave propagation is complicated under fault planes (usually footwall zone) which cause an unreliable seismic image of the zone. This zone of poor illumination is called fault shadow. Seismic imaging algorithms that doesn't take into account lateral velocity variations above imaging points fail to provide correct image under fault planes.

Lateral lithology variations: Lithological variations within rock units can cause strong lateral velocity variations which can be associated with relatively simple structures. Seismic wave propagation is complex in such settings despite the relatively simple structure (Alaei, 2005).

3. Applications of seismic modeling

Seismic modeling is useful in a wide range of applications in exploration and earthquake seismology. It plays an important role in almost all aspects of exploration seismology such as seismic data acquisition, processing, interpretation, and reservoir characterization. It increases the reliability of seismic data analysis.

Applications in seismic acquisition: In seismic acquisition, seismic forward modeling reduces the risk in seismic exploration by providing quantitative information to design better 3D surveys (e.g. Gjøystdal et al., 2007; Laurain et al., 2004; Robertsson et al., 2007). In complex geological settings seismic forward modeling can be used to test different acquisition parameters and subsurface models to achieve the optimum data collection strategy. Illumination problems of target horizons have been addressed using 2D and 3D modeling studies. The results of illumination studies can be directly applied to survey layout design. There are two categories of illumination studies used for the feasibility purposes including global approach that provides information over the whole target interface and local approach that gives information at one point in time and space (Laurain et al., 2004). Subsalt imaging has been a challenge for exploration seismology for many years and the application of seismic modeling has considerably improved the acquisition survey design for subsalt imaging (e.g. Regone, 2007). The modeling studies showed that wide angle azimuth acquisition surveys provide better illumination from subsalt structures. Seismic modeling studies have been carried out to improve seismic data acquisition over complex geologic settings of fold and thrust belts (Alaei, 2005). Fig. 5 shows raypaths from one shot record of a complex faulted anticline setting from the Zagros fold thrust belt. The acquisition geometry includes an off-end source–receiver array. In off-end source-receiver array, the seismic source is at one side of the array and receivers are deployed at the other side of the array. Complex structural settings cause poor coverage of raypaths at deeper

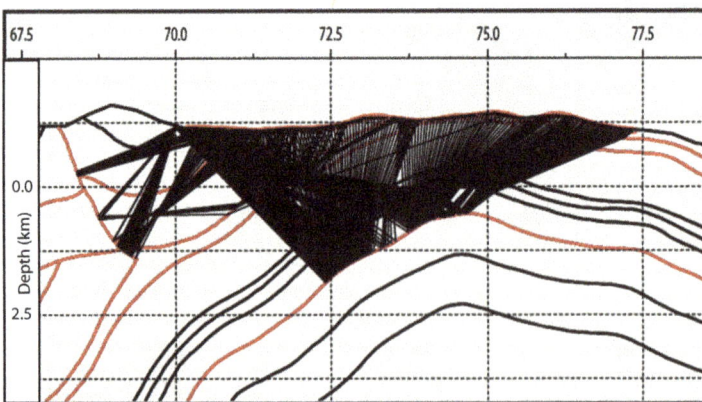

Fig. 5. Raypaths from a single shot gather ray tracing with 7km offset from a source located at x=70km. Structural complexity caused poor subsurface coverage between x=70 and x=72km

levels. The modeling shows that the large offset may partially improve the target illumination but there are still large areas of the subsurface with poor coverage. Although the 2D seismic surveys are the dominant acquisition pattern over mountainous terrains of complex geological settings, the example shown in fig. 5 clearly indicates that 2D seismic acquisition fails to provide good quality images from the subsurface and instead alternative methods such as 3D seismic acquisition can be used. However no single technology can improve the image as much as detailed analysis of survey parameters through seismic modeling.

Applications in seismic processing: Seismic modeling has been used to test different processing algorithms and flows. An important role of seismic modeling is to calibrate migration methods (Gray et al., 2001).

It can be used to optimize the processing sequences particularly in complex situations. Because of the important role of seismic modeling in seismic processing a number of synthetic models have been generated and widely used to test processing sequences. Some examples are the SEG/ EAGE 3D salt/overthrust model (Aminzadeh et al., 1997), Marmousi 2D model (Versteeg, 1994), The Society of Exploration Geophysicists Advanced Modeling Program (SEAM) (Pangman, 2007), Husky model (Stork et al., 1995; W.J. Wu et al., 1998) and Spratt Foothills model (Lines et al., 2001). Some of these models were used to test new imaging algorithms (e.g. R.S. Wu et al., 2008). Several seismic processing techniques such as multiple removal, velocity estimation (e.g. Chen & Du, 2010), migration (Moser & Howard, 2008), and seismic inversion (Jang et al., 2009; Hu et al., 2011) have been tested using the Marmousi synthetic data. SEG/EAGE 3D salt and overthrust models and associated synthetic seismic data have been used to test different migration velocity estimation and seismic imaging methods (e.g. Operto et al., 2000; Xu et al., 2004). The SEG/EAGE salt model is similar to the salt features of Gulf of Mexico and the overthrust model is similar to structures of thrust belts from South America.

Applications in seismic interpretation: Seismic forward modeling can be used to relate the response of an interpreted geologic model to real data. One application is the development of geological models to investigate the structural and stratigraphic problems faced during the seismic interpretation (Chopra & Sayers, 2009). It can be used to check the validity of interpretation particularly in complex situations. Seismic image data quality of complex geological settings is often poor that the reliance on structural styles in complex geological settings is necessary in view of the fact that the quality of the seismic images of such settings is poor. Parts of the stratal geometry maybe clearly shown while other parts show either a lack or a confusing overabundance of reflection signals (Lingrey, 1991). Seismic modeling can be used to investigate the validity of models representing different structural styles and find the best match with the real seismic data (Alaei, 2006; Alaei & Petersen, 2007; Lingrey, 1991; Morse et al., 1991).

4. Model building approaches

The integration of different data types for model definition in space and time is increasing. The model building methods can be divided into two categories: Interface based methods (e.g. Alaei, 2005) and grid based methods (e.g. Mallet, 2002). The model type can considerably influence the quality of the seismic realizations from the model. Fagin (1991) suggested a range of questions to avoid errors caused by constructing improper models. These questions are:

Should the model be 2D o 3D? How large should the model be? How many and which surfaces should the model contain? Where should the model properties (interval velocity and density) be obtained? How should the model properties vary between model interfaces? How much complexity (structural or stratigraphic) should be portrayed in the model?

Sideswipes (structural features that lies off the 2D profile) can not be simulated using 2D models. However, the 3D modeling has the capability of simulating sideswipes. Seismic response of 3D models can be viewed in different directions including time slice and mapped on the geological surfaces of the model. The model size depends on the modeling objectives. Target interface size, and depth are some of the main factors controlling the model size. It can be very large to study regional structural settings (e.g. Alaei, 2005) or small scale to investigate numerical simulations of petrophysical properties of rocks (e.g. Saenger et al., 2007).

The process starts with building the geometry of model and followed by propagating different properties such as velocity and density within different units of the model. Geometry of model is composed of stratigraphic surfaces (horizons) and faults irrespective of modeling approach. Examples of horizons are top and base of reservoir rocks, unconformities, top and base of salt, and surfaces that correspond to significant velocity variations. Faults are structural surfaces that juxtapose rocks of different properties and cause seismic wave scattering. These components shall be selected based on geological and modeling objectives. Modeling objectives that have to be included in addition to geological objectives are those which satisfy seismic wave propagation through the model. For example if there is significant velocity variation above target horizon (overburden), additional surfaces or interfaces should be included to properly simulate the seismic wave propagation through the variable velocity overburden. Layers representing velocity inversions such as thrust faults and base of salt bodies are important for modeling as they cause defocusing of seismic waves.

4.1 Interface based modeling

Interface based model building approach starts with defining the model dimensions and is followed by selecting horizons and faults of the model. The structure is constructed by interfaces (curves or lines). The curves are composed of points in depth or time domain. A minimum number of points are required to build an interface using an interpolation algorithm such as spiline method. Some of the seismic realization methods (e.g. Ray tracing methods) needs continuous second derivative. For Ray tracing methods variations in the interface geometry should be small relative to the dominant wavelength in the seismic signal. Curvature radius of the interfaces is an attribute that can be used to define a threshold for the interface smoothness. The minimum curvature radius of model interfaces should be larger than the dominant wavelength in the seismic signal. Curves representing horizons should usually be long enough to cross the model lateral boundaries. Horizons can either cross the model lateral boundaries or other horizons above or below (for example unconformities). The intersection of interfaces with either each other or model boundaries is necessary for defining blocks between the interfaces (solid model). The cross cutting horizons add to the complexity of the model. In complex models it is useful to start building the large scale architecture of the model first and then add more details into the model. The area bounded by interfaces (horizons or faults) and model boundaries corresponds to layers or blocks which include seismic properties (P and S wave velocities and density). Fig. 6 illustrates a model with 14 interfaces and corresponding blocks. There are two possibilities to define faults in an interface-based model (Fagin, 1991). They can be modeled as separate surfaces that cut the stratigraphic surfaces, or they can be represented as offsets in the modeled horizons. Although defining the fault planes as surfaces cutting the stratigraphic units is difficult, it allows us to follow the reflections from the fault plane.

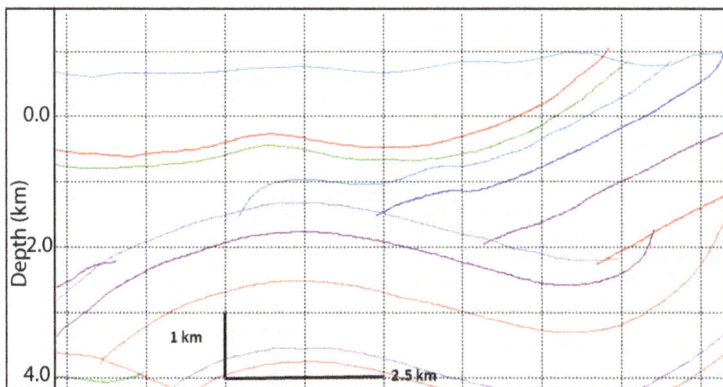

Fig. 6. Interface-based model building approach applied to a faulted anticline. 14 interfaces shown in the figure. The shallowest interface represents the topographic surface

When the geometry of the model is constructed, seismic forward realizations require properties to be assigned to each of the model layers. These properties include P and S wave velocities and density and can be constant or vary within model layers. The variation can be horizontal or vertical. The representation of properties within each layer reflects geological settings. For example in a siliciclastic sequence properties vary with depth representing compaction trends. The sources of velocities and densities are well data and reflection seismic data.

It is useful to provide information about velocity before we describe the sources of velocity data for the modeling purpose. Seismology in general and exploration seismology in particular is overflowing with velocities (Margrave, 2003). To name a few, there are instantaneous velocity, average velocity, interval velocity, root mean square (rms) velocity, migration velocity, stacking velocity, phase velocity, and group velocity. The type of velocity that is used for seismic forward modeling is the interval velocity which is simply derived by dividing the thickness of a particular layer by the travel time through the layer.

Sources of interval velocity are sonic wire line logs, checkshot surveys, and Vertical Seismic Profile (VSP) data. The thickness of the time intervals varies from 1 to 3 feet in sonic logs to hundreds of meters in checkshot surveys. Checkshot data provide travel times from source that is usually located at the land or sea surface to receivers located in the borehole and can be used to estimate interval velocity. VSP data acquired in the same way as checkshot data but includes closely (and usually evenly) spaced measurement points. VSP data can be considered as high resolution checkshot data that unlike the checkshot survey that uses only the first break data uses the entire trace information.

The interval velocity data can be derived from seismic prestack gathers. When the subsurface layers are horizontal and velocity varies more in vertical direction, reflections from interfaces are described by hyperbolas (e.g. Binodi, 2006). The change in receiver to source (offset) distance causes a delay in reflection arrival time known as moveout. For a multilayer subsurface the travel time at an offset x is (Dix, 1955):

$$T_x^2 = T_0^2 + \frac{4x^2}{V_{rms}^2} \tag{1}$$

Where T_0 is the reflection travel time at zero offset and T_x is the reflection travel time at offset x. The Vrms formula for n layers is:

$$V_{rms}(t_n)^2 = \frac{\sum_{i=1}^{n} v(t_i)^2 \Delta t_i}{\sum_{i=1}^{n} \Delta t_i} \tag{2}$$

where $V(t_i)$ is the interval velocity of layer i and t_i is the time thickness of layer i. The denominator of the formula corresponds to the total two way travel time to the base of the nth layer. The interval velocity is the one that can be directly used in modeling studies. The equation 3 can be solved for interval velocity (Dix, 1955),

$$V_{int(i)} = \frac{V_{rms}^2{}_{(i)}t_{(i)} - V_{rms}^2{}_{(i-1)}t_{(i-1)}}{t_{(i)} - t_{(i-1)}} \tag{3}$$

The velocity that is measured from seismic gathers is moveout or stacking velocity (V_{NMO}) which under certain conditions (stratified flat isotropic settings) is equivalent to rms velocity. However in complex geological settings with dipping layers and lateral velocity variations velocities measured from the seismic gathers can not be directly used to estimate interval velocity through Dix equation. Levin (1971) provided the following equation to account for the dip using V_{NMO}:

$$\cos(\theta) = \frac{V_{NMO}}{V_{rms}} \tag{4}$$

where θ is the dip angle. When the geological model is too complex and lateral velocity variations is too strong equation 3 can no longer provide accurate estimate of interval velocity and advanced model based methods must be used to estimate interval velocity. Reflection tomography is one of these methods that estimate interval velocity by using an inversion procedure to fit modeled travel times to measured travel times. Fig. 7 (Alaei, 2005) shows a regional 2D model from Zagros fold and thrust belt southwest of Iran that is 81 by 17km. The model is built using interface-based model building approach.

4.2 Grid or cell-based modeling

Constructing cell-based geological models has received a lot of attention in the past decades. In the model building process of the subsurface, model elements including faults and horizons are modeled as triangulated surfaces (Mallet, 2002). In Discrete Model (Mallet,

2008) geological model is represented by a set of points called 'nodes' that are linked to their neighbors. The nodes together with the linked neighbors generate a gird. Each node of the grid is associated with both coordinates (x,y,z) and values of physical properties (such as velocity or density).

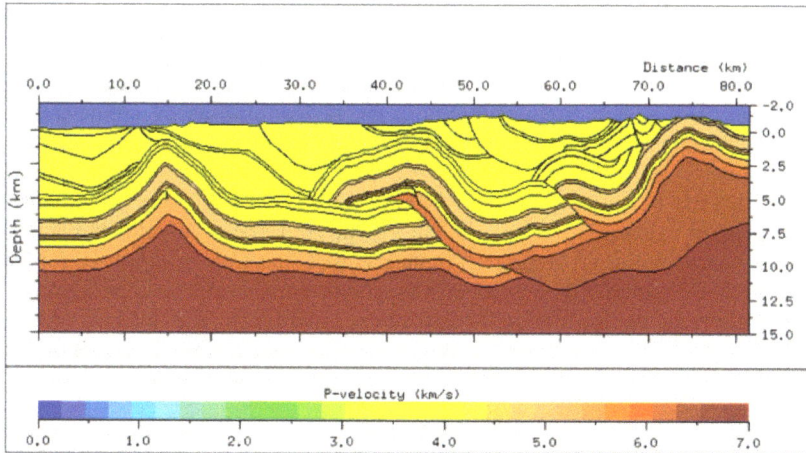

Fig. 7. Geometry of a regional complex 2D model (81x17km) from Zagros fold and thrust belt together with P wave velocity distribution. Well velocity data used to define velocities of model blocks (Alaei, 2005)

A strategy for modeling clastic reservoirs was explained by Bryant & Flint (1993). It includes five major steps: (1) definition of the space occupied by the modeled interval; (2) recognition of geological units within the model space; (3) assignment of geometries to the units; (4) arrangement of the units within the model space (architecture); (5) assignment of properties to the units. Two common approaches for the third step, assignment of geometry and orientation, are proposed. 1. Modeling of discrete objects such as shale in sand or sand in shale. 2. Modeling based on continuous variation. This is based on a Boolean method. Both methods use cell-based systems.

Patel & McMechan (2003) used well log data and control horizons to build a gridded model from physical properties such as seismic velocity. Inverse distance weighting or linear interpolation has been used to extend the well log information into the 2D model. The geometry of the control horizons is used to control the spatial extent of the properties. To obtain data for building any model with this method it is required to provide sufficient wells to sample every element in the model and enough control horizons to define the lateral extent of the structures.

Petersen (1999) proposed a modeling approach – compound modeling – to construct geological models. The compound model is composed of compound cells and each cell occupies an area. Different physical property distributions are assigned to each cell. The properties can vary within each cell. The model is consistent with geological evolution since the final distribution of properties emulates geological processes over time. In complex structural models where the sequence of events is important this feature of compound modeling will make it possible to differentiate between different stages of deformation (faulting, folding or fault-related folding, erosion etc.). If, for example, the result of a geological process such as a deformation phase, orogenic event or a sedimentation-related process is overprinted by the result of another process, the properties belonging to the latest stage (in 'time') replace the previous one for a specific position in 'space'. This ensures the time and space consistency of the geological model. The property distributions involve ranking. So, in the case of several property functions for one position in space, the one with highest priority derived from ranking will be selected. The geometry is controlled by curves of parametric description and properties by 1D functions of depth. Some characteristics of the curves are: (1) made of isolated points (x, z); (2) continuous by spline interpolation (spline means that the curve is continuous to second order); (3) x and z are functions of a common parameter, so that the curve may take any shape. The property distribution in space relates directly to space by property cells. The characteristics of property cells are the curve, the property values and curve orientation. The compound model is transformed into a grid using corresponding setting parameters that have been applied for different compound cells of the model. The grid point positions along x and z can be set according to the model requirements for the grid realization (gridding). The internal geometry of geological units gives some information about the deposition and post-deposition history of the units.

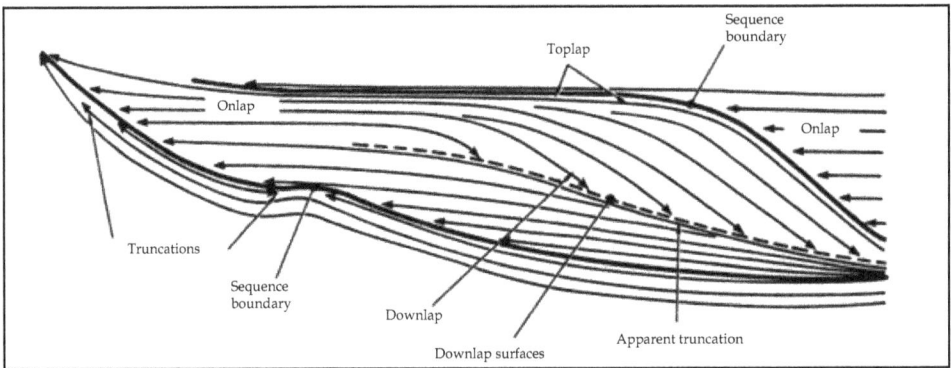

Fig. 8. Seismic patterns of a stratigraphic sequence. The reflection terminations at different locations of the unit indicate different geological processes (Vail, 1987)

Any modeling attempt without taking into consideration such details will not represent the real geology. Sedimentary bodies are rarely equi-dimensional, so proper modeling requires knowledge of the orientation of the geological units. Fig. 8 shows an example of the importance of internal orientation of geological units in modeling. If one just models the whole unit shown in fig. 8 as one block without attention to internal structure and orientation, it will not represent the real situation. Therefore, a successful modeling approach is the one that can include such geometrical details in the model so that the output will be geologically consistent. It is possible with curves and the hierarchical approach in compound modeling to build any kind of internal geometry and orientation such as truncations, onlap, downlap, and complex small-scale faulting and folding inside the geological units. Fig. 9 (Alaei & Petersen, 2007) shows the regional 2D Zagros model shown in fig. 7 that is constructed using Compound modeling approach. The model includes regional as well as small scale structural and stratigraphic details. The velocity model is based on the integration of different available data, including check shot data from 10 wells. All available density logs from the wells used in the model and for deeper layers constant values has been used.

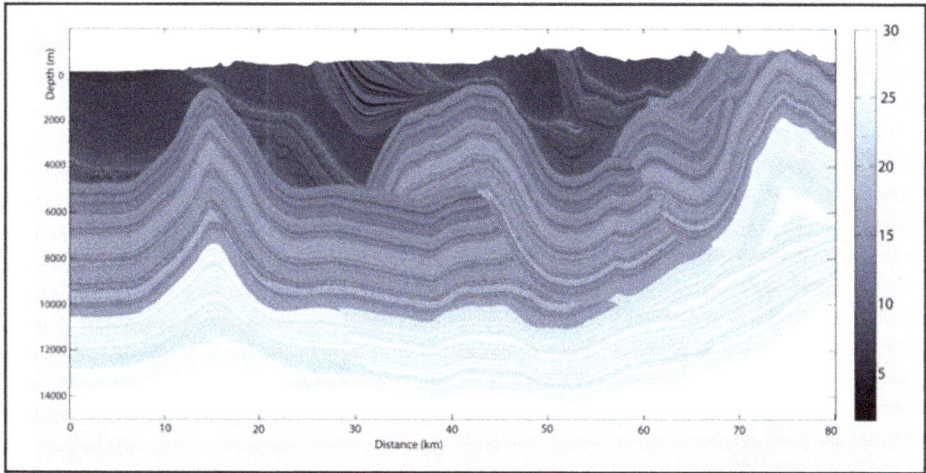

Fig. 9. 2D regional model (80x17km) from Zagros fold and thrust belt. It includes regional structural elements as well as small-scale stratigraphic detail. The color represents the scaled acoustic impedance (Alaei, 2006)

5. Seismic forward realizations

Seismic forward realizations can be carried out following the construction of model geometry and populating proper seismic properties. The goal is to predict seismic response of subsurface model recorded on a group of receivers. Seismic modeling methods can be classified into three main categories (Carcione et al., 2002): i)asymptotic, ii) integral-equation and iii) direct methods.

5.1 Asymptotic methods

Asymptotic methods (ray tracing methods) have been frequently used in seismic modeling and imaging. They do not take into account the full wavefield (e.g. Červený, 2001). In these methods, the wavefield is considered as a series of certain events, with characteristic travel time and associated amplitude. Raypaths are traced either by solving a certain differential equation that can be extracted from seismic wave equation (girded models) or by using analytic results within layers and explicit Snell's law calculations (interface based models). Raypaths are unbent in a constant velocity layer, bend at velocity interfaces (in accordance with Snell's law), and reflect at an angle equal to incidence angle at impedance interfaces. Snell's law is the relation that governs the transmission and reflection of raypaths at velocity interfaces and is used to calculate the raypath bending at velocity interfaces,

$$\frac{v_i}{Sin(\alpha)} = \frac{v_{i+1}}{Sin(\theta)} \tag{5}$$

V_i and V_{i+1} are the velocities at medium i and i+1 and α and θ are incidence and transmission angles respectively. Rays follow the geometrical rule of transmission/reflection (Snell law) also called geometric rays. Rays that follow the law of edge diffraction at a point are called diffracted rays (e.g. Klem-Musatov et al., 2008; Kravtsov & Orlov, 1993).

For a constant density variable velocity scalar wave equation

$$\nabla^2 P = \frac{1}{v(\overrightarrow{x})^2} \frac{\partial^2 P}{\partial t^2} \tag{6}$$

an approximate high frequency solution can be written as

$$P(x,t) = P(x_i) e^{[-i\omega(t-T_x)]} \quad \omega >> 0 \tag{7}$$

P and T are functions of position and are smooth scalar functions. If we take the derivate of the equation and considering the high frequency assumption we get,

$$(\nabla T)^2 = \frac{1}{v(\overrightarrow{x})^2} \tag{8}$$

and

$$2\nabla P.\nabla T + P\nabla^2 T = 0 \tag{9}$$

These equations are basic equations in the asymptotic methods and are called eikonal and transport equation (e.g. Carcione et al., 2002; Červený, 2001). The eikonal equation is a non-linear partial differential equation of first order for arrival time, T. The transport equation represents a linear partial differential equation of first order in $P(x)$. The eikonal equation describes the travel time behavior of seismic waves under high frequency condition (Bleistein et al., 2000). Kinematic ray tracing includes travel time computation of the rays and only requires seismic velocity of geological model while amplitude calculation (dynamic ray tracing) requires both velocity and density of the model to be defined.

There are different ray tracing modes depending on the source and receiver arrays (acquisition mode) and seismic modeling objectives which can be categorized into two main groups of zero offset and offset methods. Offset ray tracing includes a series of seismic traces recorded with different receivers but same source. Different source-receiver arrays can be used such as split spread (source in the middle of the receivers) and off-end (source at one side of receivers) arrays depending on the position of the source with respect to the receivers. This ray tracing mode simulates the same way that real seismic data are acquired and has been widely used to test different processing stages that are carried out on prestack common shot gather data. Processing steps such as dip filtering, and some prestack migration algorithms can be tested using the offset ray tracing of geological models (Fagin, 1991). Seismic response of a single source and receivers is called shot gather record. Shot gather ray tracing is one of the most common geometries used to simulate prestack seismic response from subsurface models. Shot gather ray tracing with off-end survey geometry was carried out at point x=49.5km of the Zagros model shown in fig. 7. Seismic response of the single shot (fig. 10) is generated by convolving travel time data derived from ray tracing with a zero phase synthetic Ricker wavelet of 35 Hz dominant frequency. The non-hyperbolic event shown in fig. 10 corresponds to the repeated layer of the anticline. Due to velocity and structural complexity, rays are bent and there are non-hyperbolic arrival times. Such non hyperbolic event will not be properly imaged if poststack seismic imaging methods are applied (Biondi, 2006).

In the zero offset ray tracing modes there is no offset between the source and receiver. Two main zero offset ray tracing modes are normal incidence ray tracing and image ray tracing.

Normal incidence ray tracing is one of the methods used in the modeling of complex geological structures (e.g. Alaei, 2005). It simulate a Common Mid Point(CMP) stack section.

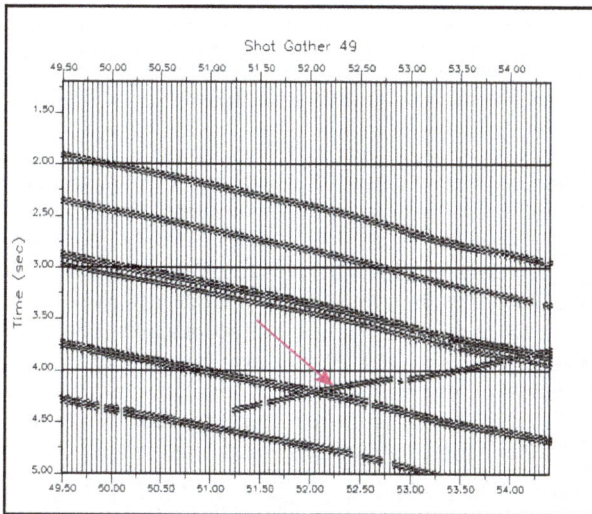

Fig. 10. Synthetic shot gather record at x=49.5km of Zagros model shown in fig. 7. Maximum offset is 4500m. The event indicated by the arrow represent a nonhyperbolic event in the shot gather record (Alaei, 2005)

CMP stack section is generated by combining (stacking) CMP gathers. CMP gathers are traces that would be recorded by a coincident source and receiver at each location and can be generated by resorting shot gather data. Stacking process attenuates random effects and improves the signal to noise ratio. However in case of complex geological settings with non hyperbolic moveout, stack section fail to give the correct image of the subsurface.

Fig. 11 shows raypaths from normal incidence ray tracing of part of the Zagros model (fig. 7 x=30 to x=70km). The source/receiver spacing was 60m which is similar to real 2D data acquired from the same area. The objective is to get the stacked unmigrated image of the complex model. The raypaths are normal to model interfaces. The raypath distribution illustrates data density available from different parts of the model interfaces and can be used to identify defocused areas. Additional detailed seismic modeling can be applied to the defocused zones.

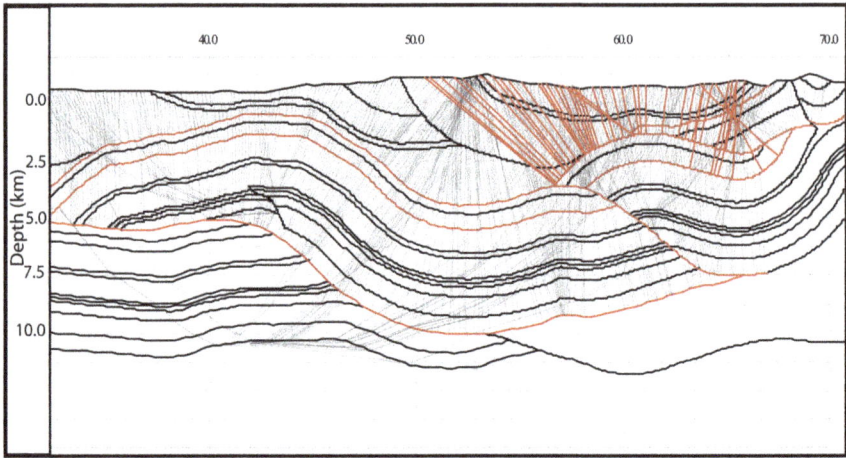

Fig. 11. Raypath image of Normal incidence ray tracing from part of the Zagros 2D model (fig. 7)

The second zero offset ray tracing mode is image ray tracing. Image ray tracing is an approximation of migrated data that can be called simulated time migration (Hubral, 1977). The raypath that represents the minimum on the time reflection surface emerges perpendicular to the recording surface. Rays are traced from points regularly distributed along the model top, and every time they hit an interface, two-way times are calculated. Reflections are considered as a continuum of diffractions and each diffraction hyperbola collapses on its least travel time peak. It can also be used to locate reflections more accurately by converting time migrated data to depth along the image rays (Thorn, et al., 1986, Johansen, et al., 2007). Imaging steeply dipping beds of complex structures is a challenge to different seismic imaging methods. Fig. 12 (left) shows 2D seismic profile from part of the Zagros model. There is an anticline in the deeper section with thick overburden. The core of the anticline is thrust faulted with flat-ramp geometry of the fault. Imaging fault plane is complex due to both structural complexity and strong velocity variations across the

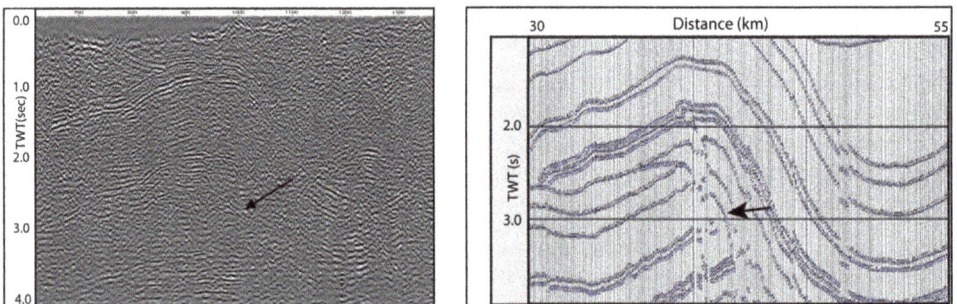

Fig. 12. 2D seismic image from faulted structure of Zagros fold and thrust belt (left). Synthetic seismic response of image ray tracing with 6km aperture. The arrow illustrates the location of steeply dipping reflector. Increasing aperture from 3 km (real data) to 6 km (synthetic data) improved the signal continuity (Alaei, 2006)

fault plane. Image ray tracing with 60m source/receiver spacing applied to the model. Imaging the steeply dipping part of the thrust plane (ramp part) is particularly important. In order to improve seismic image from the complex thrust faulted structure, a range of different migration aperture have been tested. Migration aperture is the range of spatial data considered in seismic migration. Fig. 12 (right) shows seismic response of 6km aperture. The aperture used in the processing of the real data was about 3km. The modeling results illustrate that the aperture used in the processing of the real data was not sufficient to image the steep flank of the structure properly. Increasing aperture during migration of real data will decrease the signal to noise ratio which can be improved using post migration noise cancellation filters.

5.2 Integral-equation methods

The second group of seismic modeling methods are integral-equation methods. Integral-equation methods of seismic modeling are based on integral representation of the seismic wavefield spreading from point sources (Huygens principal). There are two forms of integral methods: volume integral and boundary integral. Integral representation of scalar seismic wave equation is (Carcione et al., 2002),

$$P(x,t) = \int_D \frac{q(x_s, t - |x - x_s| / c_0)}{4\pi |x - x_s|} dx_s \qquad (10)$$

Where D is the region in space where the source term q is present and x is position vector. C_0 is the wave speed of sound. Boundary integral methods have been used to investigate the scattering of elastic waves by cracks and cavities (Bouchon, 1987; Rodrıguez-Castellanos et al., 2006; Bouchon & Sánchez-Sesma, 2007) and hydrofractures (Liu et al., 1997). Integral equation methods have been used to model wavefield scattering caused by small scale cracks or inclusions (Muijres et al., 1998; Herman & Mulder, 2002). Herman & Mulder (2002) used the integral method with two crack boundary conditions to a homogenous model containing 4000 cracks. Transmitted pressure of two different boundary conditions including compliant crack and rigid crack is illustrated in fig. 13. The compliant crack is characterized by zero pressure, and the rigid crack by zero normal particle velocity. The velocity in the model is 3000 m.s^{-1}.

5.3 Direct methods

The last category of seismic forward modeling methods are the direct methods that involve numerical solution of wave equation. Direct methods such as Finite Difference (FD) (Alterman & Karal, 1968; Claerbout, 1985) and finite element (De Basabe & Sen, 2009) require the model to be discretized into a finite number of points and therefore sometimes are called grid methods. The methods also called full wave equation method since it implicitly provides the full wave field. They have the ability to accurately model seismic waves in arbitrary heterogeneous media.

FD method is a numerical method for solving differential equations that can be applied to seismic wave equation to calculate displacement at any point in geological models. Seismic wavefield is computed at each grid point by approximating derivatives of the wave equation with finite difference formulas and solving the resulting difference equation

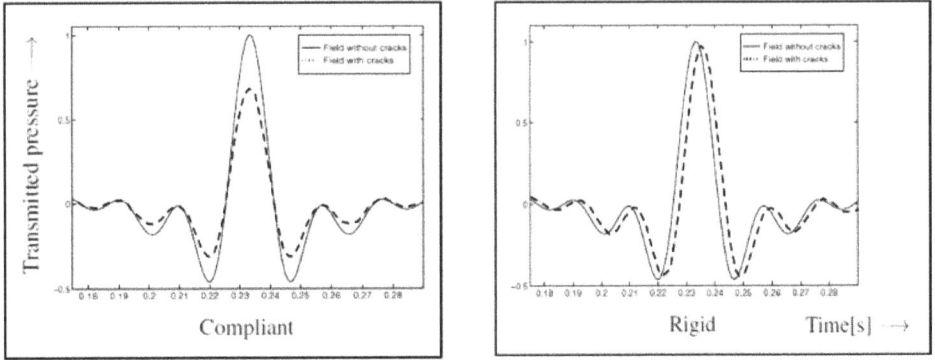

Fig. 13. Transmitted pressure for two different choices of the crack boundary condition. The velocity of the embedding equals 3,000 m/s, the crack half width 1 m and the number of cracks 4,000 (Herman & Mulder, 2002)

recursively. It includes Taylor series expansions of functions near the grid point. Explicit finite difference methods involve the estimation of wavefield at present time using the wavefield at past times. In implicit FD methods, the present values of the wavefield depend on past and future values. The mathematical formulation of finite difference seismic modeling can be found in several articles (Carcione et al., 2002; Marfurt, 1984; Margrave, 2003; Moczo et al., 2007). Let consider function U(x) with continuous first derivative. Forward, backward, and center-difference equations of the function U(x) are

$$\frac{dU(x)}{dx} = \frac{U(x + \Delta x) - U(x)}{\Delta x} \tag{11}$$

$$\frac{dU(x)}{dx} = \frac{U(x) - U(x - \Delta x)}{\Delta x} \tag{12}$$

$$\frac{dU(x)}{dx} = \frac{U(x + \Delta x) - U(x - \Delta x)}{2\Delta x} \tag{13}$$

The forward and backward-FD equations are first order approximations to the first derivative and the difference in the value of first derivative and the right hand side in equations 11 and 12 is the truncation error. Finite difference operators can be used to predict a function. For example if we know the function U(x) and its derivative at a point x_0, the function at $x_0 + \Delta x$ can be derived from equation

$$U(x + \Delta x) = U(x) + \frac{d(Ux)}{dx} \Delta x \tag{14}$$

It is easy to find an approximation to a derivative using Taylor series and the equation can be considered as a truncated Taylor series. An approximation for the second derivative can be derived by applying equations 11 and 12 and a frequently used operator is

$$\frac{d^2 U(x)}{dx^2} = \frac{U(x + \Delta x) - 2U(x) - U(x - \Delta x)}{\Delta x^2} \tag{15}$$

The equation is a centered approximation of the second derivative that is frequently used in grid-displacement FD methods.

As stated earlier the FD methods require that the geological(computational) domain is characterized by a set of discrete space-time grids. There are three different types of grids depending on the spatial distribution of functions including displacement/particle velocity and stress tensor components in the grids. In the conventional grids all functions are approximated at the same grid positions. In a partly-staggered-grid the displacement components are located at one grid position and the stress tensor components are located at other grid positions. In a staggered grid, each function (displacement and/or particle-velocity component and each shear stress-tensor component) has its own grid position. FD methods have been widely used for seismic modeling of different geological settings (e.g. Alaei & Petersen, 2007; Regone, 2007).

The Seismic Unix implementation of acoustic finite difference modeling (Cohen & Stockwell 2002) was used to generate two shots from different parts of the Zagros model (fig.9). This program uses an explicit second-order differencing method. The source was a 30 Hz Ricker wavelet. Split-spread source/receiver array was used as survey geometry with a maximum offset of 6 km.

The first shot is located on the flank of the anticline located at x=20 km of the model (fig. 9) with a thick overburden section. One of the objectives of the modeling is to investigate the effect of thick overburden on the deeper target section. Fig. 14 (left) shows the real seismic data around the first synthetic shot which is a 2D poststack time migrated image. It shows packages of reflectors at some locations and disturbed or overabundance of reflectors in the remaining part of the section, typical in complex fold belts with irregular topography. The shallow picked horizon on the seismic section (fig. 14) is located in the thick overburden. The marked area below the second picked horizon is a complex area that is not imaged as good as the upper part. Fig. 14 (right) shows the common shot gather for the first shot. It is clear that the seismic wavefield in the model is complex. The same picked events in the seismic section (fig. 14) have been picked in the shot gather. This single shot that is located on the flank of

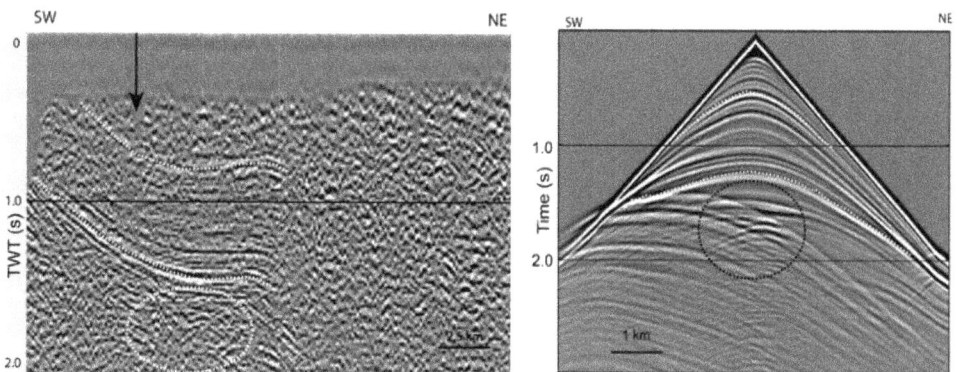

Fig. 14. 2D poststack time migrated seismic image from Zagros fold and thrust belt (left). The marked area in the deeper part is complex and has poor quality image. The right figure illustrates the shot gather. The source position is shown by arrow (Alaei & Petersen, 2007)

the anticline shows how the complexity of the overburden geology affects the seismic response. The types of events that can be seen in the shot record (particularly the diffractions and nonhyperbolic events inside the marked area in fig. 14 (right) are significant challenges for pre- or poststack time-processing methods. Advanced seismic velocity analysis and migration methods are required to image these types of complexities.

The second shot is located at X=63 km (fig. 9). The main purpose is to investigate the seismic wavefield at the top of a faulted anticline with a complex overburden. Fig. 15 (left) shows the real seismic data around the shot and the arrow shows the location of the shot. The first shallow picked horizon is in the overburden. The other picked horizons are all top reservoir rock units but faulted and repeated at the top of the structure. The overburden deepens and thickens significantly towards the southwest flank of the structure. Fig. 15 (right) shows the common shot gather. This shot is from the same location as the shot gather record illustrated in fig. 10 using ray based method. There is much more detail imaged with the FD modeling (fig. 15). The shallowest picked horizon is strong in the far offsets and weaker in the near offsets. It represents the high impedance contrast in the overburden. The picked events between 1 s and 2 s are faulted and folded top reservoir rock unit. The arrow shows the faulted leg of top reservoir. It is not easy to explain the complexity of the seismic image in the real data in the overburden section of the southwest flank of the structure using this synthetic shot.

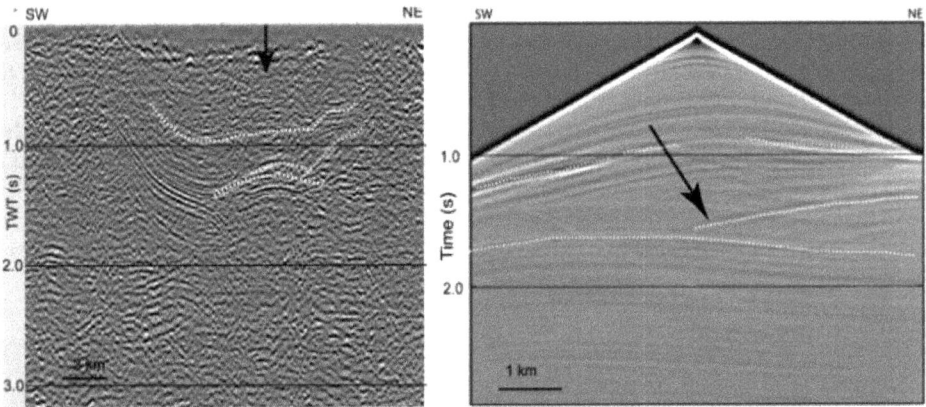

Fig. 15. Real seismic data around the shot at x=63 km of fig. 9 (left). The first shallow picked horizon is in the overburden. The other picked horizons are all faulted (repeated) top reservoir. Common shot gather of the second shot (right). The two picked events between 1s and 2s are faulted and folded top reservoir. The arrow shows the faulted leg that has been confirmed by drilling result (Alaei & Petersen, 2007)

The last FD modeling example is from another faulted anticline from Zagros fold and thrust belt. The model (fig. 16a) is built using Compound modeling and a single shot has been generated using explicit second-order differencing method (Cohen & Stockwell, 2002). To show the real complexity of the seismic wavefield, a snapshot of the single shot is illustrated in fig. 16b. Different wave modes including first arrival reflections and diffractions are illustrated.

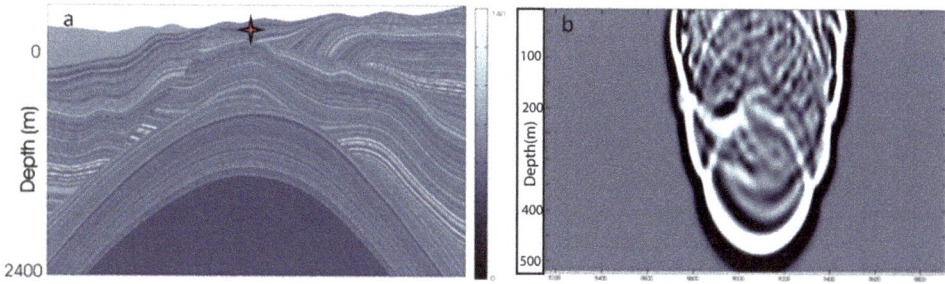

Fig. 16. a) Geological model of a complex faulted fold from Zagros fold belt. The slowness values are scaled. b) Snapshot from a shot located at the centre of the model. Complexity of the wavefield illustrated by several diffracted events

6. Summary

The main goal of this chapter is to introduce seismic forward modeling as a powerful tool to investigate the seismic wave propagation in different geological settings with a special reference to complex geological structures. Seismic forward modeling is the computation of seismic response of a geologic model and has been widely used in both earthquake and exploration seismology. The source of complexities of seismic wave transmission and reflection in subsurface have been explained. I provided some applications of seismic modeling in exploration seismology including its applications in seismic data acquisition, processing, and interpretation of complex structures. Interface and grid based model building approaches presented with examples from complex structures of Zagros fold and thrust belt. Three different seismic forward realizations including asymptotic (ray tracing methods), integral, and direct (e.g. finite difference algorithms) methods have been presented.

7. References

Al-Ali, M. N., & Verschuur, D. J. 2006, An integrated method for resolving the seismic complex near-surface problem. *Geophysical Prospecting*, 54,pp. 739-750, 1365-2478

Alaei, B. 2005. Seismic forward modeling of two fault related folds from Dezful embayment of the Iranian Zagros Mountains. *Journal of Seismic Exploration*, 14, pp. 13–30

Alaei, B., 2006. Seismic Depth Imaging of Complex Structures, an example from Zagros fold thrust belt, Iran. *PhD thesis*, University of Bergen

Alaei, B. & Pajchel, J. 2006. Single arrival Kirchhoff prestack depth migration of complex faulted folds from the Zagros mountains, Iran. *CSEG Recorder*, 31, pp. 41–48

Alaei, B. & Petersen, S. A. 2007. Geological modeling and finite difference forward realization of a regional section from the Zagros fold-and-thrust belt. *Petroleum Geoscience*, Vol. 13, pp. 241–251

Albertin, U., Woodward, M., Kapoor, J., Chang, W., Charles, S., Nichols, D., Kitchenside, P., &Mao, W. 2001. Depth imaging example and methodology in the Gulf of Mexico. *The Leading Edge*, 20, pp. 498-513

Alterman, Z. S., & Karal, F.C. Jr. 1968. Propagation of elastic waves in layered media by finite difference methods. *Bull. Seism. Soc. Am.*, vol. 58, pp.367-398

Aminzadeh, F., Brac, J., & Kunz, T. 1997. *SEG/EAGE 3D Modeling Series No 1*. Society of Exploration Geophysicists and the European Association of Geoscientists and Engineers.

Bevc D. 1997. Flooding the topography: Wave-equation datuming of land data with rugged acquisition topography. *Geophysics*, vol.62, no. 5, pp.1558–1569

Biondi, L. B. 2006. *3D Seismic Imaging*. SEG

Bleistein, N., Cohen, J. K., & Stockwell, Jr. W. 2000. *Mathematics of Multidimensional Seismic Imaging, Migration, and Inversion*. Springer Verlag

Bouchon, M. 1987. Diffraction of elastic waves by cracks or cavities using the discrete wave number method, *J. Acoust. Soc. Am.* Vol. 81, pp.1671–1676

Bouchon, M., & Sánchez-Sesma, J. S. 2007. Boundary Integral Equations and Boundary Elements Methods in Elastodynamics. *Advances in Geophysics,* Vol 48, pp. 157-189

Bryant, I.D. & Flint, S.S. 1993. Quantitative clastic reservoir modeling: problems and perspectives. *Special Publications of the International Association of Sedimentologists,* 15, 3–20.

Carcione, J.M., Herman, G.C. & ten Kroode, A.P.E. 2002. Seismic Modeling. *Geophysics, vol. 67*, no. 4, pp.1304–1325

Červený, V. 2001. *Seismic Ray Theory*. Cambridge University Press

Chen, J. B., & Du, S. Y. 2010. Kinematic characteristics and the influence of reference velocities of phase-shift-plus-interpolation and extended-split-step-Fourier migration methods. *Geophysical Prospecting*, vol. 58, pp.429–439

Cohen, J.K. & Stockwell, J.W. Jr. 2002. *CWP/SU: Seismic Unix Release 37: a free package for seismic research and processing.* Center for Wave Phenomena, Colorado School of Mines.

Claerbout, J. F. 1985. *Imaging the Earth's interior,* Blackwell Scientific Publications

Cox M. 1999. *Static Corrections for Seismic Reflection Surveys*. Society of Exploration Geophysicists

De Basabe J, D., & Sen, M. K. 2009. New developments in the finite-element method for seismic modeling, *The Leading Edge,* vol, 28, no. 5, pp. 562-567

Dix, C. H. 1955. Seismic velocities from surface measurements. *Geophysics,* vol. 20, no. 1, pp. 68-86

Ellison, A. 1993. Modeling, philosophy and limitation. *Computing & Control Engineering Journal,* 4, 190–192.

Fagin, S.W. 1991. *Seismic Modeling of Geologic Structures: Applications to Exploration Problems.* Geophysical Development series, 2. Society of Exploration Geophysicists.

Gjøystdal, H., Reinhardsen, J.E., & Åstebøl, K. 1985. Computer representation of complex 3-D geological structures using a new 'solid modeling' technique. *Geophysical Prospecting,* vol. 33,pp. 1195-1211

Gjøystdal, H., Iversen, E., Lecomte, I., Kaschwich, T., Drottning, A., & Mispel, J. 2007. Improved applicability of ray tracing in seismic acquisition, imaging, and interpretation. *Geophysics,* 75, SM261–271

Gray, H.S., Etgen, J., Dellinger, J. & Whitmore, D. 2001. Seismic migration problems and solutions. *Geophysics,* 66, 1622–1640

Herman, G. C. & Mulder, W. A. 2002. Scattering by small-scale inclusions and cracks. *64th EAGE Technical Conference,* Florence, Italy, E-10

Hu, W., Abubakar, A., Habashy, T. M., & Liu, J. 2011. Preconditioned non-linear conjugate gradient method for frequency domain full-waveform seismic inversion. *Geophysical Prospecting*, vol. 59, pp. 477–491

Huang, Y., Lin, D., Bai, B., Roby, S., & Ricardez, C. 2010. Challenges in presalt depth imaging of the deepwater Santos Basin, Brazil. *The Leading Edge,* 29, 820–825.

Hubral, P., 1977. Time migration – Some ray theoretical aspects, *Geophysical Prospecting,* vol. 25, pp.738-745

Jang, U., Min, D. J., & Shin, C. 2009. Comparison of scaling methods for waveform inversion. *Geophysical Prospecting*, vol. 57, pp. 49–59

Johansen, B. S., Granberg, E., Mellere, D., Arntsen, B., & Olsen T. 2007. Decoupling of seismic reflectors and stratigraphic timelines: A modeling study of Tertiary strata from Svalbard. *Geophysics*, vol. 72, no. 5, pp. SM273–SM280

Khattri, K., & Gir, R. 1976. A study of the seismic signatures of sedimentation models using synthetic seismograms. *Geophysical Prospecting*, vol. 24, pp. 454-477

Klem-Musatov, K., Aizenberg, A. M., Pajchel, J., Helle, H. B. 2008. *Edge and Tip Diffractions: Theory and Applications in Seismic Prospecting*, SEG

Kravtsov, Y. A., & Orlov, Y. I. 1993. *Caustics, Catastrophes, and Wave Fields*, Springer-Verlag

Krebes, E.S. 2004. Seismic Forward Modeling. *CSEG Recorder*, 30, 28–39.

Laurain, R., Gelius, L. J., Vinje, V., & Lecomte, I. 2004. A review of illumination studies. *Journal of Seismic Exploration*, 13, pp.17–34

Levin, F. K. 1971. Apparent velocity from dipping interface reflections? *Geophysics*, vol. 36, no. 3, pp. 510-516

Lines, L.R., Gray, S. H., & Lawton, D. C. 2000. *Depth imaging of Foothills seismic data.* Canadian Society of Exploration Geophysicists

Lingrey, S. 1991. Seismic modeling of an Imbricate Thrust Structure from the Foothills of the Canadian Rocky Mountains. In Fagin, S.W., ed., *Seismic Modeling of Geologic Structures Applications to exploration Problems,* , Society of Exploration Geophysicists, pp.111-125

Liu, E., Crampin, S., & Hudson, J. A. 1997. Diffraction of seismic waves by cracks with application to hydraulic fracturing. *Geophysics*, vol. 62, no. 1, pp. 253–265

Mallet, J-L. 2002. *Geomodeling.* Oxford University Press

Mallet, J-L. 2008. *Numerical Earth Models.* EAGE Publications BV

Marfurt, K. J. 1984. Accuracy of the finite-difference and finite-element modeling of the scalar and elastic wave equations. *Geophysics*, vol. 49, no. 5, pp.533-549

Margrave, G. 2003. *Numerical methods of exploration seismology with algorithms in Matlab.* http://www.crewes.org/ResearchLinks/FreeSoftware/

Marsden, D. 1993. Static corrections- a review. *The Leading Edge*, 12, pp. 43-49

May, B., & Hron, F. 1978. Synthetic seismic sections of typical traps. *Geophysics*, vol. 43, no. 6, pp. 1119-1147

Moczo, P., Robertsson, O. J. A., & Eisner, L. 2007. The Finite-Difference Time-Domain Method for Modeling of Seismic Wave Propagation. *Advances in Geophysics*, Vol. 48, pp. 421-516

Morse, P.F., Purnell, G.W. & Medwedeff, D.A. 1991. Seismic modeling of Fault-related Folds. *In:* Fagin, S.W. (ed.) *Seismic Modeling of Geologic Structures: Applications to Exploration Problems.* Society of Exploration Geophysicists, pp. 127–152

Moser, T. J., & Howard, C. B. 2008. Diffraction imaging in depth. *Geophysical Prospecting*, vol. 56, pp. 627–641

Muijres, A. J. H., Herman, G. C., & Bussink, P. G. J. 1998. Acoustic wave propagation in two-dimensional media containing small-scale heterogeneities. *Wave Motion*, vol. 27, pp. 137–154.

Operto, M. S., Xuz, S., & Lambare, G. 2000. Can we quantitatively image complex structures with rays? *Geophysics*, vol. 65, no. 4, pp. 1223-1238

Pangman, P. 2007. SEG Advanced Modeling Program. *The Leading Edge*, 27, 718-721.

Patel, M. D., & McMechan, G. A. 2003. Building 2-D stratigraphic and structure models from well log data and control horizons. *Computers and Geosciences*, 29, pp. 557–567

Petersen, S.A. 1999. Compound modeling, a geological approach to the construction of shared earth models. *61st EAGE Technical Conference*, Helsinki

Ray, A., Pfau, G., & Chen, R. 2004. Importance of ray-trace modeling in the discovery of the Thunder Horse North Filed, Gulf of Mexico. *The Leading Edge*, 23,pp. 68-70

Regone, C. J. 2007. Using finite –difference modeling to design wide-azimuth surveys for improved subsalt imaging. *Geophysics*, vol. 72, no. 5, pp. SM231–SM239

Reshef, M. 1991. Depth migration from irregular surfaces with depth extrapolation methods. *Geophysics*, vol. 56, no. 1, pp. 119-122

Robertsson, J. O. A., Bednar, B., Blanch, J., Kostov, C., & van Manen, D. 2007. Introduction to the supplement on seismic modeling with applications to acquisition, processing, and interpretation. *Geophysics*, vol. 72, no. 5, pp. SM1–SM4

Rodriguez-Castellanos, A., Sanchez-Sesma, F. J., Luzon, F., & Martin, R. 2006. Multiple Scattering of Elastic Waves by Subsurface Fractures and Cavities. *Bulletin of the Seismological Society of America*, Vol. 96, no. 4A, pp.1359-1374

Saenger, E. H., Ciz, R., Krüger, O. S., Schmalholz, S. M., Gurevich, B., & Shapiro, S. A. 2007. Finite-difference modeling of wave propagation on microscale: A snapshot of the work in progress. *Geophysics*, vol, 72, no. 5, pp.SM293-SM300

Sayers, C., & Chopra, S. 2009. Introduction to special section: Seismic modeling. *The Leading Edge*, 28, pp. 528-529

Seitchick, A., Jurick, D., Bridge, A., Brietzke , R., Beeney, K., Codd, J., Hoxha, F., Pignol, C., & Kessler, D. 2009. The Tempest Project – Addressing challenges in deepwater Gulf of Mexico depth imaging through geologic models and numerical simulation. *The Leading Edge*; 28, pp. 546-553

Sheriff, R. 2004. *Encyclopedic Dictionary of Applied Geophysics* (4th edn). Society of Exploration Geophysicists

Shtivelman, V., & Canning, A. 1988. Datum correction by wave equation extrapolation. *Geophysics*, vol. 53, no. 10, pp.1311-1322

Stork, C., Welsh, C. & Skuce, A. 1995. Demonstration of processing and model building methods on a real complex structure data set. Paper presented at SEG workshop no. 6, Houston.

Thorn, S.A. & Jones, T.P. 1986. Application of Image Ray Tracing in the Southern North Sea Gas Fields. *Geological Society of London Special Publication*, No. 23, pp.169-186

Vail, P.R. 1987. Sequence stratigraphy interpretation using sequence stratigraphy. *In:* Bally, A.W. (ed.) *Atlas of Seismic Stratigraphy, 1.* AAPG Studies in Geology, 27, pp.1–14

Versteeg, R. 1994. The Marmousi experience: Velocity model determination on a synthetic complex data set. *The Leading Edge*, 13, 927–936.

Vinje, V., Åstebøl, K., Iversen, E., & Gjøystdal, H. 1999. 3-D ray modeling by wavefront construction in open models. *Geophysics*, vol. 64, no. 6, pp. 1912-1919

Wu, R., S., Wang, Y., & Luo M. 2008. Beamlet migration using local cosine basis. *Geophysics*, vol. 73, no. 5, pp. S207–S217

Wu, W. J., Lines, L., Burton, A., Lu, H., X., Zhu, J., Jamison, W., & Bording, R. P. 1998. Prestack depth migration of an Alberta foothills data set – The Husky experience. *The Leading Edge*, 17, pp. 635-638

Xu, S., Lambare, G., & Calandra, H. 2004. Fast migration/inversion with multivalued ray fields: Part 2 – Applications to the 3D SEG/EAGE salt model. *Geophysics*, vol. 69, no. 5, pp. 1320-1328

Yang, K. Zheng, H-M. Wang, L., Liu, Y-Z., Jiang F. Cheng, J-B., & Ma Z-T. 2009. Application of an integrated wave-equation datuming scheme to overthrust data: A case history from the Chinese foothills. Geophysics, vol. 74, no.5, pp. B153–B165

Modelling Seismic Wave Propagation for Geophysical Imaging

Jean Virieux[1]* Vincent Etienne[2†] and Victor Cruz-Atienza et al.[3‡]
[1]*ISTerre, Université Joseph Fourier, Grenoble*
[2]*GeoAzur, Centre National de la Recherche Scientifique, Institut de Recherche pour le développement*
[3]*Instituto de Geofísica, Departamento de Sismologia, Universidad Nacional Autónoma de México*
[1,2]*France*
[3]*Mexico*

1. Introduction

The Earth is an heterogeneous complex media from the mineral composition scale ($\simeq 10^{-6}m$) to the global scale ($\simeq 10^{6}m$). The reconstruction of its structure is a quite challenging problem because sampling methodologies are mainly indirect as potential methods (Günther et al., 2006; Rücker et al., 2006), diffusive methods (Cognon, 1971; Druskin & Knizhnerman, 1988; Goldman & Stover, 1983; Hohmann, 1988; Kuo & Cho, 1980; Oristaglio & Hohmann, 1984) or propagation methods (Alterman & Karal, 1968; Bolt & Smith, 1976; Dablain, 1986; Kelly et al., 1976; Levander, 1988; Marfurt, 1984; Virieux, 1986). Seismic waves belong to the last category. We shall concentrate in this chapter on the forward problem which will be at the heart of any inverse problem for imaging the Earth. The forward problem is dedicated to the estimation of seismic wavefields when one knows the medium properties while the inverse problem is devoted to the estimation of medium properties from recorded seismic wavefields.

The Earth is a translucid structure for seismic waves. As we mainly record seismic signals at the free surface, we need to consider effects of this free surface which may have a complex topography. High heterogeneities in the upper crust must be considered as well and essentially in the weathering layer zone which complicates dramatically the waveform and makes the focusing of the image more challenging.

Among the main methods for the computation of seismic wavefields, we shall describe some of them which are able to estime the entire seismic signal considering different approximations as acoustic or elastic, isotropic or anisotropic, and attenuating effects. Because we are interested in seismic imaging, one has to consider methods which should be efficient especially for the many-sources problem as thousands of sources are required for imaging. These sources could be active sources as explosions or earthquakes. We assume that their

---

*Romain Brossier, Emmanuel Chaljub, Olivier Coutant and Stéphane Garambois
†Diego Mercerat, Vincent Prieux, Stéphane Operto and Alessandra Ribodetti
‡Josué Tago

distribution are known spatially as punctual sources and that the source time function is the signal we need to reconstruct aside the medium properties.

Asymptotic methods based on the high frequency ansatz (see (Virieux & Lambaré, 2007) for references or textbooks (Červený, 2001; Chapman, 2004)) and spectral methods based on a spatial and time Fourier transformations (Aki & Richards, 2002; Cagniard, 1962; de Hoop, 1960; Wheeler & Sternberg, 1968) are efficient methods which are difficult to control: whispering galeries for flat layers are efficiently considered using spectral methods. These two methods may be used either for local interpretation of specific phases or as efficient alternatives when media is expected to be simple. They could be used as well for scattering inverse problems. In the general heterogeneous case, we have to deal with volumetric methods where the medium properties are described through a volume while seismic wave fields satisfy locally partial differential equations. Although one may consider boundaries as the free surface or the water/solid interface, we may consider that variations of the medium properties are continuous at the scale of the wavelength which we want to reconstruct: the best resolution we could expect is half the wavelength (Williamson & Worthington, 1993). Therefore a volumetric grid discretization is preferred where numerical expressions of boundary conditions should be mostly implicit through properties variations.

A quite popular method because of this apparent simplicity is the finite difference method where partial derivatives are transformed into finite difference expressions as soon as the medium has been discretized into nodes: discrete equations should be exactly verified. We shall consider first this method as it is an excellent introduction to numerical methods and related specific features. We will consider both time and frequency approaches as they have quite different behaviours when considering seismic imaging strategies.

Applications will enhance the different properties of this numerical tool and the caveats we must avoid for the various types of propagation we need.

Another well-known approach is the finite element method where partial differential equations are asked to be fulfilled in a average way (to be defined) inside elements paving the entire medium. We shall concentrate into the discontinuous Galerkin method as it allows to mix acoustic and elastic wave propagation into a same formalism: this particular method shares many features of finite element formalism when describing an element, but differs by the way these elements interact each other. We avoid the description of the continuous finite element method for compactness and differences will be pointed out when necessary. Again, we shall discuss both time-domain and frequency-domain approaches.

Applications will illustrate the different capabilities of this technique and we shall illustrate what are advantages and drawbacks compared to finite difference methods while specific features will be identified compared to continuous finite element methods.

We shall conclude on the strategy for seismic imaging when comparing precision of solutions and numerical efforts for both volumetric methods.

2. Equations of seismic wave propagation

In a heterogeneous continuum medium, seismic waves verify partial differential equations locally. Integral equations may provide an alternative for the evolution of seismic fields either in the entire domain or at the scale of an elementary element of a given mesh describing the medium structure.

Fundamental laws of dynamics require the conservation of linear and angular momentum in a Galilean reference frame. In the continuum, a force applied across a surface, oriented by the unit normal $\mathbf{n}$ at a given position $\mathbf{x} = (x, y, z)$ in a Cartesian coordinate system (O, x, y, z), by one side of the material on the other side defines the traction vector $t_i = \sigma_{ij} n_j$ where the second-rank stress tensor σ has been introduced. The conservation of the angular momentum makes the stress tensor symmetrical $\sigma_{ij} = \sigma_{ji}$. We shall introduce as well a volumetric force per unit mass at the current position denoted as $\mathbf{f} = (f_x, f_y, f_z)$. The conservation of linear momentum allows one to write the acceleration of the displacement motion $\mathbf{u}(\mathbf{x})$ of a given particle at the current position as

$$\rho(\mathbf{x}) \frac{\partial^2 u_i}{\partial t^2} = \frac{\partial \sigma_{ik}}{\partial x_k} + \rho(\mathbf{x}) f_i, \tag{1}$$

where the density is denoted by $\rho(\mathbf{x})$.

Aside the translation and the rotation transformations preserving the distances inside the body we consider, the deformation of the continuum body is described by defining a strain tensor ϵ expressed as

$$\epsilon_{ij} = \frac{1}{2} \left(\frac{\partial u_j}{\partial x_i} + \frac{\partial u_i}{\partial x_j} \right). \tag{2}$$

The symmetrical definition of the deformation ensures that no rigid-body rotations are included. The particle motion is decomposed into a translation, a rotation and a deformation: the two formers transformations preserve distances inside the solid body while the third one does not preserve distances, inducing stress variations inside the solid body. In the framework of linear elasticity, there is a general linear relation between the strain and stress tensors by introducing fourth-rank tensor c_{ijkl} defined as follows

$$\sigma_{ij} = c_{ijkl} \epsilon_{kl}. \tag{3}$$

Because of symmetry properties of stress and strain tensors, we have only 36 independent parameters among the 81 elastic coefficients while the positive strain energy leads to a further reduction to 21 independent parameters for a general anisotropic medium. For the particular case of isotropic media, we end up with two coefficients which can be the Lamé coefficients λ and μ. The second one is known also as the rigidity coefficient as it characterizes the mechanical shear mode of deformation. The following expression of elastic coefficients,

$$c_{ijkl} = \lambda \delta_{ij} \delta_{kl} + \mu (\delta_{ik} \delta_{jl} + \delta_{il} \delta_{jk}), \tag{4}$$

with the Kronecker convention for δ_{ij} gives the simplified expression linking the stress tensor to the deformation tensor for isotropic media as

$$\sigma_{ij} = \lambda \epsilon_{kk} \delta_{ij} + 2\mu \epsilon_{ij}. \tag{5}$$

One may prefer the inverse of the relation (5) where the deformation tensor is expressed from the stress tensor by the introduction of the Young modulus E. Still, we have two independent coefficients. By injecting the relation (5) into the fundamental relation of dynamics (1), we end up with the so-called elastic wave propagation system, which is an hyperbolic system of second order, where only the displacement u has to be found. This system can be written as

$$\frac{\partial^2 u_x}{\partial t^2} = \frac{1}{\rho}\left[(\lambda + 2\mu)\frac{\partial^2 u_x}{\partial x^2} + (\lambda + \mu)\left(\frac{\partial^2 u_y}{\partial x \partial y} + \frac{\partial^2 u_z}{\partial x \partial z}\right) + \right.$$

$$\left. +\mu\left(\frac{\partial^2 u_x}{\partial y^2} + \frac{\partial^2 u_x}{\partial z^2}\right)\right]$$

$$\frac{\partial^2 u_y}{\partial t^2} = \frac{1}{\rho}\left[(\lambda + 2\mu)\frac{\partial^2 u_y}{\partial y^2} + (\lambda + \mu)\left(\frac{\partial^2 u_x}{\partial x \partial y} + \frac{\partial^2 u_z}{\partial y \partial z}\right) + \right.$$

$$\left. +\mu\left(\frac{\partial^2 u_y}{\partial x^2} + \frac{\partial^2 u_y}{\partial z^2}\right)\right] \tag{6}$$

$$\frac{\partial^2 u_z}{\partial t^2} = \frac{1}{\rho}\left[(\lambda + 2\mu)\frac{\partial^2 u_z}{\partial z^2} + (\lambda + \mu)\left(\frac{\partial^2 u_x}{\partial x \partial z} + \frac{\partial^2 u_y}{\partial y \partial z}\right) + \right.$$

$$\left. +\mu\left(\frac{\partial^2 u_z}{\partial x^2} + \frac{\partial^2 u_z}{\partial y^2}\right)\right],$$

where we have neglected spatial variations of Lamé coefficients. Therefore, we must reconstruct over time the three components of the displacement or equivalently of the velocity or the acceleration. Choosing the stress is a matter of mechanical behaviour in a similar way for seismic instruments which record one of these fields.

For heterogeneous media, spatial differential rules for Lamé coefficients have to be designed. We shall see how to avoid this definition in the continuum by first considering hyperbolic system of first-order equations, keeping stress field. More generally, any hyperbolic equation with n-order derivatives could be transformed in a hyperbolic system with only first derivatives by adding additional unknown fields. This mathematical transformation comes naturally for the elastodynamic case by selecting the velocity field v and the stress field σ as fields we want to reconstruct. In a compact form, this first-order system in particle velocity and stresses is the following

$$\rho\frac{\partial v_i}{\partial t} = \sigma_{ij,j} + \rho f_i \tag{7a}$$

$$\frac{\partial \sigma_{ij}}{\partial t} = \lambda v_{k,k}\delta_{ij} + \mu(v_{i,j} + v_{j,i}), \tag{7b}$$

with $i, j = x, y, z$. We may consider other dual quantities as (displacement, integrated stress) or (acceleration, stress rate) as long as the medium is at rest before the dynamic evolution. Let us underline that time partial derivatives are on the left-hand side and that spatial variations and derivations are on the right-and side.

Using simple linear algebra manipulations, alternative equivalent expressions may deserve investigation: the three components σ_{ii} could be linearly combined for three alternative components considering the trace $\sigma_1 = trace(\sigma)/3$, the x-coordinate deviatoric stress $\sigma_2 = (2\sigma_{xx} - \sigma_{yy} - \sigma_{zz})/3$ and the y-coordinate deviatoric stress $\sigma_3 = (-\sigma_{xx} + 2\sigma_{yy} - \sigma_{zz})/3$ which allows to separate partial spatial derivatives in the right hand side and material properties in the left hand side. The system (7) becomes

$$\rho\frac{\partial v_i}{\partial t} = \sigma_{ij,j} + \rho f_i$$

$$\frac{3}{3\lambda + 2\mu}\frac{\partial \sigma_1}{\partial t} = v_{i,i}$$

$$\frac{3}{2\mu}\frac{\partial \sigma_2}{\partial t} = \left(3\frac{\partial v_x}{\partial x} - v_{i,i}\right) \qquad (8)$$

$$\frac{3}{2\mu}\frac{\partial \sigma_3}{\partial t} = \left(3\frac{\partial v_y}{\partial y} - v_{i,i}\right)$$

$$\frac{1}{\mu}\frac{\partial \sigma_{ij}}{\partial t} = v_{i,j} + v_{j,i}$$

which could be useful when we move from differential formulation to integral formulation over elementary volumes. Partial differential operators only in the right-hand side of the system (8) are separated from spatial variations of model parameters on the left-hand side as a diagonal matrix $\Lambda = \left(\frac{3}{3\lambda+2\mu}, \frac{3}{2\mu}, \frac{3}{2\mu}, \frac{1}{\mu}, \frac{1}{\mu}, \frac{1}{\mu}\right)$. Similar strategies could be applied for 2D geometries.

Finally, for easing discussions on the numerical implementation, let us write both the 1D scalar second-order acoustic wave equation in the time domain as

$$\rho(x)\frac{\partial^2 u(x,t)}{\partial t^2} = \frac{\partial}{\partial x}E(x)\frac{\partial u(x,t)}{\partial x}, \qquad (9)$$

or, in frequency domain,

$$\omega^2\rho(x)u(x,\omega) + \frac{\partial}{\partial x}E(x)\frac{\partial u(x,\omega)}{\partial x} = 0, \qquad (10)$$

away from sources where one can see the importance of the mixed operator $\partial_x E(x)\partial_x$. We have introduced the Young modulus E related to unidirectional compression/delation motion. The 1D vectorial first-order acoustic wave equation can be written as

$$\rho(x)\frac{\partial v(x,t)}{\partial t} = \frac{\partial \sigma(x,t)}{\partial x}$$

$$\frac{\partial \sigma(x,t)}{\partial t} = E(x)\frac{\partial v(x,t)}{\partial x}, \qquad (11)$$

from which one can deduce immediatly the system of equations in the frequency domain

$$-i\omega\rho(x)v(x,\omega) = \frac{\partial\sigma(x,\omega)}{\partial x}$$

$$-i\omega\sigma(x,\omega) = E(x)\frac{\partial v(x,\omega)}{\partial x}. \tag{12}$$

Please note that the mixed operator does not appear explicitly. By discretizing this system and by eliminating the stress discrete values, one can go back to an equation involving only the velocity: a natural and systematic procedure for discretizing the mixed operator as proposed by Luo & Schuster (1990).

For an isotropic medium, two types of waves - compressional and shear waves - are propagating at two different velocities v_p and v_s. These velocities can be expressed as

$$v_p = \sqrt{\frac{\lambda + 2\mu}{\rho}} \quad \text{and} \quad v_s = \sqrt{\frac{\mu}{\rho}}, \tag{13}$$

except for the 1D medium where only compression/dilatation motion could take place. The displacement induced by these two different waves is such that compressive waves u^P verify $\nabla \times u_i^P = 0$ and shear waves u^S verify $\nabla \cdot u_i^S = 0$. Applying these operators to the numerical displacement will separate it into these two wavefields.

2.1 Time-domain or frequency-domain approaches

These systems of equations could be solved numerically in the time domain or in the frequency domain depending on applications. For seismic imaging, the forward problem has to be solved for each source and at each iteration of the optimisation problem. The time approach has a computational complexity increasing linearly with the number of sources while precomputation could be achieved in the frequency domain before modelling the propagation of each source. Let us write a compact form in order to emphasize the time/frequency domains approaches. The elastodynamic equations are expressed as the following system of second-order hyperbolic equations,

$$\mathbf{M}(\mathbf{x})\frac{\partial^2 \mathbf{w}(\mathbf{x},t)}{\partial t^2} = \mathbf{S}(\mathbf{x})\mathbf{w}(\mathbf{x},t) + \mathbf{s}(\mathbf{x},t), \tag{14}$$

where $\mathbf{M}$ and $\mathbf{S}$ are the mass and the stiffness matrices (Marfurt, 1984). The source term is denoted by $\mathbf{s}$ and the seismic wavefield by $\mathbf{w}$. In the acoustic approximation, $\mathbf{w}$ generally represents pressure, while in the elastic case, $\mathbf{w}$ generally represents horizontal and vertical particle displacements. The time is denoted by t and the spatial coordinates by $\mathbf{x}$. Equation (14) is generally solved with an explicit time-marching algorithm: the value of the wavefield at a time step $(n+1)$ at a spatial position x is inferred from the value of the wavefields at previous time steps (Dablain, 1986; Tal-Ezer et al., 1990). Implicit time-marching algorithms are avoided as they require solving a linear system (Marfurt, 1984; Mufti, 1985). If both velocity and stress wavefields are helpful, the system of second-order equations can be recast as a first-order hyperbolic velocity-stress system by incorporating the necessary auxiliary variables (Virieux, 1986). The time-marching approach could gain in efficiency if one consider local time steps related to the coarsity of the spatial grid (Titarev & Toro, 2002): this leads to a quite challenging load balancing program between processors when doing parallel programming as most processors are waiting for the one which is doing the maximum of number crunching as illustrated for the ADER scheme (Dumbser & Käser, 2006). Adapting the distribution of the number of nodes to each processor depending on the expected complexity of mathematical operations is still an open problem. Other integration schemes as the Runge-Kutta scheme or the Stormer/Verlet symplectic scheme (Hairer et al., 2002) could be used as well.

Seismic imaging requires the cross-correlation in time domain or the product in frequency domain of the incidents field of one source and the backpropagated residues from the receivers for this source. In order to do so, one has to save at each point of the medium the incident field from the source which could be a time series or one complex number. The storage when considering a time-domain approach could be an issue: a possible strategy is storing only few time snapshots for recomputing the incident field on the fly (Symes, 2007) at intermediate times. An additional advantage is that the attenuation effect could be introduced as well. in the time-domain approach, the complexity increases linearly with the number of sources.

In the frequency domain, the wave equation reduces to a system of linear equations, the right-hand side of which is the source, and the solution of which is the seismic wavefield. This system can be written compactly as

$$\mathbf{B}(\mathbf{x}, \omega)\mathbf{w}(\mathbf{x}, \omega) = \mathbf{s}(\mathbf{x}, \omega), \tag{15}$$

where $\mathbf{B}$ is the so-called impedance matrix (Marfurt, 1984). The sparse complex-valued matrix $\mathbf{B}$ has a symmetric pattern, although is not symmetric because of absorbing boundary conditions (Hustedt et al., 2004; Operto et al., 2007). The fourier transform is defined with the following convention

$$f(\omega) = \int_{-\infty}^{+\infty} f(t)e^{i\omega t}dt.$$

Solving the system of equations (15) can be performed through a decomposition of the matrix $\mathbf{B}$, such as lower and upper (LU) triangular decomposition, which leads to the so-called direct-solver techniques. The advantage of the direct-solver approach is that, once the decomposition is performed, equation (15) is efficiently solved for multiple sources using forward and backward substitutions (Marfurt, 1984). This approach has been shown to be efficient for 2D forward problems (Hustedt et al., 2004; Jo et al., 1996; Stekl & Pratt, 1998). However, the time and memory complexities of the LU factorization, and its limited scalability on large-scale distributed memory platforms, prevents the use of the direct-solver approach for large-scale 3D problems (*i.e.*, problems involving more than ten millions of unknowns) (Operto et al., 2007).

Iterative solvers provide an alternative approach for solving the time-harmonic wave equation (Erlangga & Herrmann, 2008; Plessix, 2007; Riyanti et al., 2006; 2007). Iterative solvers are currently implemented with Krylov-subspace methods (Saad, 2003) that are preconditioned by the solution of the dampened time-harmonic wave equation. The solution of the dampened wave equation is computed with one cycle of a multigrid. The main advantage of the iterative approach is the low memory requirement, while the main drawback results from the difficulty to design an efficient preconditioner, because the impedance matrix is indefinite. To our knowledge, the extension to elastic wave equations still needs to be investigated. As for the time-domain approach, the time complexity of the iterative approach increases linearly with the number of sources or, equivalently, of right-hand sides, in contrast to the direct-solver approach.

An intermediate approach between the direct and the iterative methods consists of a hybrid direct-iterative approach that is based on a domain decomposition method and the Schur complement system (Saad, 2003; Sourbier et al., 2011): the iterative solver is used to solve the reduced Schur complement system, the solution of which is the wavefield at interface nodes between subdomains. The direct solver is used to factorize local impedance matrices that are assembled on each subdomain. Briefly, the hybrid approach provides a compromise in terms of memory saving and multi-source-simulation efficiency between the direct and the iterative approaches.

The last possible approach to compute monochromatic wavefields is to perform the modeling in the time domain and extract the frequency-domain solution, either by discrete Fourier transform in the loop over the time steps (Sirgue et al., 2008) or by phase-sensitivity detection once the steady-state regime has been reached (Nihei & Li, 2007). An arbitrary number of frequencies can be extracted within the loop over time steps at a minimal extra cost. Time windowing can easily be applied, which is not the case when the modeling is performed in the frequency domain. Time windowing allows the extraction of specific arrivals (early arrivals, reflections, PS converted waves) for the full waveform inversion (FWI), which is often useful to mitigate the nonlinearity of the inversion by judicious data preconditioning (Brossier et al., 2009; Sears et al., 2008).

Among all of these possible approaches, the iterative-solver approach has theoretically the best time complexity (here, *complexity* denotes how the computational cost of an algorithm grows with the size of the computational domain) if the number of iterations is independent of the frequency (Erlangga & Herrmann, 2008). In practice, the number of iterations generally increases linearly with frequency. In this case, the time complexity of the time-domain approach and the iterative-solver approach are equivalent (Plessix, 2007).

For one-frequency modeling, the reader is referred to those articles (Plessix, 2007; 2009; Virieux et al., 2009) for more detailed complexity analysis of seismic modeling based on different numerical approaches. A discussion on the pros and cons of time-domain versus frequency-domain seismic modeling relating to what it is required for full waveform inversion is also provided in Vigh & Starr (2008) and Warner et al. (2008).

2.2 Boundary conditions

In seismic exploration, two boundary conditions are implemented for wave modeling: absorbing boundary conditions to mimic an infinite medium and free surface conditions on the top side of the computational domain to represent the air-solid or air-water interfaces which have the highest impedance contrast. For internal boundaries, we assume that effects are well described by variations of the physical properties of the medium: the so-called implicit formulation (Kelly et al., 1976; Kummer & Behle, 1982).

2.2.1 PML absorbing boundary conditions

For simulations in an infinite medium, an absorbing boundary condition needs to be applied at the edges of the numerical model. An efficient way to mimic such an infinite medium can be achieved with Perfectly-Matched Layers (PML), which has been initially developed by Berenger (1994) for electromagnetics, and adapted for elastodynamics by Chew & Liu (1996); Festa & Vilotte (2005). PMLs are anisotropic absorbing layers that are added at the periphery of the numerical model. The classical PML formulation is based on splitting of the elastodynamic equations. A new kind of PML, known as CPML, does not require split terms. The CPML originated from Roden & Gedney (2000) for electromagnetics was applied by Komatitsch & Martin (2007) and Drossaert & Giannopoulos (2007) to the elastodynamic system. CPML is based on an idea of Kuzuoglu & Mittra (1996), who has obtained a strictly causal form of PML by adding some parameters in the standard damping function of Berenger (1994), which enhanced the absorption of waves arriving at the boundaries of the model with grazing incidence angles.

In the frequency domain, the implementation of PMLs consists of expressing the wave equation in a new system of complex-valued coordinates $\tilde{x}$ defined by (e.g., Chew & Weedon, 1994):

$$\frac{\partial}{\partial \tilde{x}} = \frac{1.}{\xi_x(x)} \frac{\partial}{\partial x}. \tag{16}$$

In the PML layers, the damped 1D acoustic wave equation could be deduced from the equation (10) as

$$\left[\omega^2 \rho(x) + \frac{1.}{\xi_x(x)} \frac{\partial}{\partial x} \frac{E(x)}{\xi_x(x)} \frac{\partial}{\partial x} \right] u(x, \omega) = -s(x, \omega), \tag{17}$$

where $\xi_x(x) = 1 + i\gamma_x(x)/\omega$ and $\gamma_x(x)$ is a 1D damping function which defines the PML damping behavior in the PML layers. In the CPML layers, the damping function $\xi_x(x)$ becomes

$$\xi_x(x) = \kappa_x + \frac{d_x}{\alpha_x + i\omega}, \tag{18}$$

with angular frequency ω and coefficients $\kappa_x \geq 1$ and $\alpha_x \geq 0$. The damping profile d_x varies from 0 at the entrance of the layer, up to a maximum real value $d_{\theta max}$ at the end (Collino & Tsogka, 2001) such that

$$d_x = d_{xmax} \left(\frac{\delta_x}{L_{cpml}} \right)^2, \tag{19}$$

and

$$d_{xmax} = -3V_p \frac{\log(R_{coeff})}{2L_{cpml}}, \tag{20}$$

with δ_x as the depth of the element barycentre inside the CPML, L_{cpml} the thickness of the absorbing layer, and R_{coeff} the theoretical reflection coefficient. Suitable expressions for κ_x, d_x and α_x are discussed in Collino & Monk (1998); Collino & Tsogka (2001); Drossaert & Giannopoulos (2007); Komatitsch & Martin (2007); Kuzuoglu & Mittra (1996); Roden & Gedney (2000). We often choose $R_{coeff} = 0.1\%$ and the variation of the coefficient α_x goes from a maximum value ($\alpha_{x_{max}} = \pi f_0$) at the entrance of the CPML, to zero at its end. If $\kappa_x = 1$ and $\alpha_x = 0$, the classical PML formulation is obtained.

One can use directly these frequency-dependent expressions when considering the frequency approach. The formulation in the time domain is slightly more involved. The spatial derivatives are replaced by

$$\partial_{\tilde{x}} \to \frac{1}{\kappa_x}\partial_x + \zeta_x * \partial_x, \tag{21}$$

with

$$\zeta_x(t) = -\frac{d_x}{\kappa_x^2}H(t)e^{-(d_x\kappa_x+\alpha_x)t}, \tag{22}$$

where $H(t)$ denotes the Heaviside distribution. Roden & Gedney (2000) have demonstrated that the time convolution in equation (21) can be performed in a recursive way using memory variables defined by

$$\psi_x = \zeta_x * \partial_x. \tag{23}$$

The function ψ_x represents a memory variable in the sense that it is updated at each time step. Komatitsch & Martin (2007) have shown that the term κ_x has a negligible effect on the absorbing abilities, and it can be set to 1. If we take $\kappa_x = 1$, we derive the equation (23) using the equation (22) as

$$\partial_t\psi_x = -d_x\partial_x - (d_x + \alpha_x)\psi_x. \tag{24}$$

One equation is generated for each spatial derivative involved in the elastodynamic system, which can be a memory-demanding task. Once they are computed at each time step, we can introduce the memory variables into the initial elastodynamic system which requires two additional variables for the 1D equations (11) with the definition of $\psi_x(v)$ and $\psi_x(\sigma)$ leading to the following system

$$\begin{aligned}
\rho(x)\frac{\partial v}{\partial t} &= \frac{\partial \sigma}{\partial x} + \psi_x(\sigma) \\
\frac{\partial \sigma}{\partial t} &= E(x)\frac{\partial v}{\partial x} + \psi_x(v) \\
\frac{\partial \psi_x(v)}{\partial t} &= -d_x(x)\frac{\partial \psi(v)}{\partial x} - (d_x(x) + \alpha_x(x))\psi(v) \\
\frac{\partial \psi_x(\sigma)}{\partial t} &= -d_x(x)\frac{\partial \psi(\sigma)}{\partial x} - (d_x(x) + \alpha_x(x))\psi(\sigma)
\end{aligned} \tag{25}$$

At the outer edge of the PML zone, one could apply any conditions as simple absorbing conditions (Clayton & Engquist, 1977) or free surface conditions (Etienne et al., 2010) as fields go to zero nearby the outer edge.

We must underline that the extension to 2D and 3D geometries is straightforward both in the frequency domain (Brossier et al., 2010; 2008) and in the time domain (Etienne et al., 2010; Komatitsch & Martin, 2007).

2.2.2 Free surface

Planar free surface boundary conditions can be simply implemented using a strong formulation or a weak formulation.

In the first case which is often met in the finite-difference methods, one requires that the stress is zero at the free surface. The free surface matches the top side of the FD grid and the stress is forced to zero on the free surface (Gottschamer & Olsen, 2001). Alternatively, the method of image can be used to implement the free surface along a virtual plane located half a grid interval above the topside of the FD grid (Virieux, 1986). The stress is forced to vanish at the free surface by using a virtual plane located half a grid interval above the free surface where the stress is forced to have opposite values to that located just below the free surface. In case of more complex topographies, one strategy is to adapt the topography to the grid structure at the expense of numerical dispersion effect (Robertsson, 1996) or to deform the underlying meshing used in the numerical method to the topography (Hestholm, 1999; Hestholm & Ruud, 1998; Tessmer et al., 1992). In the first case, because of stair-case approximation, a local fine sampling is required (Hayashi et al., 2001).

Owing to the weak formulation used in finite-element methods, the free surface boundary condition are naturally implemented by zeroing test functions on these boundaries which follow edges of grid elements (Zienkiewicz & Taylor, 1967). This approach could be used as well for finite-difference methods through the summation-by-parts (SBP) operators based on energy minimization combined with Simultaneous Approximation Term (SAT) formulation based on a boundary value penalty method (Taflove & Hagness, 2000). In this case, boundaries on which stress should be zero are not requested to follow any grid discretisation.

Finally, one may consider an immersed boundary approach where the free surface boundary is not related to the discretisation of the medium as promoted by LeVeque (1986). Extensive applications have been proposed by Lombard & Piraux (2004); Lombard et al. (2008) where grid discretisation does not influence the application of boundary conditions. This approach might be seen as an extension of the method of images following extrapolation techniques above the free surface at any a priori order of precision.

2.3 Source implementation

There are different ways of exciting the numerical grid by the source term. The simplest one is the direct contribution of the source term in the discrete partial differential equations: for example, we may just increment by the source term after each time step or we may consider the right-hand side source term for solving the linear system in the frequency domain. Depending on the numerical approach, it is necessary to consider specific influences coming from the discretisation as we shall see for numerical methods we consider.

In order to avoid singularities of solutions nearby the source, on can use the injection technique as proposed in the pioneering work of Alterman & Karal (1968) where a specific box around the source is defined. Inside the box, only the scattering field is computed. The incident field is estimated at the edges of the box and it is substracted when propagations are estimated inside the box and added when propagations are estimated outside the box. A more general framework is proposed by Opršal et al. (2009) related to boundary integral approaches.

3. Finite-difference methods: solving the equations through a strong formulation

We shall first consider the discretization based on simple and intuitive approaches as finite-difference methods for solving these partial differential equations while focusing on techniques useful for seismic imaging which means a significant number of forward problems for many sources in the same medium. We shall identify features which might be interesting for seismic imaging. If these approaches are intuitive for solving differential equations, the numerical implementation of boundary conditions and source excitation is less obvious and requires specific strategies as we shall see.

3.1 Spatial-domain finite-difference approximations

Whatever is our strategy for the reconstruction of the wave field u, one has to discretize it. We may be very satisfied by considering a set of discrete values $(u_1, u_2, ..., u_{I-1}, u_I)$ along one direction at a given specific time t_n which can be discretized as well. Therefore, a simple way of solving this first-order differential system is by making finite difference approximations of spatial derivatives.

Still considering a 1D geometry, the partial operator $(\partial/\partial x)$ could be deduced from a Taylor expansion using Lagrange polynomial. A quite fashionable symmetrical estimation using a centered finite difference approximation is expressed as

$$\frac{\partial u_i^n}{\partial x} = \frac{u_{i+1}^n - u_{i-1}^n}{2\Delta x} + O\left[\Delta x^2\right], \tag{26}$$

which is a three-nodes stencil as three nodes are involved: two for the derivative estimation and one for the updating. Let us mention that the discrete derivative is shifted with respect to discrete values of the field. Because of the very specific antisymmetrical structure of our first-order hyperbolic system where time evolution of velocity requires only stress derivatives (and vice versa), we may consider centered approximations both in space and in time. This will lead to a leap-frog structure or a red/black pattern. Of course, we have truncation errors expressed by the function $O(\Delta x^n)$ which depends on the power n of the spatial stepping and by the function $O(\Delta t^k)$ on the power k of the time stepping.

We may require a greater precision of the derivative operator by using more points for this partial derivative approximation and a very popular centered finite difference approximation of the first derivative is the following expression

$$\frac{\partial u_i^n}{\partial x} = \frac{c_1\left(u_{i+\frac{1}{2}}^n - u_{i-\frac{1}{2}}^n\right) + c_2\left(u_{i+\frac{3}{2}}^n - u_{i-\frac{3}{2}}^n\right)}{\Delta x} + O\left[\Delta x^4\right], \tag{27}$$

where $c_1 = 9/8$ et $c_2 = -1/24$ (Levander, 1988). This fourth-order stencil is compact enough (few discrete points inside the stencil) for numerical efficiency while having a small local truncation error. This stencil is a five-nodes stencil. Let us underline that centered approximations lead to have field quantities not at the same position in the numerical grid as derivative approximations (figure 1). In other words, stress and velocity components should be specified on different positions of the spatial grid. If we still consider a full grid where stress and velocities are known at the same position, this stencil could be recast as a seven-nodes

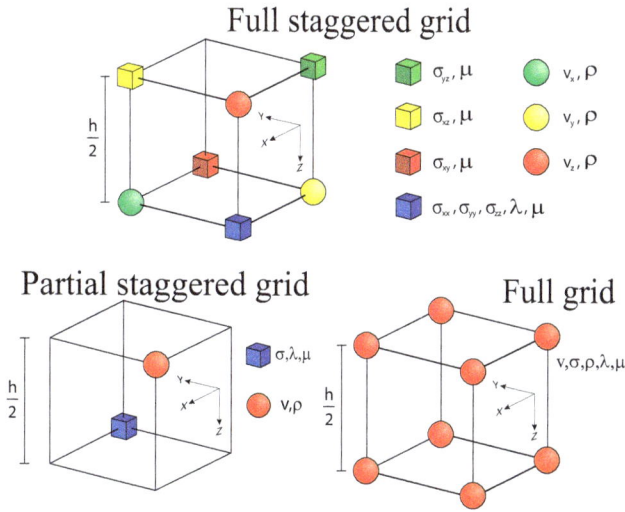

Fig. 1. Cell structure for the full staggered grid (top), the partial staggered grid (bottom left) and the full grid (bottom right, Stress tensor is denoted by σ, particle velocity by v and the density by ρ as well as Lamé coefficient by λ et μ. The regular grid step is denoted by h. Time stepping is missing

stencil given by

$$\frac{\partial u_i^n}{\partial x} = \frac{d_1 \left(u_{i+1}^n - u_{i-1}^n\right) + d_2 \left(u_{i+2}^n - u_{i-2}^n\right) + d_3 \left(u_{i+3}^n - u_{i-3}^n\right)}{\Delta x'}, \tag{28}$$

where $\Delta x' = \Delta x/2$ and where we have following specific coefficients $d_1 = c_1$, $d_2 = 0$. and $d_3 = c_2$. The fourth-order scheme would require the following theoretical coefficients $d_1 = 15/20$, $d_2 = -3/20$ and $d_3 = 1/60$. For fourth-order stencils, the two sub-grids are not entirely decoupled and are weakly coupled leading to a dispersion behaviour as if the discretization is Δx. Let us remind that these sub-grids are completely decoupled when considering second-order stencils, leading to the staggered structure. Therefore, solving partial differential equations in the staggered grid structure has a less accurate resolution but improves significantly the efficiency of the method than solving equations in the full grid even with dispersion-relation-preserving stencils (Tam & Webb, 1993). The memory saving can be easily seen when comparing nodes for staggered grid and nodes for full grid (figure 1)

When dealing with 2D and 3D geometries, we may exploit the extra freedom and estimate derivatives along the direction x from nodes shifted by half the grid step in x but also by half the grid step in y (and eventually in z). This leads to another compact stencil as shown in the figure 1 where all components of the velocity are discretized in one location while all components of the stress field are discretized half the diagonal of the grid as proposed by Saenger et al. (2000). This grid is still partially staggered and could be named as a partial grid.

These standard and partial staggered structures are sub-grids of the full grid as shown in the figure 1 which is used in aeroacoustics (Tam & Webb, 1993).

These different discretizations related to various stencils may lead to preferential directions of propagation. Numerical anisotropy effect is observed even when considering isotropic wave propagation. The figure 2 shows error variations in velocities with respect to angles of propagation for the standard grid and the partial grid: one can see that the anisotropy behavior is completely different with a rotation shift of 45^0. In 2D, the two stencils provide the same anisotropic error while the partial grid has a slightly improved numerical anisotropic performance (percentage differences go from 3 % down to 2 % in 3D geometries). Of course, the spatial sampling is such that the error should be negligible and few percentages is considered to be acceptable except nearby the source.

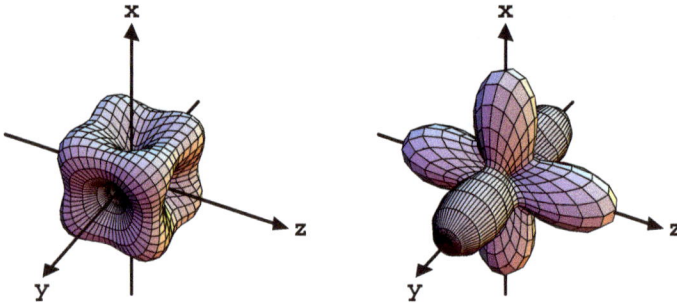

Fig. 2. Numerical anisotropic errors when considering finite difference stencils related to *partial staggered grid* (left panel) and *standard staggered grid* (right panel) Saenger et al. (2000).

Other spatial interpolations are possible. Previous discrete expressions are based on Lagrange interpolations while other interpolations are possible such as Chebychev or Laguerre polynomial or Fourier interpolations (Kosloff & Baysal, 1982; Kosloff et al., 1990; Mikhailenko et al., 2003). Interpolation basis could be local (Lagrange) or global(Fourier) ones based on equally spaced nodes or judiciously distributed nodes for keeping interpolation errors as small as possible: this will have a dramatic impact on the accuracy of the numerical estimation of the derivative and, therefore, on the resolution of partial differential equations. We should stress that local stencils should be preferred for seismic imaging for efficiency in the computation of the forward modeling.

3.2 Time-domain finite-difference approximations

Similarly, one may consider finite difference approximation for time derivatives which can be illustrated on the simple scalar wave equation. A widely used strategy is again the centered differences through the expression

$$\frac{\partial u_i^n}{\partial t} = \frac{u_i^{n+1} - u_i^{n-1}}{2\Delta t} + O\left[\Delta t^2\right]. \tag{29}$$

For understanding how the procedure of computing new values in time is performing, let us consider the simple 1D second-order scalar wave equation for displacement u. This equation

$$\frac{\partial^2 u}{\partial t^2} = c^2 \frac{\partial^2 u}{\partial x^2}, \tag{30}$$

could be discretized through these finite difference approximations

$$\frac{u_i^{n+1} - 2u_i^n + u_i^{n-1}}{(\Delta t)^2} \approx c^2 \left[\frac{u_{i+1}^n - 2u_i^n + u_{i-1}^n}{(\Delta x)^2} \right]. \tag{31}$$

The next value at the discrete time $n + 1$ comes from older values known at time n and time $n - 1$ through the expression

$$u_i^{n+1} \approx (c\Delta t)^2 \left[\frac{u_{i+1}^n - 2u_i^n + u_{i-1}^n}{(\Delta x)^2} \right] + 2u_i^n - u_i^{n-1}. \tag{32}$$

A more compact notation of this equation as

$$u_i^{n+1} = 2(1 - S^2)u_i^n + S^2(u_{i+1}^n + u_{i-1}^n) - u_i^{n-1} \tag{33}$$

shows the quantity

$$S = \frac{c\Delta t}{\Delta x},$$

known as the Courant number in the literature. This quantity is quite important for understanding the numerical dispersion and stability of finite difference schemes. The related stencil on the spatio-temporal grid as shown in the left panel of the figure 3 clearly illustrates that the value at time $n + 1$ is explicitly computed from values at time $n - 1$ and time n. In this explicit formulation, the selection of the time step Δt should verify that the Courant number is lower than 1 for any point of the medium.

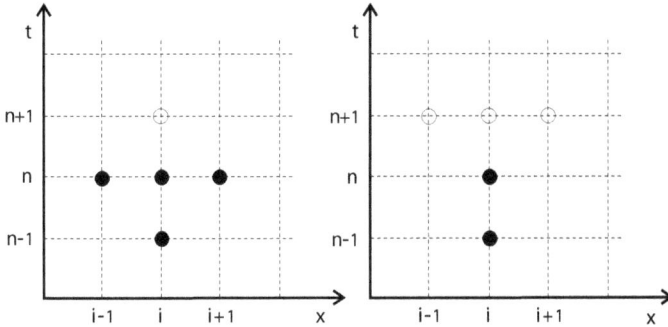

Fig. 3. Space/time finite difference stencil for an explicit scheme on the left and for an implicit scheme in a 1D configuration: black circles are known values from which the white circle is estimated.

On the contrary, we may consider spatial derivatives at time $n + 1$. This leads us to an implicit scheme where more than one value at time $n + 1$ is present in the discretisation. The equation is now

$$\frac{u_i^{n+1} - 2u_i^n + u_i^{n-1}}{(\Delta t)^2} = c^2 \left[\frac{u_{i+1}^{n+1} - 2u_i^{n+1} + u_{i-1}^{n+1}}{(\Delta x)^2} \right] \tag{34}$$

which can be described by the Courant number S through the equation

$$(1 + 2S^2)u_i^{n+1} - S^2(u_{i+1}^{n+1} + u_{i-1}^{n+1}) = 2u_i^n - u_i^{n-1}. \tag{35}$$

The right panel of the figure 3 illustrates the structure of the stencil and that three unknowns have to be estimated through the single equation (35). By considering different spatial nodes, we may find these three unknowns by solving a linear system. The Courant number could take any value for time integration as long as discrete sampling is correctly performed.

Other implicit stencils might be designed by averaging the spatial derivatives over the three times $n - 1$, n and $n + 1$. We may as well average the time derivative over the three positions $i - 1$, i and $i + 1$. This lead to another linear system to be solved. These weighting strategies could be designed for reducing numerical noise as numerical dispersion and/or anisotropy: a road for further improvements.

As discretisation in space and time goes to zero, one expects the solution to be more precise but cumulative rounding errors should prevent to have too small values. In expressions (26) and (29), truncation error $O[\Delta x^2]$ goes to zero as the square of the discrete increment. We shall say that this is a second-order precision scheme both in space and in time. One consider often the stencil with the fourth-order precision in space and second-order precision in time, denoted as $O\left[\Delta x^4, \Delta t^2\right]$, as an optimal one for finite-differences simulations.

3.3 Frequency-domain finite-difference approximations

The second-order acoustic equation (10) provides a generalization of the Helmholtz equation. In exploration seismology, the source is generally a local point source corresponding to an explosion or a vertical force.

Attenuation effects of arbitrary complexity can be easily implemented in equations (10) and (12) using complex-valued wave speeds in the expression of the bulk modulus, thanks to the correspondence principle transforming time convolution into products in the frequency domain: in the frequency domain, one has to replace elastic coefficients by corresponding viscoelastic complex moduli for considering visco-elastic behaviors (Bland, 1960). For example, according to the Kolsky-Futterman model (Futterman, 1962; Kolsky, 1956), the complex wave speed $\bar{c}$ is given by

$$\bar{c} = c\left[\left(1 + \frac{1}{\pi Q}|log(\omega/\omega_r)|\right) + i\frac{sgn(\omega)}{2Q}\right]^{-1}, \tag{36}$$

where the P wave speed is denoted by $c = \sqrt{E/\rho}$, the attenuation factor by Q and a reference frequency by ω_r. The function sgn gives the sign of the function.

Since the relationship between the wavefields and the source terms is linear in the first-order and second-order wave equations, one can explicitly expressed the matrix structure of equations (10) and (12) through the compact expression,

$$[M + S]\, u = Bu = s, \tag{37}$$

where $\mathbf{M}$ is the mass matrix, $\mathbf{S}$ is the complex stiffness/damping matrix. This expression holds as well in 2D and 3D geometries. The dimension of the square matrix $\mathbf{B}$ is the number of nodes in the computational domain multiplied by the number of wavefield components. System (37) could be solved using a sparse direct solver. A direct solver performs first a LU decomposition of $\mathbf{B}$ followed by forward and backward substitutions for the solutions (Duff et al., 1986) as shown by the following equations:

$$\mathbf{Bu} = (\mathbf{LU})\,\mathbf{u} = \mathbf{s} \tag{38}$$

$$\mathbf{Ly} = \mathbf{s}; \quad \mathbf{Uu} = \mathbf{y} \tag{39}$$

Exploration seismology requires to perform seismic modeling for a large number of sources, typically, up to few thousands for 3D acquisition. The use of direct solver is the efficient computation of the solutions of the system (37) for multiple sources. Combining different stencils for constructing a compact and accurate stencil can follow strategies developped for acoustic and elastic wave propagation (Jo et al., 1996; Operto et al., 2007; Stekl & Pratt, 1998). The numerical anisotropy is dramatically reduced

The mass matrix $\mathbf{M}$ is a diagonal matrix although never explicitly constructed when considering explicit time integration. In the frequency domain formulation, we may spread out the distribution of mass matrix over neighboring nodes in order to increase the precision without increasing the computer cost as we have to solve a linear system in all cases. This strategy is opposite to the finite element approach where often the mass matrix is lumped into a diagonal matrix for explicit time integration (Marfurt, 1984). For a frequency formulation, considering the mass matrix as a non-diagonal matrix does not harm the solver. The weights of distribution are obtained through minimization of the phase velocity dispersion in an infinite homogeneous medium (Brossier et al., 2010; Jo et al., 1996): the numerical dispersion is dramatically reduced.

3.4 PML absorbing boundary condition implementation

Implementation of PML conditions in the frequency domain is straightforward using unsplit variables while, in the time domain, we need to introduce additional variables for handling the convolution through memory variables or to use split unphysical field variables (Cruz-Atienza, 2006). These additional variables are only necessary in the boundary layers following the figure 4

We first consider an infinite homogeneous medium which is embedded into a cubic box of a $16\ km$ size and a grid stepping of $h = 100\ m$. The thickness of the PML layer is $1\ km$ leading to $nsp = ten$ nodes inside the PML zone. The P-wave velocity is $4000\ m/s$ while the S-wave velocity is $2300\ m/s$ and the density $2500\ kg/m^3$. The figure 5 shows various time sections of the 3D volume for the vertical particle velocity where one can see that the explosive wavefront is entering the PML zone at the time $2.8\ s$. The last two snapshots shows the vanishing of the wavefront with completely negligible residues at the final time (the decrease of the elastic energy is better than $0.2\ \%$ for ten nodes and could reach $0.03\ \%$ for twenty nodes).

When we have discontinuous interfaces crossing the PML zone, we may expect difficulties coming from various angles of propagation waves (Chew & Liu, 1996; Festa & Nielsen, 2003; Marcinkovich & Olsen, 2003). Therefore, a simple heterogeneous medium is considered

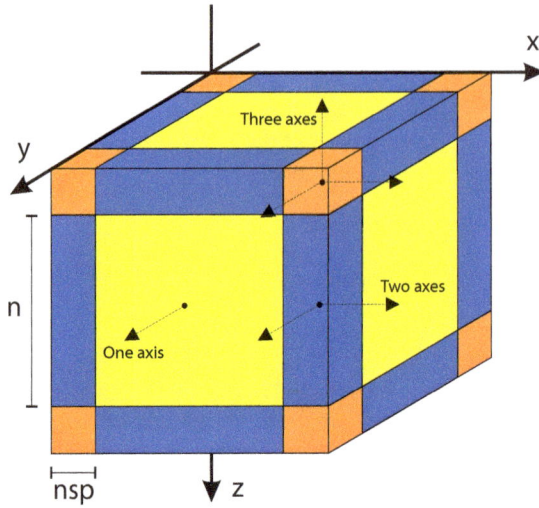

Fig. 4. Three kinds of PML boundary layers should be considered where only one coordinate is involved (yellow zone), two coordinates are involved (blue zone) and three coordinates are involved (red zone). Internally, standard elastodynamic equations are solved

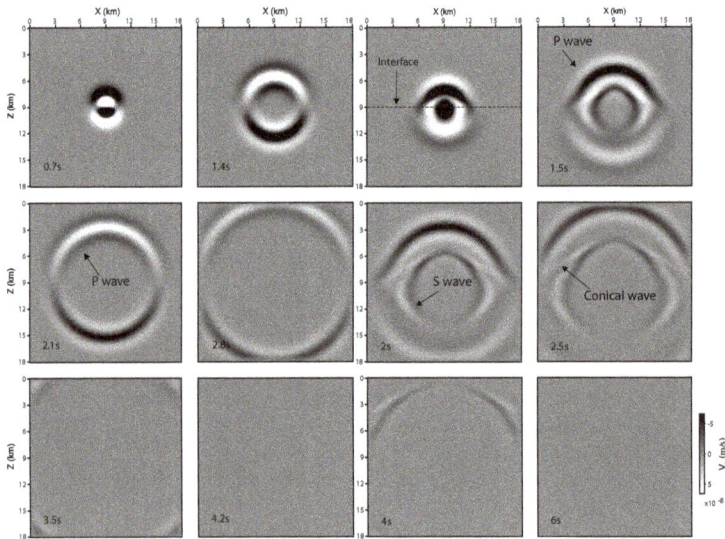

Fig. 5. Snapshots for y=0 of the vertical particle velocity at different times for an explosive source: on the left for an homogeneous infinite medium and on the right for an heterogeneous medium. Please note the vanishing of the seismic waves, thanks to the PML absorption

with two layers where physical parameters are (v_p, v_s, ρ)=(4330 m/s, 2500 m/s, 2156 kg/m^3) and (v_p, v_s, ρ)=(6000 m/s, 4330 m/s, 2690 kg/m^3). The figure 5 shows that, in spite of the complexities of waves generated at the horizontal flat interface, the PML layer succeeds to absorb seismic energy with a residual energy of 0.3 % in this case, still far better than standard paraxial absorbing boundary conditions (Clayton & Engquist, 1977).

3.5 Source and receiver implementation on coarse grids

Seismic imaging by full waveform inversion is initiated at an initial frequency as small as possible to mitigate the non linearity of the inverse problem resulting from the use of local optimization approach such as gradient methods. The starting frequency for modeling in exploration seismics can be as small as 2 Hz which can lead to grid intervals as large as 200 m. In this framework, accurate implementation of point source at arbitrary position in a coarse grid is critical. One method has been proposed by Hicks (2002) where the point source is approximated by a windowed Sinc function. The Sinc function is defined by

$$sinc(x) = \frac{sin(\pi x)}{\pi x}, \tag{40}$$

where $x = (x_g - x_s)$, x_g denotes the position of the grid nodes and x_s denotes the position of the source. The Sinc function is tapered with a Kaiser function to limit its spatial support (Hicks, 2002) . For multidimensional simulations, the interpolation function is built by tensor product construction of 1D windowed Sinc functions. If the source positions matches the position of one grid node, the Sinc function reduces to a Dirac function at the source position and no approximation is used for the source positioning. If the spatial support of the Sinc function intersects a free surface, part of the Sinc function located above the free surface is mirrored into the computational domain with a reverse sign following the method of image. Vertical force can be implemented in a straightforward way by replacing the Sinc function by its vertical derivative. The same interpolation function can be used for the extraction of the pressure wavefield at arbitrary receiver positions. The accuracy of the method of Hicks (2002) is illustrated in Figure 6a which shows a 3.75 Hz monochromatic wavefield computed in a homogeneous half space. The wave speed is 1500 m/s and the density is 1000 kg/m^3. The grid interval is 100 m. The free surface is half a grid interval above the top of the FD grid and the method of image is used to implement the free surface boundary condition. The source is in the middle of the FD cell at 2 km depth. The receiver line is oriented in the Y direction. Receivers are in the middle of the FD cell in the horizontal plane and at a depth of 6 m just below the free surface. Comparison between the numerical and the analytical solutions at the receiver positions are first shown when the source is positioned at the closest grid point and the numerical solutions are extracted at the closest grid point (Figure 6b). The amplitude of the numerical solution is strongly overestimated because the numerical solution is extracted at a depth of 50 m below free surface (where the pressure vanishes) instead of 6 m. Second, a significant phase shift between numerical and analytical solutions results from the approximate positioning of the sources and receivers. In contrast, a good agreement between the numerical and analytical solutions both in terms of amplitude and phase is shown in Figure 6c where the source and receiver positioning is implemented with the windowed Sinc interpolation.

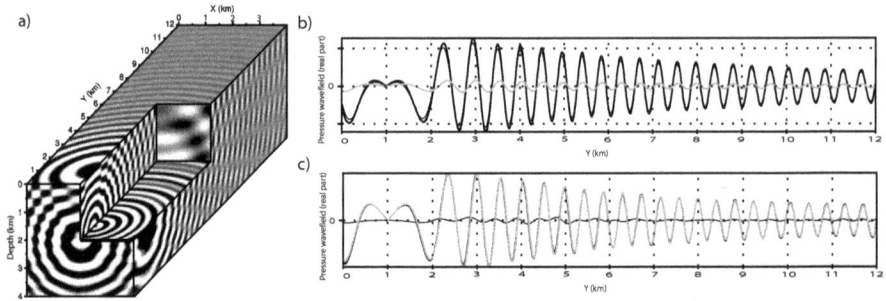

Fig. 6. a) Real part of a $3.75Hz$ monochromatic wavefield in a homogeneous half space. (b) Comparison between numerical (black) and analytical (gray) solutions at receiver positions when the closest grid point is used for both the source implementation and the extraction of the solution at the receiver positions on a coarse FD grid. (c) Same comparison between numerical (black) and analytical (gray) solutions at receiver positions when the Sinc interpolation with 4 coefficients is used for both the source implementation and the extraction of the solution at the receiver positions on a coarse FD grid.

4. Realistic examples for acoustic and elastic propagations using FD formulations

We shall provide two simple examples of seismic modeling using finite-differences methods both in the frequency and time approaches. The first example concerns seismic exploration problem where the acoustic approximation is often used while the second one is related to seismic risk mitigation where free surface effects including elastic propagation are quite important.

4.1 3D EAGE/SEG salt model

The salt model is a constant density acoustic model covering an area of 13.5 km × 13.5 km × 4.2 km (Aminzadeh et al., 1997)(Figure 7). The salt model is representative of a Gulf Coast salt structure which contains salt sill, different faults, sand bodies and lenses. The salt model is discretized with 20 m cubic cells, representing an uniform mesh of 676 x 676 x 210 nodes. The minimum and maximum velocities in the Salt model are 1500 m/s and 4482 m/s respectively. We performed a simulation for a frequency of 7.33 Hz and for one source located at $x = 3600$ m, $y = 3.600$ m and $z = 100$ m. The original model is resampled with a grid interval of 50 m corresponding to 4 grid points per minimum wavelength. The dimension of the resampled grid is 270 x 270 x 84 which represents 8.18 millions of unknowns after addition of the PML layers. Results of simulations performed with either in the frequency domain or in the time domain are compared in Figure 7. The time duration of the simulation is 15 s.

We obtain a good agreement between the two solutions (Figure 7d) although we show a small phase shift between the two solutions at offsets greater than 5000 m. This phase shift results from the propagation in the high-velocity salt body. The direct-solver modeling is performed on 48 MPI process using 2 threads and 15 Gbytes of memory per MPI process. The memory and the elapsed time for the LU decomposition were 402 Gbytes and 2863 s, respectively. The elapsed time for the solution step for one right-hand side (RHS) is 1.4 s when we process 16 RHS at a time during the solution step in MUMPS. The elapsed time for one time-domain simulation on 16 processors is 211 s. The frequency-domain approach is more than one order

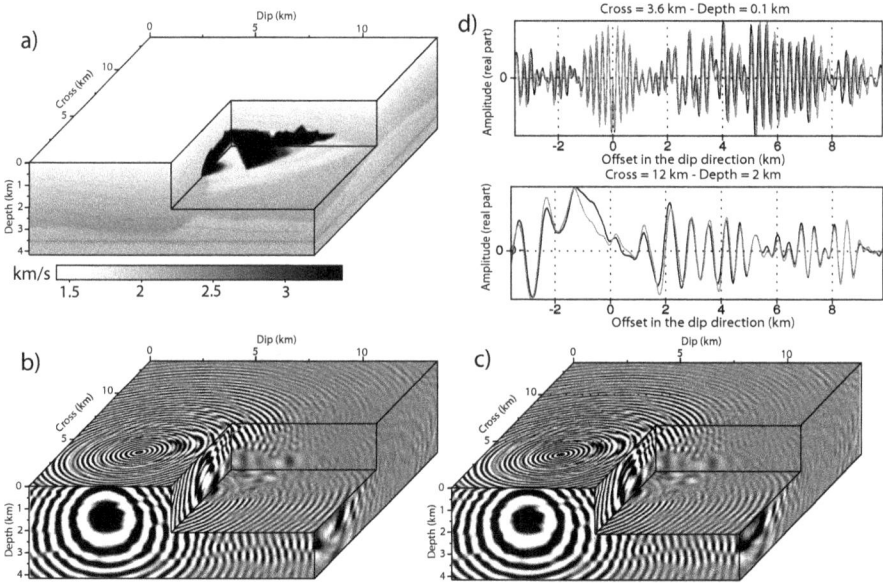

Fig. 7. (a) Salt velocity model. (b-c) 7.33-Hz monochromatic wavefield (real part) computed with a finite-different formulation in the frequency domain (b) and in the time domain (c). (d) Direct comparison between frequency-domain (gray) and time-domain (black) solutions. The receiver line in the dip direction is: (top) at 100 m depth and at 3600 m depth in the cross direction. The amplitudes are corrected for 3D geometrical spreading. (bottom) at 2500 m depth and at 15000 m in the cross direction.

of magnitude faster than the time-domain one when a large number of RHS members (2000) and a small number of processors (48) are used (Table 1). For a number of processors equal to the number of RHS members, the two approaches have the same cost. Of note, in the latter configuration ($N_P = N_{rhs}$), the cost of the two methods is almost equal in the case of the salt model (0.94 h versus 0.816 h).

Over the last decades, simulations of wave propagation in complex media have been efficiently tackled with finite-difference methods (FDMs) and applied with success to numerous physical problems (Graves, 1996; Moczo et al., 2007). Nevertheless, FDMs suffer from some critical issues that are inherent to the underlying Cartesian grid, such as parasite diffractions in cases where the boundaries have a complex topography. To reduce these artefacts, the discretisation should be fine enough to reduce the 'stair-case' effect at the free surface. For instance, a second-order rotated FDM requires up to 60 grid points per wavelength to compute an accurate seismic wavefield in elastic media with a complex topography (Bohlen & Saenger, 2006). Such constraints on the discretisation drastically restrict the possible field of realistic applications. Some interesting combinations of FDMs and finite-element methods (FEMs) might overcome these limitations (Galis et al., 2008). The idea is to use an unstructured FEM scheme to represent both the topography and the shallow part of the medium, and to adopt for the rest of the model a classical FDM regular grid. For the same reasons as the issues related to the topography, uniform grids are not suitable for highly

Model	Method	Pre. (hr)	Sol. (hr)	Total (hr)	Pre. (hr)	Sol. (hr)	Total (hr)
Salt	Time	0	39	39	0	0.94	0.94
Salt	Frequency	0.8	0.78	1.58	0.80	0.016	0.816

Table 1. Comparison between time-domain and frequency-domain modeling for 32 (left) and 2000 (right) processors. The number of sources is 2000. *Pre.* denotes the elapsed time for the source-independent task during seismic modeling (i.e., the LU factorization in the frequency-domain approach). *Sol.* denotes the elapsed time for multi-RHS solutions during seismic modeling (i.e., the substitutions in the frequency-domain approach).

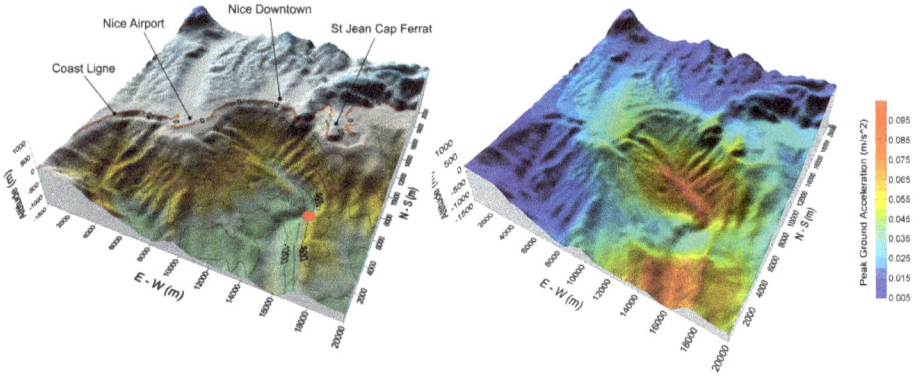

Fig. 8. On the left, the French Riviera medium with complex topography and bathymetry: an hypothetical earthquake of magnitude 4.5 is at a depth of 10 km below the epicenter shown by a red ball. The simulation medium is 20 km by 20 km by 15 km. On the right, the related peak ground acceleration (PGA). Please note that the acceleration is always lower than one tenth of the Earth acceleration g

heterogeneous media, since the grid size is determined by the shortest wavelength. Except in some circumstances, like mixing grids (Aoi & Fujiwara, 1999) or using non uniform Cartesian grids (Pitarka, 1999) in the case of a low velocity layer, it is almost impossible to locally adapt the grid size to the medium properties in the general case. From this point of view, FEMs are appealing, since they can use unstructured grids or meshes. Due to ever-increasing computational power, these kinds of methods have been the focus of a lot of interest and have been used intensively in seismology (Aagaard et al., 2001; Akcelik et al., 2003; Ichimura et al., 2007).

4.2 PGA estimation in the French Riviera

Peak ground acceleration (PGA) are estimated using empirical attenuation laws calibrated through databases of seismic records of various areas: these laws should be adapted to each area around the world and European moderate earthquakes require a specific calibration (Berge-Thierry et al., 2003). Aside these attenuation laws, numerical tools as finite-differences time-domain methods allows the deterministic estimation of the peak ground acceleration (PGA) in specific areas of interest once the medium is known and the source specified.

Small areas as the French Riviera where a complex topography and bathymetric makes the simulation difficult. We would like to illustrate the procedure of time-domain simulation on this specific example (Cruz-Atienza et al., 2007). The Figure 8 shows a very simple model surrounding the city of Nice: the box is 20 km by 20 km by 15 km in depth. The P-wave velocity is 5700 m/s while the S-wave velocity is 3300 m/s and the density 2600 km/m^3. The water is characterized by a P-wave velocity of 1530 m/s while the density is about 1030 km/m^3. The grid step is 50 m and the time integration step is 0.005 s.

The numerical simulation of an hypothetical earthquake of magnitude 4.5 at a depth of 10 km in the Mediterranean Sea provides us a deterministic estimation of the PGA as shown in the Figure 8. This small source is characterized upto a frequency of 3 Hz and we select a source time function with this expected spectral content.

Successful applications have been proposed in the Los Angeles basin and is improved as we increase our knowledge about the medium of propagation and about the source location and its characterization. The PGA is estimated everywhere and one can see that increase of the PGA is observed at the sea bottom and nearby the coast. One can show that the amplification of PGA is decreased when considering the water layer at the expense of a longer duration of the seismic signal.

Of course, various simulations should be performed using different models of the medium and for various source scenarii. These simulations could help to assess the variability of the acceleration for possible potential earthquakes and may be used for the mitigation of seismic risks. The importance of constraining the model structure should be emphasized and we can accumulate this knowledge through various and different initiatives performed for a more accurate reconstruction of the velocity structure (Rollet et al., 2002). One tool is the seismic imaging procedure we have underlined in this chapter.

5. Finite-elements discontinuous Galerkin methods: a weak formulation

Finite element methods, often more intensive in computer resources, introduce naturally boundary conditions in an explicit manner. Therefore, we expect improved accurate solutions with this numerical approach at the expense of computer requirements. The system of equations (14) in time has now a non-diagonal mass matrix while the system of equations (15) has a impedance matrix particularly ill-conditioned in 3D geometry taking into account its dimensionality. Therefore, for 2D geometries, the frequency formulation is still a quite feasible option while time domain approaches are there appealing when considering 3D geometries. Due to ever-increasing computational power, finite element methods using unstructured meshes have been the focus of increased interest and have been used extensively in seismology (Aagaard et al., 2001; Akcelik et al., 2003; Ichimura et al., 2007).

Usually, the approximation order remains low, due to the prohibitive computational cost related to a non-diagonal mass matrix. However, this high computational cost can be avoided by mass lumping, a standard technique that replaces the large linear system by a diagonal matrix (Chin-Joe-Kong et al., 1999; Marfurt, 1984) and leads to an explicit time integration. Another class of FEMs that relies on the Gauss-Lobatto-Legendre quadrature points has removed these limitations, and allows for spectral convergence with high approximation orders. This high-order FEM, called the spectral element method (SEM) (Komatitsch & Vilotte, 1998; Seriani & Priolo, 1994) has been applied to large-scale geological models up to the global scale (Chaljub et al., 2007; Komatitsch et al., 2008). The major limitation of SEM is the exclusive use of hexahedral meshes, which makes the design of an optimal mesh cumbersome in contrast to the flexibility offered by tetrahedral meshes. With tetrahedral meshes (Frey & George, 2008), it is possible to fit almost perfectly complex topographies or geological discontinuities and the mesh width can be adapted locally to the medium properties (h-adaptivity). The extension of the SEM to tetrahedral elements represents ongoing work, while some studies have been done in two dimensions on triangular meshes (Mercerat et al., 2006; Pasquetti & Rapetti, 2006). On the other hand, another kind of FEM has been proven to give accurate results on tetrahedral meshes: the Discontinuous Galerkin finite-element method (DG-FEM) in combination with the arbitrary high-order derivatives (ADER) time integration (Dumbser & Käser, 2006). Originally, DG-FEM has been developed for the neutron transport equation (Reed & Hill, 1973). It has been applied to a wide range of applications such as electromagnetics (Cockburn et al., 2004), aeroacoustics (Toulopoulos & Ekaterinaris, 2006) and plasma physics (Jacobs & Hesthaven, 2006), just to cite a few examples. This method relies on the exchange of numerical fluxes between adjacent elements. Contrary to classical FEMs, no continuity of the basis functions is imposed between elements and, therefore, the method supports discontinuities in the seismic wavefield as in the case of a fluid/solid interface. In such cases, the DG-FEM allows the same equation to be used for both the elastic and the acoustic media, and it does not require any explicit conditions on the interface (Käser & Dumbser, 2008), which is, on the contrary, mandatory for continuous formulations, like the SEM (Chaljub et al., 2003). Moreover, the DG-FEM is completely local, which means that elements do not share their nodal values, contrary to conventional continuous FEM. Local operators make the method suitable for parallelisation and allow for the mixing of different approximation orders (p-adaptivity).

5.1 3D finite-element discontinuous Galerkin method in the time domain

Time domain approaches are quite attractive when considering explicit time integration. However, in most studies, the DG-FEM is generally used with high approximation orders. We present a low-order DG-FEM formulation with the convolutional perfectly matched layer (CPML) absorbing boundary condition (Komatitsch & Martin, 2007; Roden & Gedney, 2000) that is suitable for large-scale three-dimensional (3D) seismic wave simulations. In this context, the DG-FEM provides major benefits.

The p-adaptivity is crucial for efficient simulations, in order to mitigate the effects of the very small elements that are generally encountered in refined tetrahedral meshes. Indeed, the p-adaptivity allows an optimised time stepping to be achieved, by adapting the approximation order according to the size of the elements and the properties of the medium. The benefit of such a numerical scheme is particularly important with strongly heterogeneous media. Due to the mathematical formulation we consider, the medium properties are assumed to be constant per element. Therefore, meshes have to be designed in such a way that this assumption is compatible with the expected accuracy. The discretization must be able to represent the geological structures fairly well, without over-sampling, while the spatial resolution of the imaging process puts constraints on the coarsest parameterisation of the medium. If we consider full waveform inversion (FWI) applications, the expected imaging resolution reaches half a wavelength, as shown by Sirgue & Pratt (2004). Therefore, following the Shannon theorem, a minimum number of four points per wavelength is required to obtain such accuracy. These reasons have motivated the development of DG-FEM with low orders. We focus on the quadratic interpolation, which yields a good compromise between accuracy, discretisation and computational cost.

5.1.1 3D time-domain elastodynamics

It is worth to provide notations for specific manipulation of equations for DG-FEM approaches. The first-order hyperbolic system (8) under the so-called pseudo-conservative form can be written following the approach of Ben Jemaa et al. (2007) as

$$\rho \partial_t \vec{v} = \sum_{\theta \in \{x,y,z\}} \partial_\theta (\mathcal{M}_\theta \vec{\sigma}) + \vec{f}$$

$$\Lambda \partial_t \vec{\sigma} = \sum_{\theta \in \{x,y,z\}} \partial_\theta (\mathcal{N}_\theta \vec{v}) + \Lambda \partial_t \vec{\sigma}_0, \tag{41}$$

with the definitions of the velocity and stress vectors as $\vec{v}^t = (v_x \ v_y \ v_z)^t$ and $\vec{\sigma} = (\sigma_1 \ \sigma_2 \ \sigma_3 \ \sigma_{xy} \ \sigma_{xz} \ \sigma_{yz})^t$. Under this pseudo-conservative form, the RHS of (41) does not include any term that relates to the physical properties. The diagonal matrix Λ has been introduced in the system (8) and its inverse is required for the computation of the stress components (equation (41)). Matrices $\mathcal{M}_\theta$ and $\mathcal{N}_\theta$ are constant real matrices (Etienne et al., 2010). The extension of the pseudo-conservative form for the visco-elastic cases could be considered with the inclusion of memory variables while the anisotropic case should be further analysed since the change of variable may depend on the physical parameters. Finally, in the equation (41), the medium density is denoted by ρ, while $\vec{f}$ and $\vec{\sigma}_0$ are the external forces and the initial stresses, respectively.

5.1.2 Spatial discretisation

Following standard strategies of finite-element methods (Zienkiewicz et al., 2005), we want to approximate the solution of the equation (41) by means of polynomial basis functions defined in volume elements. The spatial discretisation is carried out with non-overlapping and conforming tetrahedra. We adopt the nodal form of the DG-FEM formulation (Hesthaven & Warburton, 2008), assuming that the stress and velocity vectors are approximated in the tetrahedral elements as follows

$$\widehat{\vec{v}}_i(\vec{x}, t) = \sum_{j=1}^{d_i} \vec{v}_{ij}(\vec{x}_j, t) \, \varphi_{ij}(\vec{x})$$

$$\widehat{\vec{\sigma}}_i(\vec{x}, t) = \sum_{j=1}^{d_i} \vec{\sigma}_{ij}(\vec{x}_j, t) \, \varphi_{ij}(\vec{x}), \tag{42}$$

where i is the index of the element, $\vec{x}$ is the spatial coordinates inside the element, and t is the time. d_i is the number of nodes or degrees of freedom (DOF) associated with the interpolating Lagrangian polynomial basis function φ_{ij} relative to the j-th node located at position $\vec{x}_j$. Vectors $\vec{v}_{ij}$ and $\vec{\sigma}_{ij}$ are the velocity and stress vectors, respectively, evaluated at

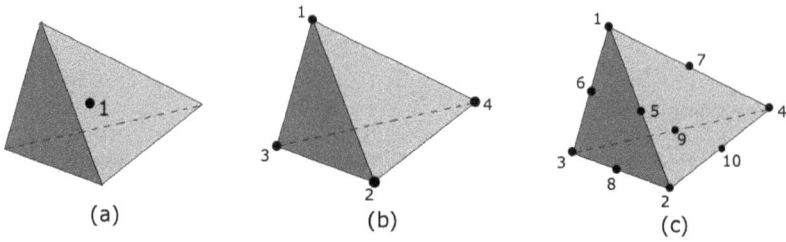

(a)   (b)   (c)

Fig. 9. (a) P_0 element with one unique DOF. (b) P_1 element with four DOF. (c) P_2 element with 10 DOF.

the j-th node of the element. Although it is not an intrinsic limitation, we have adopted here the same set of basis functions for the interpolation of the velocity and the stress components. In the following, the notation P_k refers to a spatial discretisation based on polynomial basis functions of degree k, and a P_k element is a tetrahedron in which a P_k scheme is applied. The number of DOF in a tetrahedral element is given by $d_i = (k+1)(k+2)(k+3)/6$. For instance, in a P_0 element (Figure 9.a), there is only one DOF (the stress and velocity are constant per element), while in a P_1 element (Figure 9.b), there are four DOF located at the four vertices of the tetrahedron (the stress and velocity are linearly interpolated). It is worth noting that the P_0 scheme corresponds to the case of the finite-volume method (Ben Jemaa et al., 2009; 2007; Brossier et al., 2008). For the quadratic approximation order P_2, one node is added at the middle of each edge of the tetrahedron, leading to a total of 10 DOF per element (Figure 9.c). The first step in the finite-element formulation is to obtain the weak form of the elastodynamic system. To do so, we multiply the equation (41) by a test function φ_{ir} and integrate the system over the volume of the element i. For the test function, we adopt the same kind of function as used for the approximation of the solution. This case corresponds to the standard Galerkin method and can be written as

$$\int_{V_i} \varphi_{ir} \rho \partial_t \vec{v} \, dV = \int_{V_i} \varphi_{ir} \sum_{\theta \in \{x,y,z\}} \partial_\theta (\mathcal{M}_\theta \vec{\sigma}) \, dV$$

$$\int_{V_i} \varphi_{ir} \Lambda \partial_t \vec{\sigma} \, dV = \int_{V_i} \varphi_{ir} \sum_{\theta \in \{x,y,z\}} \partial_\theta (\mathcal{N}_\theta \vec{v}) \, dV \qquad \forall r \in [1, d_i], \tag{43}$$

where the volume of the tetrahedral element i is denoted by V_i. For the purpose of clarity, we have omitted the external forces and stresses in the equation (43). Standard manipulations of finite-elements methods (integration by parts, Green theorem for fluxes along boundary surfaces) are performed as well as an evaluation of centered flux scheme for its non-dissipative property (Ben Jemaa et al., 2007; Delcourte et al., 2009; Remaki, 2000). Moreover, we assume constant physical properties per element. We define the tensorial product $\otimes$ as the Kronecker product of two matrices **A** and **B** given by

$$
\mathbf{A} \otimes \mathbf{B} =
\begin{bmatrix}
a_{11}\mathbf{B} & \cdots & a_{1m}\mathbf{B} \\
\cdot & \cdot & \cdot \\
\cdot & \cdot & \cdot \\
\cdot & \cdot & \cdot \\
a_{n1}\mathbf{B} & \cdots & a_{nm}\mathbf{B}
\end{bmatrix},
\tag{44}
$$

where $(n \times m)$ denotes the dimensions of the matrix **A**. We obtain the expression

$$
\rho_i(\mathcal{I}_3 \otimes \mathcal{K}_i)\partial_t \vec{v}_i = - \sum_{\theta \in \{x,y,z\}} (\mathcal{M}_\theta \otimes \mathcal{E}_{i\theta})\vec{\sigma}_i + \frac{1}{2} \sum_{k \in N_i} \left[(\mathcal{P}_{ik} \otimes \mathcal{F}_{ik})\vec{\sigma}_i + (\mathcal{P}_{ik} \otimes \mathcal{G}_{ik})\vec{\sigma}_k \right]
$$

$$
(\Lambda_i \otimes \mathcal{K}_i)\partial_t \vec{\sigma}_i = - \sum_{\theta \in \{x,y,z\}} (\mathcal{N}_\theta \otimes \mathcal{E}_{i\theta})\vec{v}_i + \frac{1}{2} \sum_{k \in N_i} \left[(\mathcal{Q}_{ik} \otimes \mathcal{F}_{ik})\vec{v}_i + (\mathcal{Q}_{ik} \otimes \mathcal{G}_{ik})\vec{v}_k \right], \tag{45}
$$

where $\mathcal{I}_3$ represents the identity matrix. In the system (45), the vectors $\vec{v}_i$ and $\vec{\sigma}_i$ should be red as the collection of all nodal values of the velocity and stress components in the element i. The system (45) indicates that the computations of the stress and velocity wavefields in one element require information from the directly neighbouring elements. This illustrates clearly the local nature of DG-FEM. The flux-related matrices $\mathcal{P}$ and $\mathcal{Q}$ are defined as follows

$$
\mathcal{P}_{ik} = \sum_{\theta \in \{x,y,z\}} n_{ik\theta}\,\mathcal{M}_\theta
$$

$$
\mathcal{Q}_{ik} = \sum_{\theta \in \{x,y,z\}} n_{ik\theta}\,\mathcal{N}_\theta,
$$

where the component along the θ axis of the unit vector $\vec{n}_{ik}$ of the face S_{ik} that points from element i to element k is denoted by $n_{ik\theta}$, while we also introduce the mass matrix, the stiffness matrix and the flux matrices with $\theta \in \{x,y,z\}$ respectively,

$$
(\mathcal{K}_i)_{rj} = \int_{V_i} \varphi_{ir}\, \varphi_{ij}\, dV \qquad j,r \in [1, d_i],
$$

$$
(\mathcal{E}_{i\theta})_{rj} = \int_{V_i} (\partial_\theta \varphi_{ir})\, \varphi_{ij}\, dV \qquad j,r \in [1, d_i],
$$

$$
(\mathcal{F}_{ik})_{rj} = \int_{S_{ik}} \varphi_{ir}\, \varphi_{ij}\, dS \qquad j,r \in [1, d_i] \tag{46}
$$

$$
(\mathcal{G}_{ik})_{rj} = \int_{S_{ik}} \varphi_{ir}\, \varphi_{kj}\, dS \qquad r \in [1, d_i] \qquad j \in [1, d_k].
$$

It is worth noting that, in the last equation of the system (46), the DOF of elements i and k appear (d_i and d_k, respectively) indicating that the approximation orders are totally decoupled from one element to another. Therefore, the DG-FEM allows for varying approximation orders in the numerical scheme. This feature is referred to as p-adaptivity. Moreover, given an approximation order, these matrices are unique for all elements (with a normalisation according to the volume or surface of the elements) and they can be computed before hand with appropriate integration quadrature rules. The memory requirement is therefore low, since only a collection of small matrices is needed according to the possible combinations of approximation orders. The maximum size of these matrices is $(d_{max} \times d_{max})$ where d_{max} is the maximum number of DOF per element and the number of matrices to store is given by the square of the number of approximation orders mixed in the numerical domain. The four matrices $\mathcal{K}_i$, $\mathcal{E}_i$, $\mathcal{F}_{ik}$ and $\mathcal{G}_{ik}$ are computed by numerical integration using Hammer quadrature (Hammer & Stroud, 1958) and explicit forms of these matrices could be found in Etienne et al. (2010) for P_0, P_1 and P_2 orders.

It should be mentioned that, in order to retrieve both the velocity and the stress components, the system (45) requires the computation of $\mathcal{K}_i^{-1}$, which can also be performed before hand. Note that, if we want to consider variations in the physical properties inside the elements, the pseudo-conservative form makes the computation of flux much easier and computationally more efficient than in the classical elastodynamic system. These properties come from the fact that, in the pseudo-conservative form, the physical properties are located in the left-hand side of the system (41). Therefore, no modification of the stiffness and flux matrices nor additional terms are needed in the system (45) to take into account the variation of properties. Only the mass matrix needs to be evaluated for each element and for each physical property according to the expression

$$(\mathcal{K}_i)_{rj} = \int_{V_i} \chi_i(\vec{x})\, \varphi_{ir}(\vec{x})\, \varphi_{ij}(\vec{x})\, dV \qquad j, r \in [1, d_i], \tag{47}$$

where $\chi_i(\vec{x})$ represents the physical property (ρ_i or one of the Λ_i components) varying inside the element.

5.1.3 Time discretisation

The time integration of the system (45) relies on the second-order explicit leap-frog scheme that allows to compute alternatively the velocity and the stress components between a half time step. The system (45) can be written as

$$\rho_i(\mathcal{I}_3 \otimes \mathcal{K}_i)\frac{\vec{v}_i^{n+\frac{1}{2}} - \vec{v}_i^{n-\frac{1}{2}}}{\Delta t} = -\sum_{\theta \in \{x,y,z\}} (\mathcal{M}_\theta \otimes \mathcal{E}_{i\theta})\vec{\sigma}_i^n + \frac{1}{2}\sum_{k \in N_i}\left[(\mathcal{P}_{ik} \otimes \mathcal{F}_{ik})\vec{\sigma}_i^n + (\mathcal{P}_{ik} \otimes \mathcal{G}_{ik})\vec{\sigma}_k^n\right]$$

$$(\Lambda_i \otimes \mathcal{K}_i)\frac{\vec{\sigma}_i^{n+1} - \vec{\sigma}_i^n}{\Delta t} = -\sum_{\theta \in \{x,y,z\}} (\mathcal{N}_\theta \otimes \mathcal{E}_{i\theta})\vec{v}_i^{n+\frac{1}{2}}$$

$$+ \frac{1}{2}\sum_{k \in N_i}\left[(\mathcal{Q}_{ik} \otimes \mathcal{F}_{ik})\vec{v}_i^{n+\frac{1}{2}} + (\mathcal{Q}_{ik} \otimes \mathcal{G}_{ik})\vec{v}_k^{n+\frac{1}{2}}\right], \tag{48}$$

where the superscript n indicates the time step. We chose to apply the definition of the time step as given by Käser et al. (2008), which links the mesh width and time step as follows

$$\Delta t < \min_i(\frac{1}{2k_i + 1} \cdot \frac{2r_i}{V_{Pi}}), \tag{49}$$

where r_i is the radius of the sphere inscribed in the element indexed by i, V_{Pi} is the P-wave velocity in the element, and k_i is the polynomial degree used in the element. Equation (49) is a heuristic stability criterion that usually works well. However, there is no mathematical proof for unstructured meshes that guarantees numerical stability.

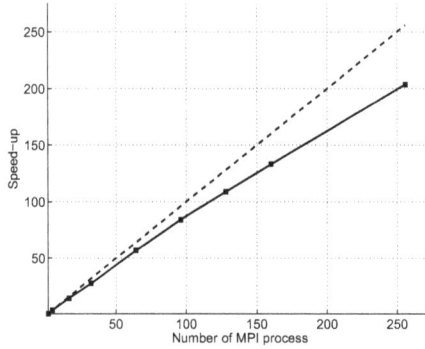

Fig. 10. Speed-up observed when the number of MPI processes is increased from 1 to 256 for modelling with a mesh of 1.8 million P_2 elements. The ideal speed-up is plotted with a dashed line, the observed speed-up with a continuous line. These values were observed on a computing platform with bi-processor quad core Opteron 2.3 GHz CPUs interconnected with Infiniband at 20 Gb/s.

5.1.4 Computational aspects

The DG-FEM is a local method, and therefore it is naturally suitable for parallel computing. In our implementation, the parallelism relies on a domain-partitioning strategy, assigning one subdomain to one CPU. This corresponds to the single program mutiple data (SPMD) architecture, which means that there is only one program and each CPU uses the same executable to work on different parts of the 3D mesh. Communication between the subdomains is performed with the message passing interface (MPI) parallel environment (Aoyama & Nakano, 1999), which allows for applications to run on distributed memory machines. For efficient load balancing among the CPUs, the mesh is divided with the partitioner METIS (Karypis & Kumar, 1998), to balance the number of elements in the subdomains, and to minimise the number of adjacent elements between the subdomains. These two criteria are crucial for the efficiency of the parallelism on large-scale numerical simulations. Figure 10 shows the observed speed-up (i.e. the ratio between the computation time with one CPU, and the computation time with N CPUs) when the number of MPI processes is increased from 1 to 256, for strong scaling calculations on a fixed mesh of 1.8 million P_2 elements. This figure shows good efficiency of the parallelism, of around 80%. In our formulation, another key point is the time step, which is common for all of the subdomains. The time step should satisfy the stability condition given in equation (49) for every element. Consequently, the element with the smallest time step imposes its time step on all of the subdomains. We should mention here a more elaborate approach with local time stepping (Dumbser et al., 2007) that allows for elements to have their own time step independent of the others. Nevertheless, the p-adaptivity offered by DG-FEM allows mitigation of the computational burden resulting from the common time step as we shall see.

5.1.5 Source excitation

We proceed with the addition of the excitation to incremental increase of each involved field component. The excitation of a point source is projected onto the nodes of the element that

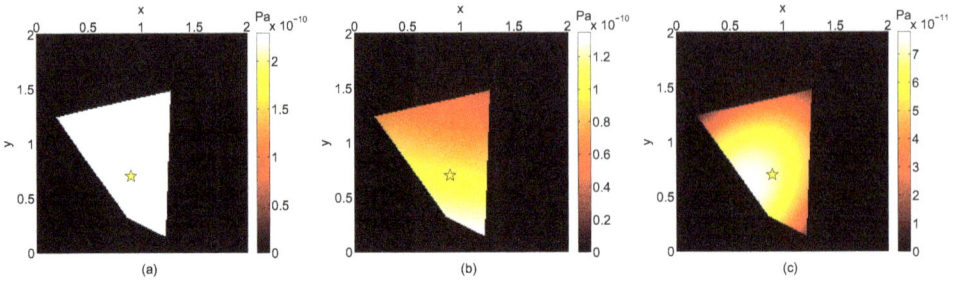

(a)   (b)   (c)

Fig. 11. (a) Cross-section of the mesh near the source position, indicated with a yellow star in the xy plane. This view represents the spatial support of the stress component in a P_0 element containing the point source. (b) Same as (a) with a P_1 element. (c) Same as (a) with a P_2 element.

contains the source as follows

$$\vec{s}_i^n = \frac{\vec{\varphi}_i(\vec{x}_s)}{\sum_{j=1}^{d_i} \varphi_{ij}(\vec{x}_s) \int_{V_i} \varphi_{ij}(\vec{x}) dV} s(t), \tag{50}$$

with $\vec{s}_i^n$ the nodal values vector associated to the excited component, $t = n\Delta t$, $\vec{x}_s$ the position of the point source and $s(t)$ the source function. Equation (50) gives the source term that should be added to the right-hand side of equation (48) for the required components. It should be noticed that this term is only applied to the element containing the source. Depending on the approximation order, the spatial support of the source varies. Figure 11.a shows that the support of a P_0 element is actually the whole volume of the element (represented on the cross-section with a homogeneous white area). In this case, no precise localisation of the source inside the element is possible due to the constant piece-wise interpolation approximation. On the other hand, in a P_1 element (Figure 11.b), the spatial support of the source is linear and allows for a rough localisation of the source. In a P_2 element (Figure 11.c), the quadratic spatial support tends to resemble the expected Dirac in space close to the source position. It should be noted that the limitations concerning source localisation also apply to the solution extraction at the receivers, according to the approximation order of the elements containing the receivers.

5.1.6 Free surface condition

Among the various approaches presented previously, we proceed by considering that the free surface follows the mesh elements. For the element faces located on the free surface, we use an explicit condition by changing the flux expression locally. This is carried out with the concept of virtual elements, which are exactly symmetric to the elements located on the free surface. Inside the virtual elements, we impose a velocity wavefield that is identical to the wavefield of the corresponding inner elements, and we impose an opposite stress wavefield on this virtual element. Thanks to the nodal formulation, the velocity is seen as continuous across the free surface, while the stress is equal to zero on the faces related to the free surface.

This is a quite natural approach similar to the one used in continuous finite-element methods where the test function is set to zero on the free surface boundary.

5.1.7 Absorbing boundary condition

We proceed through some simulations of wave propagation in a homogeneous, isotropic and purely elastic medium for an illustration of CPML conditions. The model size is 8 km × 8 km × 8 km, and the medium properties are: V_P = 4000 m/s, V_S = 2310 m/s and ρ = 2000 kg/m^3. An explosive source is placed at coordinates (x_s= 2000 m, y_s = 2000 m, z_s = 4000 m) and a line of receivers is located at coordinates (3000 m $\leq x_r \leq$ 6000 m, y_r = 2000 m, z_r = 4000 m) with 500 m between receivers. The conditions of the tests are particularly severe, since the source and the receivers are located close to the CPMLs (at a distance of 250 m), thus favouring grazing waves. The source signature is a Ricker wavelet with a dominant frequency of 3 Hz and a maximum frequency of about 7.5 Hz. Due to the explosive source, only P-wave is generated and the minimum wavelength is about 533 m. The mesh contains 945,477 tedrahedra with an average edge of 175 m, making a discretisation of about 3 elements per λ_{min}. Figures 12.c and 12.d show the results obtained with the P_2 interpolation and CPMLs of 10-elements width (L_{cpml} = 1750 m) at all edges of the model. With the standard scale, no reflection can be seen from the CPMLs. When the amplitude is magnified by a factor of 100, some spurious reflections are visible. This observation is in agreement with the theoretical reflection coefficient (R_{coeff} = 0.1%) in equation (20).

As shown by Collino & Tsogka (2001), the thickness of the absorbing layer plays an important role in the absorption efficiency. In Figures 12.a and 12.b, the same test was performed with CPMLs of 5-elements width (L_{cpml} = 875 m) at all edges of the model. Compared to Figures 12.c and 12.d, the amplitude of the reflections have the same order of magnitude. Nevertheless, in the upper and left parts of the model, some areas with a strong amplitude appear close to the edges. These numerical instabilities arise at the outer edges of the CPMLs, and they expand over the complete model during the simulations.

Instabilities of PML in long time simulations have been studied in electromagnetics (Abarbanel et al., 2002; Bécache et al., 2004). For elastodynamics, remedies have been proposed by Meza-Fajardo & Papageorgiou (2008) for an isotropic medium with standard PML. These authors proposed the application of an additional damping in the PML, onto the directions parallel to the layer, leading to a multiaxial PML (M-PML) which does not follow strictly the matching property of PML in the continuum and which has a less efficient absorption power. Through various numerical tests, Etienne et al. (2010) has shown that instabilities could be delayed outside the time window of simulation when considering extended M-PML from CPML.

5.1.8 Saving computation time and memory

Table 2 gives the computation times for updating the velocity and stress wavefields in one element for one time step, for different approximation orders, without or with the update of the CPML memory variables (i.e. elements located outside or inside the CPMLs). These computation times illustrate the significant increase with respect to the approximation order, and they allow an evaluation of the additional costs of the CPML memory variables computation from 40% to 60%. The effects of this additionnal cost have to be analysed in

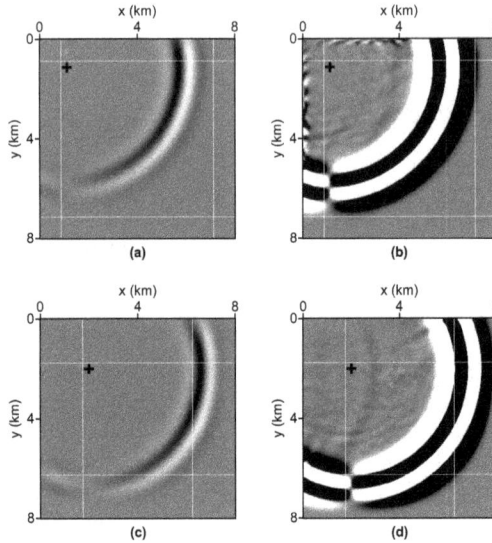

Fig. 12. Snapshots at 1.6 s of the velocity component v_x in the plane xy that contains the source location. CPMLs of 10-elements width are applied at all edges of the model. The modelling was carried out with P_2 interpolation. White lines, the limits of the CPMLs; black cross, the position of the source. (a) Real amplitude. (b) Amplitude magnified by a factor of 100. (c) & (d) Same as (a) & (b) with CPMLs of 5-elements width.

Approximation order	Element outside CPML	Element inside CPML
P_0	2.6 μs	3.6 μs
P_1	5.0 μs	8.3 μs
P_2	21.1 μs	29.9 μs

Table 2. Computation times for updating the velocity and stress wavefields in one element for one time step. These values correspond to average computation times for a computing platform with bi-processor quad core Opteron 2.3 GHz CPUs interconnected with Infiniband 20 at Gb/s.

the context of a domain-partitioning strategy. The mesh is divided into subdomains, using a partitioner. Figure 13.a shows the layout of the subdomains that were obtained with the partitioner METIS (Karypis & Kumar, 1998) along the xy plane used in the previous validation tests. The mesh was divided into 32 partitions, although only a few of these are visible on the cross-section in Figure 13.a. We used an unweighted partitioning, meaning that each partition contains approximately the same number of elements.

The subdomains, partially located in the CPMLs, contain different numbers of CPML elements. In large simulations, some subdomains are totally located inside the CPMLs, and some others outside the CPMLs. In such a case, the extra computation costs of the subdomains located in the absorbing layers penalise the whole simulation. Indeed, most of the subdomains spend 40% to 60% of the time just waiting for the subdomains located in the CPMLs to

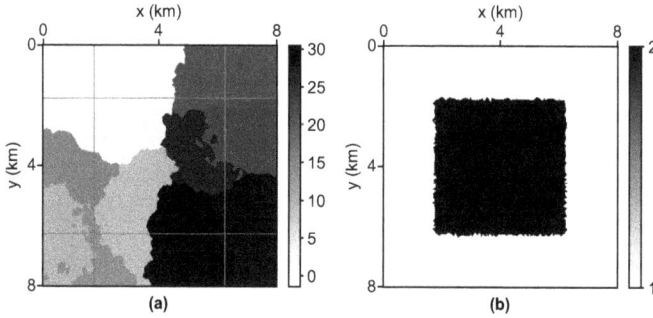

Fig. 13. (a) Layout of the subdomains obtained with the partitioner METIS (Karypis & Kumar, 1998) along the xy plane that contains the source location. Grey lines, the limits of the CPMLs. The mesh was divided into 32 partitions, although only a few of these are visible on this cross-section. (b) View of the approximation order per element along the same plane. Black, the P_2 elements; white, the P_1 elements.

complete the computations at each time step. For a better load balancing, we propose to benefit from the p-adaptivity of DG-FEM, using lower approximation orders in the CPMLs. Indeed, inside the absorbing layers, we do not need a specific accuracy, and consequently the approximation order can be decreased. Table 2 indicates that such a mixed numerical scheme is advantageous, since the computation time required for a P_0 or P_1 element located in the CPML is shorter than the computation time of a standard P_2 element. Figure 13.b shows the approximation order per element when P_1 is used in the CPMLs and P_2 in the rest of the medium. We should note here that the interface between these two areas is not strictly aligned to a cartesian axis, and has some irregularities due to the shape of the tetrahedra. Although it is possible to constrain the alignment of the element faces parallel to the CPML limits, we did not observe significant differences in the absorption efficiency whether the faces are aligned or not.

Figure 14.a shows the seismograms computed when the modelling was carried out with P_2 inside the medium and P_1 in the CPMLs. Absorbing layers of 10-elements width are applied at all edges of the model. For comparison, Figure 14.b shows the results obtained with P_2 inside the medium and P_0 in the CPMLs. In this case, the spurious reflections have significant amplitudes, preventing any use of these seismograms. On the other hand, the seismograms computed with the mixed scheme P_2/P_1 show weak artefacts, and are reasonably comparable with the seismograms obtained with complete P_2 modelling. Therefore, taking into account that the computation time and the memory consumption of the P_2/P_1 simulation are nearly half of those required with the full P_2 modelling, we can conclude that this mixed numerical scheme is of interest. It should be noticed that it is possible to adopt a weighted partitioning approach to overcome partly load balancing issues We should also stress that the saving in CPU time and memory provided with this kind of low-cost absorbing boundary condition is crucial for large 3D simulations, and this becomes a must in the context of 3D seismic imaging applications that require a lot of forward problems, such as FWI.

P2/P1   P2/P0

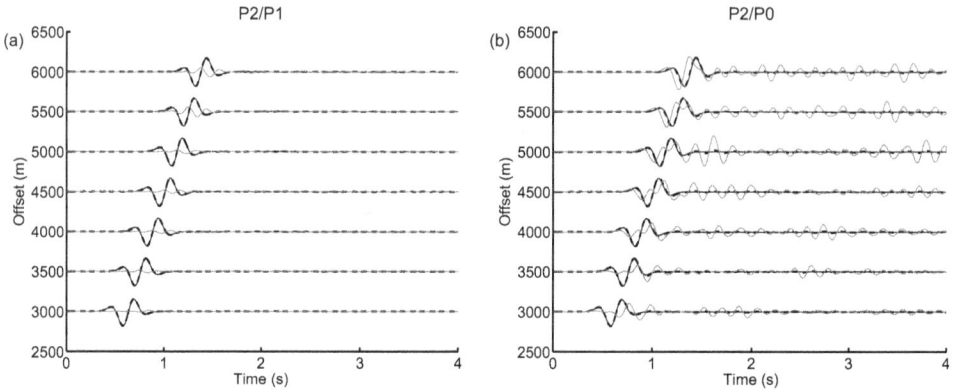

Fig. 14. (a) Seismograms of the velocity component v_x. The amplitude of each seismogram is normalised. The modelling is done with P_1 in the CPMLs and P_2 inside the medium. Black continuous line, numerical solution in large model without reflection in the time window; dashed line, numerical solution with 10-elements width CPMLs; grey line, residuals magnified by a factor of 10. (b) Same as (a) except the modelling is done with P_0 in the CPMLs and P_2 inside the medium.

5.1.9 Accuracy of DG-FEM with tetrahedral meshes

There are a variety of studies in the literature concerning the dispersive and dissipative properties of DG-FEM with reference to wave-propagation problems. Let us quote few examples: Ainsworth et al. (2006) provided a theoretical study for the 1D case; Basabe et al. (2008) analysed the effects of basis functions on 2D periodic and regular quadrilateral meshes; and Käser et al. (2008) discussed the convergence of the DG-FEM combined with ADER time integration and 3D tetrahedral meshes. More related to our particular concern here, Delcourte et al. (2009) provided a convergence analysis of the DG-FEM with a centred flux scheme and tetrahedral meshes for elastodynamics. They demonstrated the sensitivity of the DG-FEM to the mesh quality, and they proved that the convergence is limited by the second-order time integration we have used in the present study, despite the order of the basis function. Specific analysis of the convergence in the scheme we have presented could be found in Etienne et al. (2010).

5.2 2D finite-element discontinuous Galerkin method in the frequency domain

On land exploration seismology, there is a need to perform elastic wave modeling in area of complex topography such as foothills and thrust belts (Figure 15) in the frequency domain. Moreover, onshore targets often exhibit weathered layers with very low wave speeds in the near surface which require a locally-refined discretisation for accurate modeling. In shallow water environment, a mesh refinement is also often required near the sea floor for accurate modeling of guided and interface waves near the sea floor. Accurate modeling of acoustic and elastic waves in presence of complex boundaries of arbitrary shape and the local adaptation of the discretisation to local features such as weathered near surface layers or sea floor

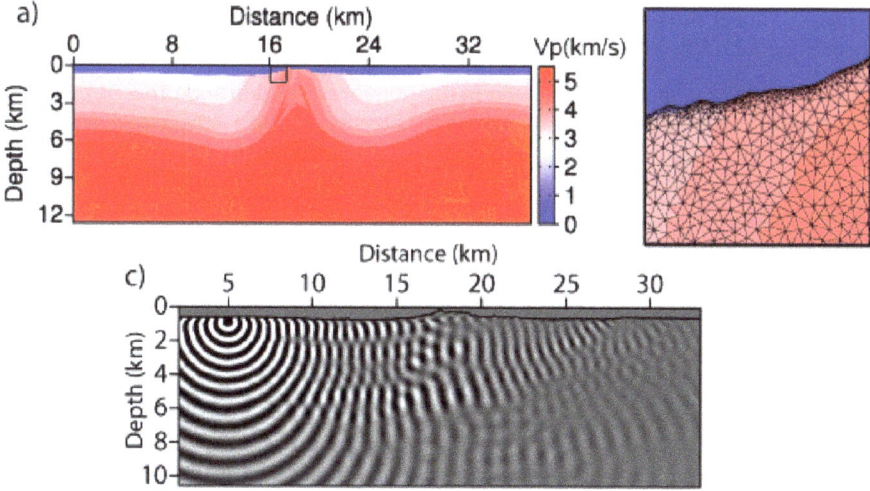

Fig. 15. Application of the DG method in seismic exploration. (a) Velocity model representative of a foothill area affected by a hilly relief and a weathered layer in the near surface. (b) Close-up of the unstructured triangular mesh locally refined near the surface. (c) Example of monochromatic pressure wavefield.

were two of our motivations behind the development of a discontinuous element method on unstructured meshes for acoustic and elastic wave modeling.

5.2.1 hp-adaptive discontinuous Galerkin discretisation

Similarly to the time formulation we adopt the nodal form of the DG formulation, assuming that the wavefield vector is approximated in triangular elements for 2D geometry which leads to the following expression,

$$\vec{u}_i(\omega, x, y, z) = \sum_{j=1}^{d_i} \vec{u}_{ij}(\omega, x_j, y_j, z_j) \varphi_{ij}(\omega, x, y, z), \tag{51}$$

where $\vec{u}$ is the wavefield vector of components such as the following vector $\vec{u} = (p, v_x, v_y, v_z)$ for acoustic propagation. The index of the element in an unstructured mesh is denoted by i. The expression $\vec{u}_i(\omega, x, y, z)$ denotes the wavefield vector in the element i and (x, y, z) are the coordinates inside the element i. In the framework of the nodal form of the DG method, φ_{ij} denotes Lagrange polynomial and d_i is the number of nodes in the element i. The position of the node j in the element i is denoted by the local coordinates (x_j, y_j, z_j).

In the frequency domain, the pseudo-conservative form (41) could be written in a 2D geometry as

$$\mathcal{M}\vec{u} = \sum_{\theta \in \{x, y, z\}} \partial_\theta \left(\mathcal{N}_\theta \vec{u} \right) + \vec{s}, \tag{52}$$

where $\mathcal{N}_\theta \vec{u}$ are linear fluxes and the source vector is denoted by $\vec{s}$. Expressions of matrices **M** and **N** could be found in Brossier et al. (2010).

The weak form of the system (52) is similar in the frequency domain and proceed by selecting a test function φ_{ir} and then an integration over the element volume V_i which gives

$$\int_{V_i} \varphi_{ir} \mathcal{M}_i \vec{u}_i \, dV = \int_{V_i} \varphi_{ir} \sum_{\theta \in \{x,y,z\}} \partial_\theta \left(\mathcal{N}_\theta \vec{u}_i \right) \, dV + \int_{V_i} \varphi_{ir} \vec{s}_i dV, \tag{53}$$

where the quantity $r \in [1, d_i]$. In the framework of Galerkin methods, we used the same function for the test function and the shape function. Similar procedures as for the 3D case and related to standard steps of the finite-element method lead to the discrete expression,

$$\left(\mathcal{M}_i \otimes \mathcal{K}_i \right) \vec{u}_i = - \sum_{\theta \in \{x,y,z\}} \left(\mathcal{N}_\theta \otimes \mathcal{E}_{i\theta} \right) \vec{u}_i + \frac{1}{2} \sum_{k \in N_i} \left[\left(\mathcal{Q}_{ik} \otimes \mathcal{F}_{ik} \right) \vec{u}_i + \left(\mathcal{Q}_{ik} \otimes \mathcal{G}_{ik} \right) \vec{u}_k \right] + \left(\mathcal{I} \otimes \mathcal{K}_i \right) \vec{s}_i$$

$$\tag{54}$$

where the mass matrix $\mathcal{K}_i$, the stiffness matrix $\mathcal{E}_i$ and the flux matrices $\mathcal{F}_i$ and $\mathcal{G}_i$ are similar to those defined for the 3D case (equation (46)). The matrix $\mathcal{Q}$ is also defined as for the 3D case (equation (46))

It is worth repeting that, in the equation (46), arbitrary polynomial order of the shape functions can be used in elements i and k indicating that the approximation orders are totally decoupled from one element to another. Therefore, the DG allows for varying approximation orders in the numerical scheme, leading to the p-adaptivity.

The equation (54) can be recast in matrix form as

$$\mathbf{B} \, \mathbf{u} = \mathbf{s}. \tag{55}$$

5.2.2 Which interpolation orders to choose?

For the shape and test functions, we used low-order Lagrangian polynomials of orders 0, 1 and 2, referred to as P_k, $k \in 0, 1, 2$ in the following (Brossier, 2009; Etienne et al., 2009). Let us remind that our motivation behind seismic modeling is to perform seismic imaging of the subsurface by full waveform inversion, the spatial resolution of which is half the propagated wavelength and that the physical properties of the medium are piecewise constant per element in our implementation of the DG method. The spatial resolution of the FWI and the piecewise constant representation of the medium direct us towards low-interpolation orders to achieve the best compromise between computational efficiency, solution accuracy and suitable discretisation of the computational domain. The P_0 interpolation (or finite volume scheme) was shown to provide sufficiently-accurate solution on 2D equilateral triangular mesh when ten cells per minimum propagated wavelength are used (Brossier et al., 2008), while 10 cells and 3 cells per propagated wavelengths provide sufficiently-accurate solutions on unstructured triangular meshes with the P_1 and the P_2 interpolation orders, respectively (Brossier, 2011). Of note, the P_0 scheme is not convergent on unstructured meshes when centered fluxes are used (Brossier et al., 2008). This prevents the use of the P_0 scheme in 3D medium where uniform tetrahedral meshes do not exist (Etienne et al., 2010). A second remark is that the finite volume scheme on square cells is equivalent to second-order accurate

	FD^{2D}	$DG^{2D}_{P_0}$	$DG^{2D}_{P_1}$	$DG^{2D}_{P_2}$
n_d	1	1	3	6
n_z	9	5-9	13-25	24-48

Table 3. Number of nodes per element (n_d) and number of non-zero coefficients per row of the impedance matrix (n_z) for the FD and DG methods. The number n_z depends on the number of wavefield components involved in the r.h.s of the first-order wave equation n_{der}.

FD stencil (Brossier et al., 2008) which is consistent with a discretisation criterion of 10 grid points per wavelength (Virieux, 1986). Use of interpolation orders greater than 2 would allow us to use coarser meshes for the same accuracy but these coarser meshes would lead to an undersampling of the subsurface model during imaging. On the other hand, use of high interpolation orders on mesh built using a criterion of 4 cells par wavelength would provide an unnecessary accuracy level for seismic imaging at the expense of the computational cost resulting from the dramatic increase of the number of unknowns in the equation (55).

The computational cost of the LU decomposition depends on the numerical bandwidth of the matrix, the dimension of the matrix (i.e., the number of rows/columns) and the number of non-zero coefficients per row (n_z). The dimension of the matrix depends in turn of the number of cell (n_{cell}), of the number of nodes per cell (n_d) and the number of wavefield components (n_{wave}) (ranging from 3 to 5 in 2D geometry). The number of nodes in a 2D triangular element is given by Hesthaven & Warburton (2008) and leads to the following expression $n_d = (k + 1)(k+2)/2$ where k denotes the interpolation order similar to what is done in the 3D geometry.

The numerical bandwidth is not significantly impacted by the interpolation order. The dimension of the matrix and the number of non-zero elements per row of the impedance matrix are respectively given by $n_{wave} \times n_d \times n_{cell}$ and $(1 + n_{neigh}) \times n_d \times n_{der} + 1$, where n_{neigh} is the number of neighbor cell (3 in 2D geometry) and n_{der} is the number of wavefield components involved in the r.h.s of the velocity-pressure wave equation, equation (52). Table 3 outlines the number of non-zero coefficients per row for the mixed-grid FD and DG methods. Increasing the interpolation order will lead to an increase of the number of non-zero coefficients per row, a decrease of the number of cells in the mesh and an increase of the number of nodes in each element. The combined impact of the 3 parameters n_z, n_{cell}, n_d on the computational cost of the DG method makes difficult the definition of the optimal discretisation of the frequency-domain DG method. The medium properties should rather drive us towards the choice of a suitable discretisation.

One must underline that the LU factorization is quite demanding in computer memory and has also some drawbacks for scalability, suggesting that nodes with high memory should be preferred at the expense of the CPU numbers.

5.2.3 Boundary conditions and source implementation

Absorbing boundary conditions are implemented with unsplitted PML in the frequency-domain DG method (Brossier, 2011) following the same approach than for the FD method: one can see that the PML implementation in the frequency is straightforward. We have found that constraining the meshing to have edges of elements in the PML zone parallel to the direction of dissipation of the waves improves the efficiency.

Free surface boundary condition is implemented with the method of image. A virtual cell is considered above the free surface with the same velocity and the opposite pressure components to those below the free surface. This allows us to fulfill the zero pressure condition at the free surface while keeping the correct numerical estimation of the particle velocity at the free surface. Using these particle velocities and pressures in the virtual cell, the pressure flux across the free surface interface vanishes, while the velocity flux is twice the value that would have been obtained by neglecting the flux contribution above the free surface. As in the FD method, this boundary condition has been implemented by modifying the impedance matrix accordingly without introducing explicitly the virtual element in the mesh. The rigid boundary condition is implemented following the same principle except that the same pressure and the opposite velocity are considered in the virtual cell.

Concerning the source excitation, the point source at arbitrary positions in the mesh is implemented by means of the Lagrange interpolation polynomials for $k \geq 1$. This means that the source excitation is performed at the nodes of the cell containing the source with appropriate weights corresponding to the projection of the physical position of the source on the polynomial basis. When the source is located in the close vicinity of a node of a triangular cell, all the weights are almost zero except that located near the source. In the case of the P_2 interpolation, a source close to the vertex of the triangular cell is problematic because the integral of the P_2 basis function over the volume of the cell is zero for nodes located at the vertex of the triangle. In this case, no source excitation will be performed (see equation (54)). To overcome this problem specific to the P_2 interpolation, one can use locally a P_1 interpolation in the element containing the source at the expense of the accuracy or distribute the source excitation over several elements or express the solution in the form of local polynomials (i.e., the so-called modal form) rather than through nodes and interpolating Lagrange polynomials (i.e., the so-called nodal form).

Another issue is the implementation of the source in P_0 equilateral mesh. If the source is excited only within the element containing the source, a checker-board pattern is superimposed on the wavefield solution. This pattern results from the fact that one cell out of two is excited in the DG formulation because the DG stencil does not embed a staggered-grid structure (the unexcited grid is not stored in staggered-grid FD methods; see Hustedt et al. (2004) for an illustration). To overcome this problem, the source can be distributed over several elements of the mesh or P_1 interpolation can be used in the area containing the sources and the receivers, while keeping P_0 interpolation in the other parts of the model (Brossier et al., 2010).

Of note, use of unstructured meshes together with the source excitation at the different nodes of the element contribute to mitigate the checker-board pattern in the in P_1 and P_2 schemes. The same procedure as for the source is used to extract the wavefield solution at arbitrary receiver positions.

6. Realistic examples for highly contrasted and strongly heterogeneous media using finite-elements methods

We shall consider two examples for the illustration of the Discontinuous Galerkin approach. The first one is related to the problem of 3D wave propagation inside an active volcano using the time-domain approach while the second one deals with the problem of 2D wave propagation above a oil reservoir using the frequency-domain approach.

6.1 The volcano *La Soufrière*

6.1.1 Characteristics of the model

La Soufrière of Guadeloupe (France) is one of nine active volcanoes of Lesser Antilles. It belongs to a recent volcanic system situated in the south part of the *Basse-Terre*. A P-wave velocity model of the volcano has been obtained by first arrival time tomography (Coutant et al., 2010). Figure 16 is the reconstructed Vp velocity model that reveals the existence of a high velocity zone below the dome of *La Soufrière*. The dimensions of the model are $1400\ m \times 1400\ m \times 1000\ m$ in xyz respectively. We consider a constant Poisson ratio of 0.25 to assess the S-wave velocity model from V_P. The velocity ranges from $660\ m/s$ to $3800\ m/s$ for V_P and $380\ m/s$ to $2200\ m/s$ for V_S. Considering a maximum frequency of $25\ Hz$, the minimum wavelength is about $15\ m$. In addition, we consider a constant density equal to $2000\ kg/m^3$. Absorbing layers of CPML type with a thickness of $300\ m$ are added at each side edge of the model as well at the bottom edge. Therefore, the complete dimensions of the numerical model are $2000\ m \times 2000\ m \times 1300\ m$.

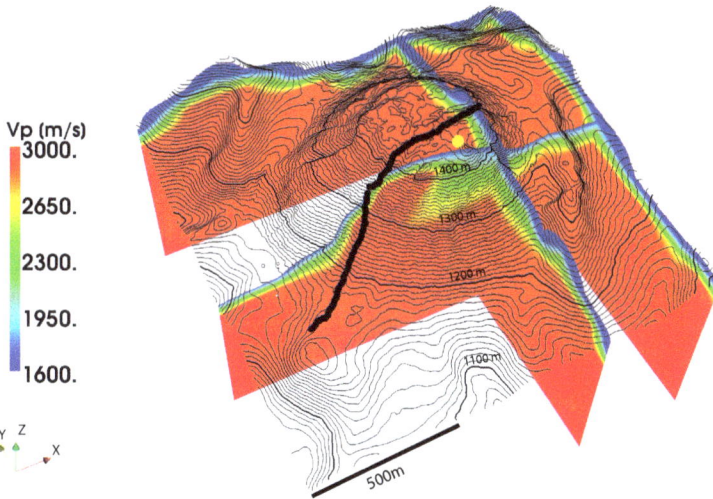

Fig. 16. Topography of the volcano *La Soufrière* with the underlying reconstructed Vp velocity structure. The position of the dynamite shot is indicated with a yellow circle and the receivers with black triangles.

6.1.2 Construction of the tetraedral mesh

The mesh has been built with the mesher TETGEN (Si, 2006) combined with an iterative *h*-refinement procedure to obtain a locally adapted mesh to the velocity field(with an average of 3 elements per minimum wavelength λ_{min}): a cross-section is shown in the left panel of the Figure 17. For building this mesh, we have started our iterative reconstruction with a uniform mesh shown in the right panel of the Figure 17. After the sixth refinement iteration, the discretization criteria are met. Areas of high velocities are correlated with the parts of the

Modelling time	5 s
Nb elements	4.6 million
Nb unknowns	414 million
Min/Max element edge	1.29 - 58.62 m
Nb time steps	37 787
Nb CPUs	512
Total memory	10 GB
Memory per CPU	14 - 23 MB
Computation time	6 h 45 min.
Time / unknown / time step	0.79 μs

Table 4. Statistics of the modelling for *La Soufrière* performed on an IBM Blue Gene machine.

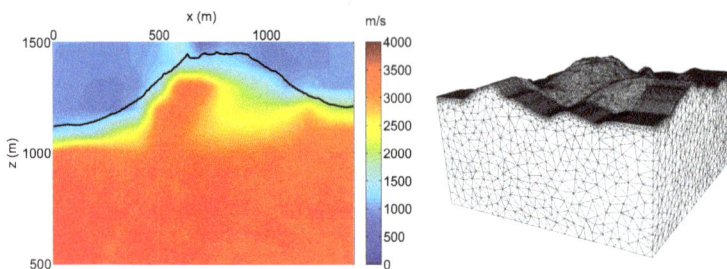

Fig. 17. On the left, cross-section of the P-wave velocity model in the plane xz in the middle of the volcano *La Soufrière*. The back line represents the topography. On the right, initial mesh of the volcano *La Soufrière* from which we deduce automatically the one used for modeling by adapting the mesh size to the local P-wave velocity. Absorbing layers of CPML type with a thickness of 300 m are added at each side edge of the model.

mesh where the elements are the largest ones. On the contrary, near the free surface, we find the finest elements.

6.1.3 Numerical result

We have performed 3D simulations with the Discontinuous Galerkin Finite-Element Method in the time domain. The computations have been performed on a Blue Gene machine with 512 processors. The statistics for these computations are given in Table 4.

The configuration of the seismic acquisition is given in Figure 16. This is a quasi-2D system with a profile according to the East-West direction, which includes 100 single-component receivers (v_z) with 10 m between receivers. The source is a shot of dynamite. For the numerical simulations, we used an explosive source with a Ricker function of dominant frequency of 10 Hz (maximum frequency 25 Hz). We present in Figure 18 a comparison between the observed and computed data. Despite significant uncertainties and approximations (source function, Poisson ratio, density, absence of attenuation, low signal to noise ratio), there are striking similarities in the data. In particular, the seismic traces exhibit well marked discontinuities related to the strong velocity contrasts and the complex topography of the volcano *La Soufrière*.

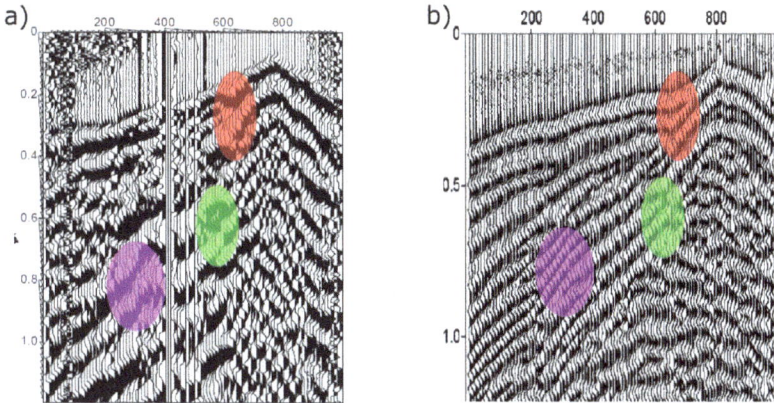

Fig. 18. (a) Recorded seismograms (component v_z). (b) Computed seismograms. Some similarities between both set of data are highlighted with color shapes.

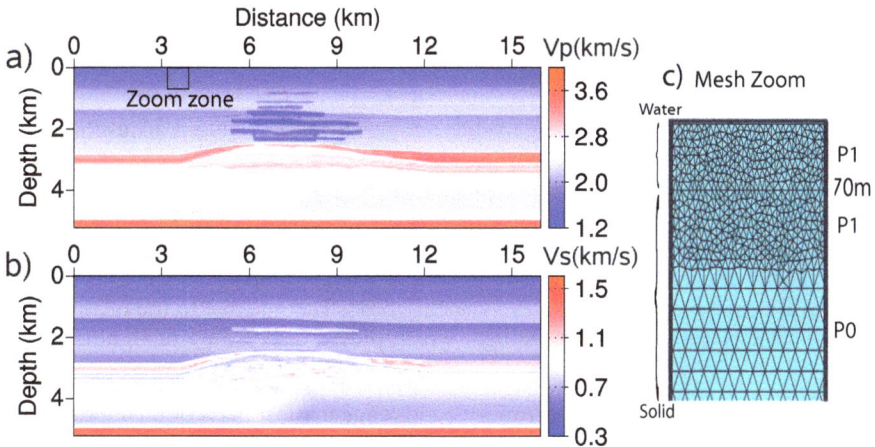

Fig. 19. The synthetic Valhall model for (a) P-wave and (b) S-wave velocities. Panel (c) represents a zoom of the shallow mesh.

6.2 Application 2D in the frequency-domain: the synthetic Valhall application

This 2D application is based on a synthetic representation of the Valhall zone in the North Sea, Norway. This model is representative of oil and gas fields in shallow water environments of the North Sea (Munns, 1985). The model is described as an heterogeneous P- and S- wave velocity model (Figure 19a-b). The water layer is only 70 m depth. The main targets are a gas cloud in the large sediment layer, and in a deeper part of the model, the trapped oil underneath the cap rock, which is formed of chalk. Gas clouds are easily identified by the low P-wave velocities, whereas their signature is much weaker in the V_S model, as gas does not affect S-waves propagation.

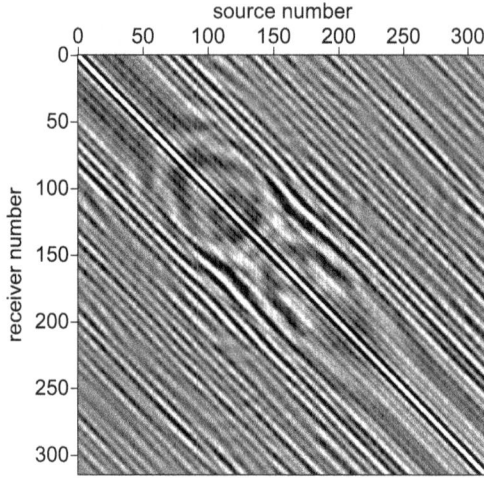

Fig. 20. Frequency-domain data for the hydrophone component at 4 Hz. The data (real-part) are plotted in the source/receiver domain.

In order to investigate seismic imaging in such environment, the selected acquisition mimics a four-component ocean-bottom cable survey (Kommedal et al., 2004), as is deployed on the field. A line of 315 explosive sources is positioned 5 m below water surface to simulated air-gun sources and a cable of 315 3-components sensors is located on the sea floor (1 hydrophone and 2 geophones). This geological setting, which is composed of a significant soft sea-bed with high Poisson'ratio due to soft and unconsolidated sediments leads to a particularly ill-posed problem for S-wave velocity reconstruction, due to the relatively small shear-wave velocity contrast at the sea bed, which prevents recording of significant P-to-S converted waves.

For the meshing of the model, the narrow velocity range in most parts of the model requires the use of a regular mesh as much as possible for computational efficiency. However, to correctly discretize the shallow-water layer and liquid-solid interface, a p-adapted mesh implemented with a mixed P_0-P_1 interpolation is chosen (Figure 19c): a refined unstructured P_1 layer of cells is used for the first 130 m of the subsurface for accurate modeling of the interface waves at the liquid/solid interface, and for accurate positioning of the sources, located 5 m below the surface, and of the receivers, located on the sea floor. Below this 130 m depth zone, a regular equilateral mesh is used in combination of a P_0 interpolation.

Figure 20 illustrates an example of frequency-domain data (real-part of the complex-value wavefield) at 4Hz. These data are plotted in the source/receiver domain for the full acquisition survey. The diagonal part of the figure represents the collocation of the source and the closest receiver, as the source moves in the acquisition. The Figure 21 illustrates a time-domain shot-gather for the 3 components of the sensors and a source located at position 4 km : the frequency-domain solutions at all the receiver positions have been computed for the single source at all the frequencies of the source spectrum, between 0 and

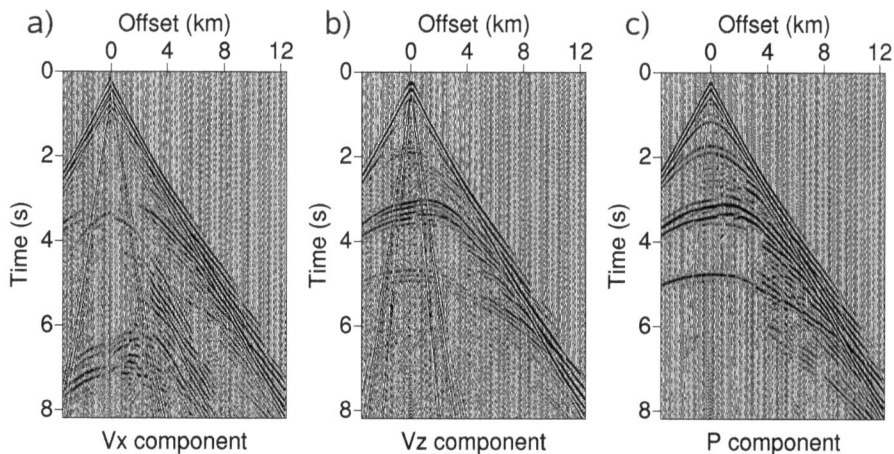

Fig. 21. Time-domain shot-gather for a source located at position 4 km. The data are computed from frequency-domain simulation and Fourier transformed in the time-domain for (a) the horizontal geophone, (b) the vertical geophone and (c) the hydrophone components.

13 Hz. These frequency-domain complex-values data have then been Fourier transform to the time-domain. The time-domain show specific properties of propagation in such environments : the hydrophone and the vertical geophones are mainly sensitive to P-wave arrivals that dominate the elastic propagation in soft-sediment zones. The horizontal geophone allows however to record some late P-to-S conversions, which could be used to image the V_S model from seismic imaging methods.

7. Conclusion

We have presented mainly two families of techniques for solving partial differential equations for elastodynamics: some finite-differences formulations in both time domain and frequency domain and some finite-element methods also in both time domain and frequency domain. Both approaches have appealing features, especially when considering seismic imaging where numerous forward problems should be performed. Such classification helps to understand the advantages and limitations of each particular method to model a specific physical phenomenon

The discretization of the strong formulation of the partial differential equations has been presented through finite-difference techniques. These approaches are easy to implement and quite flexible. They are currently the methods of choice for large-scale modelling and inversion in exploration geophysics, especially in the marine environment. They may however demand a very fine discretization when the earth model contains large contrasts; and accurately modelling the responses around a sharp interface is quite challenging. We have introduced various perspectives as summation-by-parts formulation or the immersed boundary approach as well as simple mesh deformation might broaden the use of finite-differences techniques by avoiding the stair-case approximation.

The weak formulation expressed in the finite-element methods has been considered under the specific family of Discontinuous Galerkin approaches. The use of test functions gives us more freedom and the integral form provides us flexibility in the meshing. However, they lead to numerical challenges: they are more difficult to implement than the finite-difference method, they are often more expensive in computational time and memory, and they are more complicated to use because the accuracy of the response depends on the quality of the meshing. Therefore, they are not intensively used for seismic imaging and are until now more oriented to seismic modeling in the final reconstructed model.

It should be noticed that attempts exist to combine the advantages of these methods in one approach for computing elastic fields, at least for specific applications. Even, one can think that decoupling the inverse problem procedure and the forward problem is possible: we can flip-flop between the two forward problem formulations inside iterations of the inverse problem.

When the modelling method serves as the kernel of an inversion algorithm, additional constrains generally appear because the gradient of the misfit functional needs to be evaluated. The choice of the modelling approach notably depends (1) on the needed accuracy, (2) the efficiency in evaluating the solution and the gradient of the misfit functional in an inversion algorithm, and (3) the simplicity of use.

Finally, the practical implementation shall probably be adapted to the data acquisition. Densely sampled acquisition in exploration geophysics with or without blending, or in lithospheric investigation with the recent deployment of sensors such as the USarray experiment challenges our modelling choice. This seems to indicate that development in modelling and associated inversion approaches remain crucial to improve our knowledge of the subsurface, notably by extracting more information from the, ever larger, recorded data sets.

8. Acknowledgements

We are gratefull to René-Édouard Plessix (SHELL) and Henri Calandra (TOTAL) for fruitful discussions. This work was partially performed using HPC resources from GENCI-[CINES/IDRIS] (Grant 2010-046091 & Grant 2010-082280). Facilities from mesocentres SIGAMM in Nice and CIMENT in Grenoble are also greatly acknowledged. Diego Mercerat received support from the European Research Council (ERC) Advanced grant 226837, and a Marie Curie Re-integration grant (project 223799).

9. References

Aagaard, B. T., Hall, J. F. & Heaton, T. H. (2001). Characterization of near-source ground motion with earthquake simulations, *Earthquake Spectra* 17: 177–207.

Abarbanel, S., Gottlieb, D. & Hesthaven, J. S. (2002). Long-time behavior of the perfectly matched layer equations in computational electromagnetics, *Journal of scientific Computing* 17: 405–422.

Ainsworth, M., Monk, P. & Muniz, W. (2006). Dispersive and Dissipative Properties of Discontinuous Galerkin Finite Element Methods for the Second-Order Wave Equation, *Journal of Scientific Computing* 27(1-3): 5–40.

Akcelik, V., Bielak, J., Biros, G., Epanomeritakis, I., Fernandez, A., Ghattas, O., Kim, E. J., Lopez, J., O'Hallaron, D., Tu, T. & Urbanic, J. (2003). High resolution forward and inverse earthquake modeling on terascale computers, *SC '03: Proceedings of the 2003 ACM/IEEE conference on Supercomputing*, IEEE Computer Society, Washington, DC, USA, p. 52.

Aki, K. & Richards, P. G. (2002). *Quantitative seismology, theory and methods, second edition*, University Science Books, Sausalito,California.

Alterman, Z. & Karal, F. C. (1968). Propagation of elastic waves in layered media by finite-difference methods, *Bulletin of the Seismological Society of America* 58: 367–398.

Aminzadeh, F., Brac, J. & Kunz, T. (1997). *3-D Salt and Overthrust models*, SEG/EAGE 3-D Modeling Series No.1.

Aoi, S. & Fujiwara, H. (1999). 3D finite-difference method using discontinuous grids, *Bulletin of the Seismological Society of America* 89: 918–930.

Aoyama, Y. & Nakano, J. (1999). *RS/6000 SP: Practical MPI Programming*, Red Book edn, IBM Corporation, Texas.

Basabe, J. D., Sen, M. & Wheeler, M. (2008). The interior penalty discontinuous galerkin method for elastic wave propagation: grid dispersion, *Geophysical Journal International* 175: 83–93.

Bécache, E., Petropoulos, P. G. & Gedney, S. G. (2004). On the long-time behavior of unsplit perfectly matched layers, *IEEE Transactions on Antennas and Propagation* 52: 1335–1342.

Ben Jemaa, M., Glinsky-Olivier, N., Cruz-Atienza, V. M. & Virieux, J. (2009). 3D Dynamic rupture simulations by a finite volume method, *Geophysical Journal International* 178: 541–560.

Ben Jemaa, M., Glinsky-Olivier, N., Cruz-Atienza, V., Virieux, J. & Piperno, S. (2007). Dynamic non-planar crack rupture by a finite volume method, *Geophysical Journal International* 172(1): 271–285.

Berenger, J.-P. (1994). A perfectly matched layer for absorption of electromagnetic waves, *Journal of Computational Physics* 114: 185–200.

Berge-Thierry, C., Cotton, F., Scotti, O., Pommera, D. & Fukushima, Y. (2003). New empirical response spectral attenuation laws for moderate european earthquakes, *Journal of Earthquake Engineering* 7(2): 193–222.

Bland, D. (1960). *The theory of linear viscoelasticity*, Pergamon Press, Oxford.

Bohlen, T. & Saenger, E. H. (2006). Accuracy of heterogeneous staggered-grid finite-difference modeling of Rayleigh waves, *Geophysics* 71: 109–115.

Bolt, B. & Smith, W. (1976). Finite element computation of seismic anomalies from bodies of arbitrary shape, *Geophysics* 41: 145–150.

Brossier, R. (2009). *Imagerie sismique à deux dimensions des milieux visco-élastiques par inversion des formes d'onde: développements méthodologiques et applications.*, PhD thesis, Université de Nice-Sophia-Antipolis.

Brossier, R. (2011). Two-dimensional frequency-domain visco-elastic full waveform inversion: Parallel algorithms, optimization and performance, *Computers & Geosciences* 37(4): 444 – 455.

Brossier, R., Etienne, V., Operto, S. & Virieux, J. (2010). Frequency-domain numerical modelling of visco-acoustic waves based on finite-difference and finite-element discontinuous galerkin methods, *in* D. W. Dissanayake (ed.), *Acoustic Waves*, SCIYO, pp. 125–158.

Brossier, R., Operto, S. & Virieux, J. (2009). Seismic imaging of complex onshore structures by 2D elastic frequency-domain full-waveform inversion, *Geophysics* 74(6): WCC63–WCC76.

Brossier, R., Operto, S. & Virieux, J. (2010). Which data residual norm for robust elastic frequency-domain full waveform inversion?, *Geophysics* 75(3): R37–R46.

Brossier, R., Virieux, J. & Operto, S. (2008). Parsimonious finite-volume frequency-domain method for 2-D P-SV-wave modelling, *Geophysical Journal International* 175(2): 541–559.

Cagniard, L. (1962). *Reflection and refraction of progressive seismic waves*, McGraw-Hill (translated from Cagniard 1932).

Červený, V. (2001). *Seismic Ray Theory*, Cambridge University Press, Cambridge.

Chaljub, E., Capdeville, Y. & Vilotte, J.-P. (2003). Solving elastodynamics in a fluid-solid heterogeneous sphere: a parallel spectral element approximation on non-conforming grids, *Journal of Computational Physics* 187: 457–491.

Chaljub, E., Komatitsch, D., Vilotte, J.-P., Capdeville, Y., Valette, B. & Festa, G. (2007). Spectral element analysis in seismology, *in* R.-S. Wu & V. Maupin (eds), *Advances in Wave Propagation in Heterogeneous Earth*, Vol. 48 of *Advances in Geophysics*, Elsevier - Academic Press, London, pp. 365–419.

Chapman, C. (2004). *Fundamentals of seismic waves propagation*, Cambridge University Press, Cambridge, England.

Chew, W. C. & Liu, Q. H. (1996). Perfectly matched layers for elastodynamics: a new absorbing boundary condition, *Journal of Computational Acoustics* 4: 341–359.

Chew, W. C. & Weedon, W. H. (1994). A 3-D perfectly matched medium from modified Maxwell's equations with stretched coordinates, *Microwave and Optical Technology Letters* 7: 599–604.

Chin-Joe-Kong, M. J. S., Mulder, W. A. & Van Veldhuizen, M. (1999). Higher-order triangular and tetrahedral finite elements with mass lumping for solving the wave equation, *Journal of Engineering Mathematics* 35: 405–426.

Clayton, R. & Engquist, B. (1977). Absorbing boundary conditions for acoustic and elastic wave equations, *Bulletin of the Seismological Society of America* 67: 1529–1540.

Cockburn, B., Li, F. & Shu, C. W. (2004). Locally divergence-free discontinuous Galerkin methods for the Maxwell equations, *Journal of Computational Physics* 194: 588–610.

Cognon, J. H. (1971). Electromagnetic and electrical modelling by finite element methods, *Geophysics* 36: 132–155.

Collino, F. & Monk, P. (1998). Optimizing the perfectly matched layer, *Computer methods in Applied Mechanics and Engineering* 164: 157–171.

Collino, F. & Tsogka, C. (2001). Application of the perfectly matched absorbing layer model to the linear elastodynamic problem in anisotropic heterogeneous media, *Geophysics* 66: 294–307.

Coutant, O., Doré, F., Nicollin, F., Beauducel, F. & Virieux, J. (2010). Seismic tomography of the Soufrière de Guadeloupe upper geothermal system, *Geophysical Research Abstracts*, Vol. 12, EGU.

Cruz-Atienza, V. (2006). *Rupture dynamique des faille non-planaires en différences finies*, PhD thesis, Université de Nice-Sophia Antipolis.

Cruz-Atienza, V., Virieux, J., Khors-Sansorny, C., Sardou, O., Gaffet, S. & Vallée, M. (2007). Estimation quantitative du PGA sur la Côte d'Azur, *Ecole Centrale Paris 1*, Association Française du Génie Parasismique (AFPS), p. 8.

Dablain, M. (1986). The application of high order differencing for the scalar wave equation, *Geophysics* 51: 54–66.

de Hoop, A. (1960). A modification of Cagniard's method for solving seismic pulse problems, *Applied Scientific Research* 8: 349–356.

Delcourte, S., Fezoui, L. & Glinsky-Olivier, N. (2009). A high-order discontinuous Galerkin method for the seismic wave propagation, *ESAIM: Proc.* 27: 70–89. URL: *http://dx.doi.org/10.1051/proc/2009020*

Drossaert, F. H. & Giannopoulos, A. (2007). A nonsplit complex frequency-shifted PML based on recursive integration for FDTD modeling of elastic waves, *Geophysics* 72(2): T9–T17.

Druskin, V. & Knizhnerman, L. (1988). A spectral semi-discrete method for the numerical solution of 3-d non-stationary problems, *electrical prospecting: Izv. Acad. Sci. USSR, Physics of Solid Earth (Russian, translated into English)* 8: 63–74.

Duff, I. S., Erisman, A. M. & Reid, J. K. (1986). *Direct methods for sparse matrices*, Clarendon Press, Oxford, U. K.

Dumbser, M. & Käser, M. (2006). An Arbitrary High Order Discontinuous Galerkin Method for Elastic Waves on Unstructured Meshes II: The Three-Dimensional Isotropic Case, *Geophysical Journal International* 167(1): 319–336.

Dumbser, M., Käser, M. & Toro, E. (2007). An Arbitrary High Order Discontinuous Galerkin Method for Elastic Waves on Unstructured Meshes V: Local Time Stepping and p-Adaptivity, *Geophysical Journal International* 171(2): 695–717.

Erlangga, Y. A. & Herrmann, F. J. (2008). An iterative multilevel method for computing wavefields in frequency-domain seismic inversion, *Expanded Abstracts*, Soc. Expl. Geophys., pp. 1956–1960.

Etienne, V., Chaljub, E., Virieux, J. & Glinsky, N. (2010). An hp-adaptive discontinuous Galerkin finite-element method for 3D elastic wave modelling, *Geophysical Journal International* 183(2): 941–962.

Etienne, V., Virieux, J. & Operto, S. (2009). A massively parallel time domain discontinuous Galerkin method for 3D elastic wave modeling, *Expanded Abstracts, 79th Annual SEG Conference & Exhibition, Houston*, Society of Exploration Geophysics.

Festa, G. & Nielsen, S. (2003). PML absorbing boundaries, *Bulletin of Seismological Society of America* 93: 891–903.

Festa, G. & Vilotte, J.-P. (2005). The newmark scheme as a velocity-stress time stag- gering: An efficient PML for spectral element simulations of elastodynamics, *Geophysical Journal International* 161(3): 789–812.

Frey, P. & George, P. (2008). *Mesh Generation*, ISTE Ltd & John Wiley Sons Inc, London (UK) & Hoboken (USA).

Futterman, W. (1962). Dispersive body waves, *Journal of Geophysics Research* 67: 5279–5291.

Galis, M., Moczo, P. & Kristek, J. (2008). A 3-D hybrid finite-difference - finite-element viscoelastic modelling of seismic wave motion, *Geophysical Journal International* 175: 153–184.

Goldman, M. & Stover, C. (1983).   Finite-difference calculations of the transient field of an axially symmetric earth for vertical magnetic dipole excitation, *Geophysics* 48: 953–963.

Gottschamer, E. & Olsen, K. B. (2001).   Accuracy of the explicit planar free-surface boundary condition implemented in a fouth-order staggered-grid velocity-stress finite-difference scheme, *Bulletin of the Seismological Society of America* 91: 617–623.

Graves, R. (1996).   Simulating seismic wave propagation in 3D elastic media using staggered-grid finite differences, *Bull. Seismol. Soc. Am.* 86: 1091–1106.

Günther, T., Rücker, C. & Spitzer, K. (2006). Three-dimensional modelling and inversion of dc resistivity data incorporating topography â€" II: Inversion, *Geophysical Journal International* 166: 506–517.

Hairer, E., Lubich, C. & Wanner, G. (2002). *Geometrical numerical integration: structure-preserving algorithms for ordinary differential equations*, Springer-Verlag.

Hammer, P. & Stroud, A. (1958).   Numerical evaluation of multiple integrals, *Mathematical Tables Other Aids Computation* 12: 272–280.

Hayashi, K., Burns, D. R. & Toks^f oz, M. (2001). Discontinuous-grid finite-difference seismic modeling including surface topography, *Bulletin of the Seismological Society of America* 96(6): 1750–1764.

Hesthaven, J. & Warburton, T. (2008). *Nodal discontinuous Galerkin methods: algorithms, analysis, and applications*, Springler.

Hestholm, S. (1999).   3-D finite-difference visocelastic wave modeling including surface topography, *Geophysical Journal International* pp. 852–878.

Hestholm, S. & Ruud, B. (1998). 3-D finite-difference elastic wave modeling including surface topography, *Geophysics* pp. 613–622.

Hicks, G. J. (2002). Arbitrary source and receiver positioning in finite-difference schemes using kaiser windowed sinc functions, *Geophysics* 67: 156–166.

Hohmann, G. W. (1988). Numerical modeling for electromagnetic methods of geophysics, *in* M. N. Nabighian (ed.), *Electromagnetic methods in applied geophysics*, Investigations in Geophysics, Society of Exploration Geophysics, pp. 313–363.

Hustedt, B., Operto, S. & Virieux, J. (2004).   Mixed-grid and staggered-grid finite difference methods for frequency domain acoustic wave modelling, *Geophysical Journal International* 157: 1269–1296.

Ichimura, T., Hori, M. & Kuwamoto, H. (2007).   Earthquake motion simulation with multi-scale finite element analysis on hybrid grid, *Bulletin of the Seismological Society of America* 97(4): 1133–1143.

Jacobs, G. & Hesthaven, J. S. (2006). High-order nodal discontinuous Galerkin particle-in-cell methods on unstructured grids, *Journal of Computational Physics* 214: 96–121.

Jo, C. H., Shin, C. & Suh, J. H. (1996). An optimal 9-point, finite-difference, frequency-space 2D scalar extrapolator, *Geophysics* 61: 529–537.

Karypis, G. & Kumar, V. (1998). *METIS - A software package for partitioning unstructured graphs, partitioning meshes and computing fill-reducing orderings of sparse matrices - Version 4.0*, University of Minnesota.

Käser, M. & Dumbser, M. (2008).   A highly accurate discontinuous Galerkin method for complex interfaces between solids and moving fluids, *Geophysics* 73(3): 23–35.

Käser, M., Hermann, V. & de la Puente, J. (2008). Quantitative accuracy analysis of the discontinuous Galerkin method for seismic wave propagation, *Geophysical Journal International* 173(2): 990–999.

Kelly, K., Ward, R., Treitel, S. & Alford, R. (1976). Synthetic seismograms - a finite-difference approach, *Geophysics* 41: 2–27.

Kolsky, H. (1956). The propagation of stress pulses in viscoelastic solids, *Philosophical Magazine* 1: 693–710.

Komatitsch, D., Labarta, J. & Michéa, D. (2008). A simulation of seismic wave propagation at high resolution in the inner core of the Earth on 2166 processors of MareNostrum, *Lecture Notes in Computer Science* 5336: 364–377.

Komatitsch, D. & Martin, R. (2007). An unsplit convolutional perfectly matched layer improved at grazing incidence for the seismic wave equation, *Geophysics* 72(5): SM155–SM167.

Komatitsch, D. & Vilotte, J. P. (1998). The spectral element method: an efficient tool to simulate the seismic response of 2D and 3D geological structures, *Bulletin of the Seismological Society of America* 88: 368–392.

Kommedal, J. H., Barkved, O. I. & Howe, D. J. (2004). Initial experience operating a permanent 4C seabed array for reservoir monitoring at Valhall, *SEG Technical Program Expanded Abstracts* 23(1): 2239–2242.
URL: *http://link.aip.org/link/?SGA/23/2239/1*

Kosloff, D. & Baysal, E. (1982). Forward modeling by a Fourier method, *Geophysics* 47: 1402–1412.

Kosloff, D., Kessler, D., Filho, A., Tessmer, E., Behle, A. & Strahilevitz, R. (1990). Solution of the equations of dynamic elasticity by a Chebychev spectral method, *Geophysics* 55: 464–473.

Kummer, B. & Behle, A. (1982). Second-order finite-difference modelling of SH-wave propagation in laterally inhomogeneous media, *Bulletin of the Seismological Society of America* 72: 793–808.

Kuo, J. & Cho, D.-H. (1980). Transient time-domain electromagnetics, *Geophysics* 45: 271–291.

Kuzuoglu, M. & Mittra, R. (1996). Frequency dependence of the constitutive parameters of causal perfectly matched anisotropic absorbers, *IEEE Microwave and Guided Wave Letters* 6: 447–449.

Levander, A. R. (1988). Fourth-order finite-difference P-SV seismograms, *Geophysics* 53(11): 1425–1436.

LeVeque, R. J. (1986). Intermediate boundary conditions for time-split methods applied to hyperbolic partial differential equations, *Mathematical Computations* 47: 37–54.

Lombard, B. & Piraux, J. (2004). Numerical treatment of two-dimensional interfaces for acoustic and elastic waves, *Journal of Computational Physics* 195: 90–116.

Lombard, B., Piraux, J., Gelis, C. & Virieux, J. (2008). Free and smooth boundaries in 2-D finite-difference schemes for transient elastic waves, *Geophysical Journal International* 172: 252–261.

Luo, Y. & Schuster, G. T. (1990). Parsimonious staggered grid finite-differencing of the wave equation, *Geophysical Research Letters* 17(2): 155–158.

Madariaga, R. (1976). Dynamics of an expanding circular fault, *Bulletin of Seismological Society of America* 66: 639–666.

Marcinkovich, C. & Olsen, K. (2003). On the implementation of perfectly matched layers in a three-dimensional fourth-order velocity-stress finite difference scheme, *Journal of Geophysical Research* 108: doi:10.1029/2002GB002235.

Marfurt, K. (1984). Accuracy of finite-difference and finite-elements modeling of the scalar and elastic wave equation, *Geophysics* 49: 533–549.

Mercerat, E. D., Vilotte, J. P. & Sanchez-Sesma, F. J. (2006). Triangular spectral element simulation of two-dimensional elastic wave propagation using unstructured triangular grids, *Geophysical Journal International* 166: 679–698.

Meza-Fajardo, K. & Papageorgiou, A. (2008). A nonconvolutional, split-field, perfectly matched layer for wave propagation in isotropic and anisotropic elastic media: Stability analysis, *Bulletin of the Seismological Society of America* 98(4): 1811–1836.

Mikhailenko, B., Mikhailov, A. & Reshetova, G. (2003). Numerical viscoelastic modelling by the spectral laguerre method, *Geophysical Prospecting* 51: 37–48.

Moczo, P., Kristek, J., Galis, M., Pazak, P. & Balazovjech, M. (2007). The finite-difference and finite-element modeling of seismic wave propagation and earthquake motion, *Acta Physica Slovaca* 52(2): 177–406.

Mufti, I. R. (1985). Seismic modeling in the implicit mode, *Geophysical Prospecting* 33: 619–656.

Munns, J. W. (1985). The Valhall field: a geological overview, *Marine and Petroleum Geology* 2: 23–43.

Nihei, K. T. & Li, X. (2007). Frequency response modelling of seismic waves using finite difference time domain with phase sensitive detection (TD-PSD), *Geophysical Journal International* 169: 1069–1078.

Operto, S., Virieux, J., Amestoy, P., L'Éxcellent, J.-Y., Giraud, L. & Ben Hadj Ali, H. (2007). 3D finite-difference frequency-domain modeling of visco-acoustic wave propagation using a massively parallel direct solver: A feasibility study, *Geophysics* 72(5): SM195–SM211.

Opršal, I., Matyska, C. & Irikura, K. (2009). The source-box wave propagation hybrid methods: general formulation and implementation, *Geophysical Journal International* 176: 555–564.

Oristaglio, M. & Hohmann, G. W. (1984). Diffusion of electromagnetic fields into a two-dimensional earth: A finite-difference approach, *Geophysics* 49: 870–894.

Pasquetti, R. & Rapetti, F. (2006). Spectral element methods on unstructured meshes: Comparisons and recent advances, *Journal of Scientific Computing* 27: 377–387.

Pitarka, A. (1999). 3D elastic finite-difference modeling of seismic motion using staggered grids with nonuniform spacing, *Bulletin of the Seismological Society of America* 89(1): 54–68.

Plessix, R. E. (2007). A Helmholtz iterative solver for 3D seismic-imaging problems, *Geophysics* 72(5): SM185–SM194.

Plessix, R. E. (2009). Three-dimensional frequency-domain full-waveform inversion with an iterative solver, *Geophysics* 74(6): WCC53–WCC61.

Reed, W. & Hill, T. (1973). Triangular mesh methods for the neuron transport equation, *Technical Report LA-UR-73-479*, Los Alamos Scientific Laboratory.

Remaki, M. (2000). A new finite volume scheme for solving Maxwell's system, *COMPEL* 19(3): 913–931.

Riyanti, C. D., Erlangga, Y. A., Plessix, R. E., Mulder, W. A., Vuik, C. & Oosterlee, C. (2006). A new iterative solver for the time-harmonic wave equation, *Geophysics* 71(E): 57–63.

Riyanti, C. D., Kononov, A., Erlangga, Y. A., Vuik, C., Oosterlee, C., Plessix, R. E. & Mulder, W. A. (2007). A parallel multigrid-based preconditioner for the 3D heterogeneous high-frequency Helmholtz equation, *Journal of Computational physics* 224: 431–448.

Robertsson, J. O. A. (1996). A numerical free-surface condition for elastic/viscoelastic finite-difference modeling in the presence of topography, *Geophysics* 61: 1921–1934.

Roden, J. A. & Gedney, S. D. (2000). Convolution PML (CPML): An efficient FDTD implementation of the CFS-PML for arbitrary media, *Microwave and Optical Technology Letters* 27(5): 334–339.

Rollet, N., Deverchère, J., Beslier, M., Guennoc, P., Réhault, J., Sosson, M. & Truffert, C. (2002). Back arc extension, tectonic inheritance and volcanism in the ligurian sea, western mediterranean, *Tectonics* 21(3).

Rücker, C., Günther, T. & Spitzer, K. (2006). Three-dimensional modeling and inversion of DC resistivity data incorporating topography - Part I: Modeling, *Geophysical Journal International* 166: 495–505.

Saad, Y. (2003). *Iterative methods for sparse linear systems*, SIAM, Philadelphia.

Saenger, E. H., Gold, N. & Shapiro, S. A. (2000). Modeling the propagation of elastic waves using a modified finite-difference grid, *Wave motion* 31: 77–92.

Sears, T., Singh, S. & Barton, P. (2008). Elastic full waveform inversion of multi-component OBC seismic data, *Geophysical Prospecting* 56(6): 843–862.

Seriani, G. & Priolo, E. (1994). Spectral element method for acoustic wave simulation in heterogeneous media, *Finite elements in analysis and design* 16: 337–348.

Si, H. (2006). *TetGen - A Quality Tetrahedral Mesh Generator and Three-Dimensional Delaunay Triangulator - Version 1.4*, University of Berlin.

Sirgue, L., Etgen, J. T. & Albertin, U. (2008). 3D Frequency Domain Waveform Inversion using Time Domain Finite Difference Methods, *Proceedings 70th EAGE, Conference and Exhibition, Roma, Italy*, p. F022.

Sirgue, L. & Pratt, R. G. (2004). Efficient waveform inversion and imaging : a strategy for selecting temporal frequencies, *Geophysics* 69(1): 231–248.

Sourbier, F., Haiddar, A., Giraud, L., Ali, H. B. H., Operto, S. & Virieux, J. (2011). Three-dimensional parallel frequency-domain visco-acoustic wave modelling based on a hybrid direct-iterative solver, *Geophysical Prospecting* Special issue: Modelling methods for geophysical imaging, 59(5), 835-856.

Stekl, I. & Pratt, R. G. (1998). Accurate viscoelastic modeling by frequency-domain finite difference using rotated operators, *Geophysics* 63: 1779–1794.

Symes, W. W. (2007). Reverse time migration with optimal checkpointing, *Geophysics* 72(5): SM213–SM221.

Taflove, A. & Hagness, C. (2000). *Computational electrodynamics: the finite-difference time-domaine method*, Artech House, London, United Kindom.

Tal-Ezer, H., Carcione, J. & Kosloff, D. (1990). An accurate and efficient scheme for wave propagation in linear viscoelastic media, *Geophysics* 55: 1366–1379.

Tam, C. K. & Webb, J. C. (1993). Dispersion-relation-preserving finite difference schemes for computational acoustics, *Journal of Computational Physics* 107: 262–281.

Tessmer, E., Kosloff, D. & Behle, A. (1992). Elastic wave propagation simulation in the presence of surface topography, *Geophysical Journal International* 108: 621–632.

Titarev, V. & Toro, E. (2002). Ader: arbitrary high order godunov approach, *SIAM Journal Scientific Computing* 17: 609–618.

Toulopoulos, I. & Ekaterinaris, J. A. (2006). High-order discontinuous Galerkin discretizations for computational aeroacoustics in complex domains, *AIAA J.* 44: 502–511.

Vigh, D. & Starr, E. W. (2008). Comparisons for Waveform Inversion, Time domain or Frequency domain?, *Extended Abstracts*, pp. 1890–1894.

Virieux, J. (1986). P-SV wave propagation in heterogeneous media, velocity-stress finite difference method, *Geophysics* 51: 889–901.

Virieux, J. & Lambaré, G. (2007). Theory and observations - body waves: ray methods and finite frequency effects, *in* B. Romanovitz & A. Diewonski (eds), *Treatise of Geophysics, volume 1: Seismology and structure of the Earth*, Elsevier.

Virieux, J., Operto, S., Ben Hadj Ali, H., Brossier, R., Etienne, V., Sourbier, F., Giraud, L. & Haidar, A. (2009). Seismic wave modeling for seismic imaging, *The Leading Edge* 28(5): 538–544.

Warner, M., Stekl, I. & Umpleby, A. (2008). 3D wavefield tomography: synthetic and field data examples, *SEG Technical Program Expanded Abstracts* 27(1): 3330–3334.

Wheeler, L. & Sternberg, E. (1968). Some theorems in classical elastodynamics, *Archive for Rational Mechanics Analysis* 31: 51–90.

Williamson, P. R. & Worthington, M. H. (1993). Resolution limits in ray tomography due to wave behavior: Numerical experiments, *Geophysics* 58: 727–735.

Yee, K. S. (1966). Numerical solution of initial boundary value problems involving Maxwell's equations in isotropic media, *IEEE Trans. Antennas and Propagation* 14: 302–307.

Zienkiewicz, O. C., Taylor, R. L. & Zhu, J. Z. (2005). *The Finite Element Method: Its Basis and Fundamentals*, Elsevier, London. 6th edition.

Zienkiewicz, O. & Taylor, R. (1967). *The Finite Element Method for Solid and Structural Mechanics*, McGraw Hill, New York.

Shear Wave Velocity Models Beneath Antarctic Margins Inverted by Genetic Algorithm for Teleseismic Receiver Functions

Masaki Kanao[1] and Takuo Shibutani[2]
[1]National Institute of Polar Research,
Research Organization of Information and Systems, Tokyo,
[2]Disaster Prevention Research Institute,
Kyoto University, Gokasho, Kyoto,
Japan

1. Introduction

Study on shear velocity structure of the crust and the uppermost mantle around Antarctic continent was started in the 1960's by using surface waves of the earthquakes occurred around Antarctic plate (Evison et al., 1960; Kovach and Press, 1961). Permanent seismic stations have been operated from the end of 1980's at Antarctic margins except for the South Pole (SPA) on the continental ice sheet. A majority of the stations were established as the Federation of Digital Seismographic Networks (FDSN; Butler and Anderson, 2008). The FDSN was composed of several national and governmental organizations such as the Global Seismographic Network (GSN) organized by the Incorporated Research Institutions for Seismology (IRIS), the Australian Government (AG), GEOSCOPE by French, Geo Forschungs Netz (GEOFON) by Germany, MEDNET by Italy, PACIFIC21 (Tsuboi, 1995) by Japan, and others.

In recent few years, surface wave tomography studies around Antarctic continent and surrounding oceans have been conducted by using the FDSN data (Roult et al., 1994; Ritzwoller et al., 2001; Danesi and Morelli, 2001; Kobayashi and Zhao, 2004). Enderby Land, particularly around the Napier Mountains, was one of the oldest Archaean cratons with a spatial extent about 500 km (Ellis, 1987; Black et al., 1987). However, surface wave analyses could not provide enough spatial resolution for detail discussion about fine crustal structure. Therefore, it is necessary to achieve smaller-scale heterogeneities in the specified area by using recently available broadband waveform data.

In this chapter, precise shear velocity models of the crust and the uppermost mantle were investigated from teleseismic receiver functions beneath several FDSN stations in Antarctica (Fig. 1; MAW; 67.6°S, 62.9°E, SYO; 69.0°S, 39.6°E, DRV; 66.7°S, 140.0°E, VNDA; 77.5°S, 161.9°E, PMSA; 64.8°S, 64.0°W). The obtained velocity models were discussed in relationship with the regional tectonics and crustal evolution of each terrain around the stations.

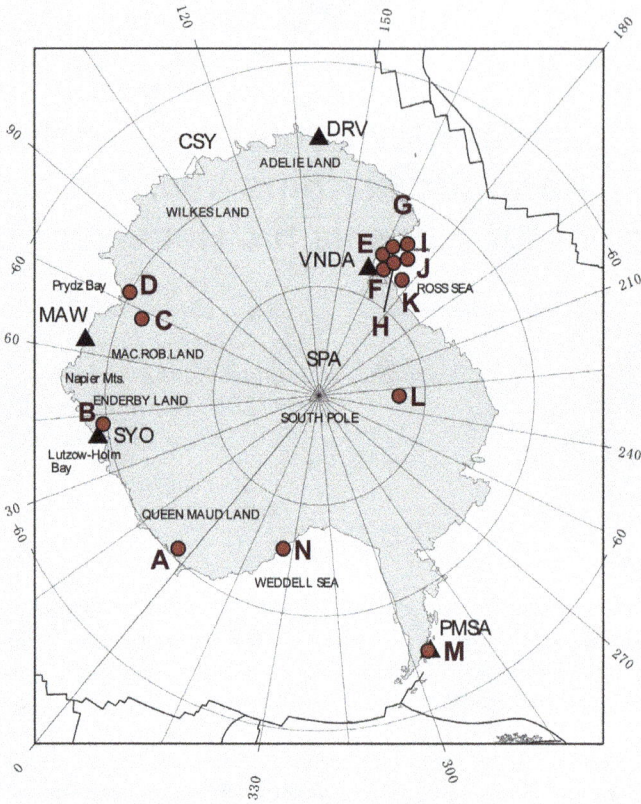

Fig. 1. Map showing the location of analyzed seismic stations by GA inversion (Solid triangles; SYO, MAW, DRV, PMSA and VNDA) in Antarctic continental margin and the related regional localities. Open triangle stations (CSY, SPA and SBA) are planned to be analyzed in future. Solid red circles are stations raveled in Von Frese et al. (1999). Alphabet numerals are location in Antarctica raveled after Von Frese et al., (1999), the same representation as in Fig. 8

2. Methodology and data

The coda parts of teleseismic P-waves contain a significant amount of information on the seismic structure in the vicinity of the recorded stations. The "receiver functions" were defined as the structural response and consisted of P-to-S converted waves and their reverberations, and were most sensitive to the shear velocity beneath the station (Fig. 2). To derive the structural response (receiver functions) beneath the recording station, the source-equalization method (Langston, 1979) was generally applied for the coda part of teleseismic P-waves. The structural response could be isolated from that of the instrument and effective source function. The followings are the procedure to produce the receiver functions.

Three components (V, R, T) of the total response at a station on a teleseismic P wave are expressed as,

$$D_V(t) = I(t) * S(t) * E_V(t)$$
$$D_R(t) = I(t) * S(t) * E_R(t) \quad\quad (1)$$
$$D_T(t) = I(t) * S(t) * E_T(t)$$

where $I(t)$ is the impulse response of the recording system, $S(t)$ is the effective seismic source function, and $E(t)$ implies the impulse response of the earth's structure.

For a steeply incident P wave, source equalization method (Langston, 1979) assumes,

$$E_V(t) = \delta(t) \quad\quad (2)$$

where $\delta(t)$ is the Dirac delta function. From (1) and (2),

$$D_V(t) = I(t) * S(t) \qu\quad\quad (3)$$

Thus the observed structual response $E_R(t)$ and $E_T(t)$ are available by deconvolving $D_V(t)$ from $D_R(t)$ and $D_T(t)$.

What are the receiver functions?

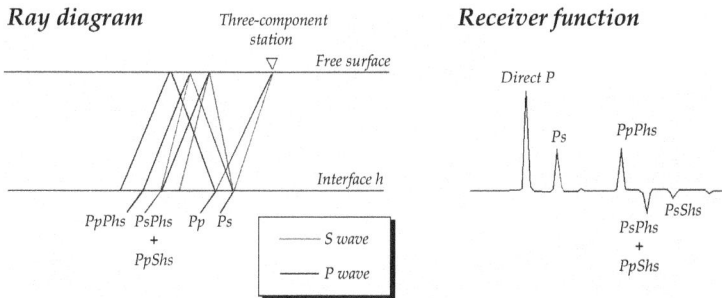

Creating receiver functions

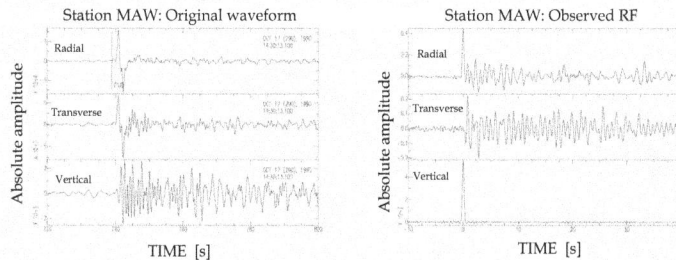

Fig. 2. (upper) The receiver functions (RF, the crustal response) are consist of P-to-S converted waves and their reverberations, which are most sensitive to the shear velocity beneath the station. (lower) the observed receiver functions $(E_R(t)$ and $E_T(t))$ can be obtained by deconvolving the original waveforms $D_V(t)$ from $D_R(t)$ and $D_T(t)$

Since the receiver functions are sensitive to *P*-to-*S* conversions through the interfaces beneath the recording station, the waveform inversion result could produce a shear velocity structure (Owens et al., 1984; Kind et al., 1995). By applying these conventional procedures, receiver functions were obtained at five permanent FDSN stations in Antarctica.

Here, we introduce the actual procedure to create receiver functions as an example for the station MAW. Before the inversion, we used the stacked receiver functions for all 20 radial components in the backazimuth group within 70°. The incoherent noise could be suppressed by stacking the waveforms, while the coherent signals were enhanced. Dataset of teleseismic waveforms for the other four stations were the same as used in the linearized inversion analyses by Kanao et al. (2002).

Inversion of the receiver functions to recover crustal and uppermost mantle structure have been widely recognized as sensitive to the starting model if a conventional linearization scheme was employed (Ammon et al., 1990). Prior to this study, a linearized time domain inversion was applied to determine the velocity model for several Antarctic stations (Kanao, 1997; Kanao et al., 2002). The starting model dependency might be excluded by employing a non-linear inversion scheme based on a Genetic Algorithm (GA; Sambridge and Drijkoningen, 1992; Shibutani et al., 1996).

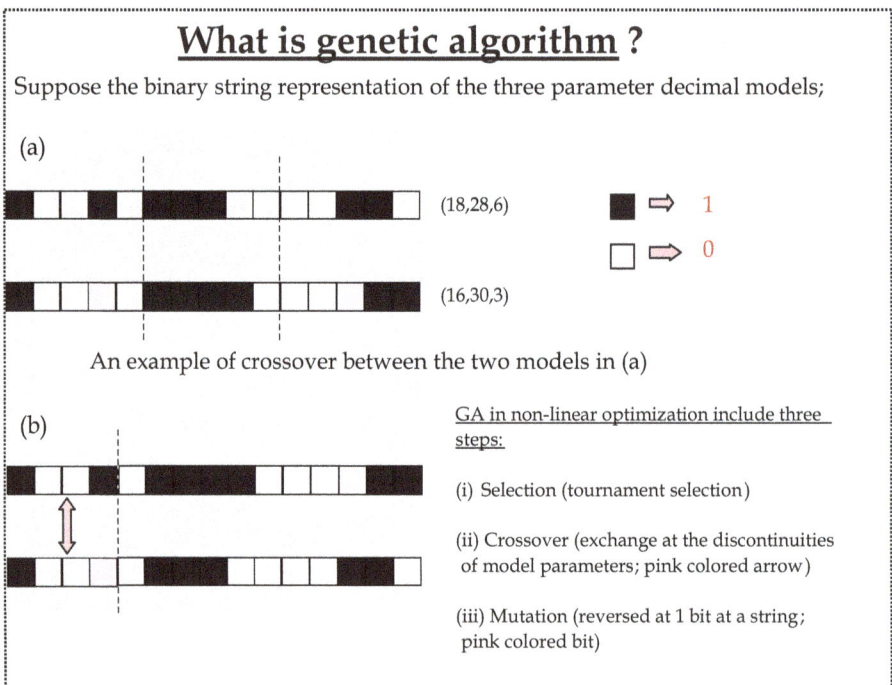

Fig. 3. Schematic illustration for Genetic Algorithm (GA). GA in non-linear optimization include three steps; (i) Selection (tounament selection), (ii) Crossover (exchange at the discontinuities of model parameters) and (iii) Mutation (reversed at 1 bit at a string), respectively. (modified after Sambridge and Drijkoningen (1992))

The GA approach makes use of a 'cloud' or 'population' of the models to minimize the dependence on a starting model; a set of 'biological' analogues are used to produce new generations of the models from previous generations, with preferential development of the models with a good fit between observed and theoretical receiver functions. Figure 3 shows a schematic illustration for GA. GA in non-linear optimization included three steps; (i) Selection (tounament selection), (ii) Crossover (exchange at the discontinuities of model parameters) and (iii) Mutation (reversed at 1 bit at a string), respectively.

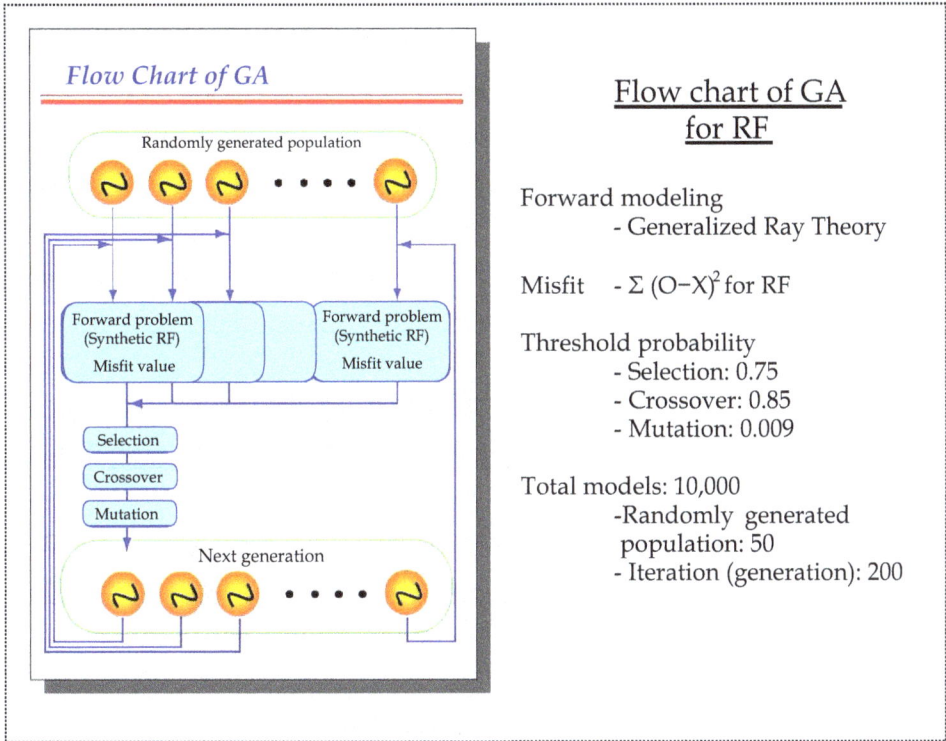

Flow Chart of GA

Randomly generated population

Forward problem (Synthetic RF)
Misfit value

Forward problem (Synthetic RF)
Misfit value

Selection
Crossover
Mutation

Next generation

Flow chart of GA for RF

Forward modeling
- Generalized Ray Theory

Misfit - $\Sigma (O-X)^2$ for RF

Threshold probability
- Selection: 0.75
- Crossover: 0.85
- Mutation: 0.009

Total models: 10,000
-Randomly generated population: 50
- Iteration (generation): 200

Fig. 4. Flow chart of Genetic Algorithm (GA) for receiver function inversion. Beginning with a randomly generated initial population and corresponding misfit values which are defined by square sum of the difference between the receiver function predicted for each model and that obtained from observed waveforms, succeeding populations are created by selection, crossover and mutation

In this study, a non-linear GA was applied for the stacked radial receiver functions of each station. Figure 4 represented a flow chart of GA for receiver function inversion. Beginning with a randomly generated initial population and corresponding misfit values, which were defined by the square sum of the difference between the receiver function predicted for each model and that obtained from observed waveforms, succeeding populations were created by selection, crossover and mutation procedures.

The approach provided a good sampling of the model space, and enabled the estimation of the shear-wave speed distribution within the crust, along with an indication of the ratio between Vp and Vs. Many models with an acceptable fit to the data were generated during the inversion, and a stable crustal model was produced by employing a weighted average of the best 1,000 models encountered in the development of GA. The weighting criterion was based on the inverse of the misfit for each model, so that the best-fitting model could have the greatest influence.

For the inversion, the crust and the uppermost mantle down to 60 km were assumed to be composed from five major layers (Table 1). The model parameters in each layer were the thickness, Vs at the upper and the lower boundaries, in addition to the Vp/Vs ratio. The Vs for each layer was constructed by linearly connecting the values at the upper and the lower boundaries, so as to give a sequence of constant velocity-gradient segments separated by velocity discontinuities. The thickness and the upper and the lower limit in each layer were defined after the averaged continental crust. 'Q_α'and 'Q_β' values were assumed to be fixed in each layer on the basis of Coda-Q inversion results after Kanao and Akamatsu (1995). A smoothness constraint in the inversion was implemented by minimizing a roughness norm of the velocity model (Ammon et al., 1990).

		Basement	Crust			Mantle
			upper	middle	lower	
Thickness	lower	0.0	5.0	5.0	5.0	5.0
(km)	upper	5.0	20.0	20.0	20.0	20.0
Vs (upper)	lower	2.90	3.10	3.40	3.70	4.00
(km/s)	upper	3.90	4.10	4.40	4.70	5.00
Vs (lower)	lower	2.90	3.10	3.40	3.70	4.00
(km/s)	upper	3.90	4.10	4.40	4.70	5.00
Vp/Vs	lower	1.65	1.65	1.65	1.65	1.70
	upper	2.00	1.80	1.80	1.80	1.90
Q_α		200	300	500	800	1360
Q_β		80	120	200	300	600

Table 1. Model parameters in GA receiver function inversion. 'Vs (upper)' and 'Vs (lower)' are the S wave velocity at the upper and the lower boundaries of each layer. The 'Lower' and the 'upper' for four variables indicate the lower and the upper bounds. The thickness and the upper and the lower limit in each layer were defined after the averaged continental crust. 'Q_α'and 'Q_β' were assumed to be fixed in each layer by referring from Coda-Q inversion results after Kanao and Akamatsu (1995)

After examining the trade-off curves between the model roughness and waveform-fit residuals, we selected the most suitable pair of the above parameters. A number of iterations up to 200 times were conducted in the inversion in order to reduce the waveform-fit residuals (misfit-values) to an acceptable value, and the most stable solutions were adopted as the final models (Fig. 5). We obtained 50 population models for the every iteration. In total, we selected 10,000 models to determine the best fitted.

The waveform fits between synthetic and observed receiver functions were generally adopted, implying the adequate inversion procedures with reasonable smoothness constrained. Figure 6 represented the synthetic radial receiver functions by assuming the

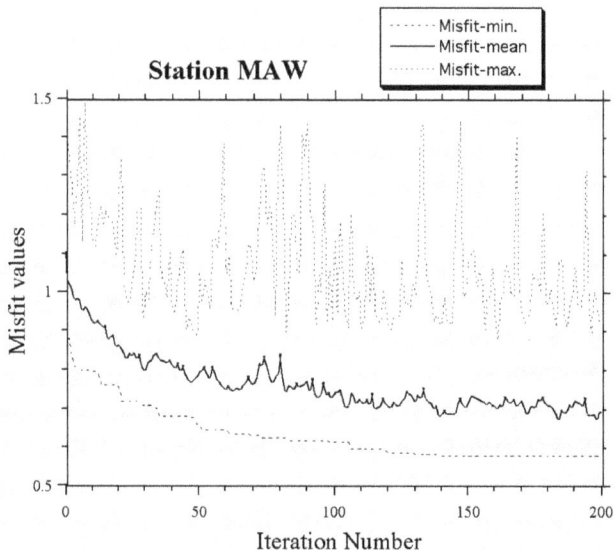

Fig. 5. Misfit values vs. the number of iteration during the GA inversion for an example of MAW. Variations in the mean, the minimum and the maximum misfit values for each population are drawn to be reached into the stable values

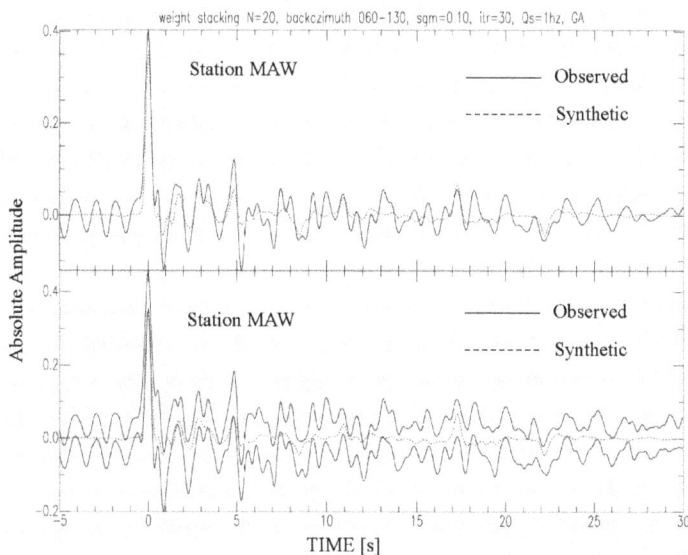

Fig. 6. Synthetic radial receiver functions by assuming the S-wave models and the Vp/Vs ratio determined by the averaged one for the best 1,000 models in the GA inversion (broken traces) compared with the observed mean (upper solid trace) and +/-1 standard error bounding (lower two solid traces) of the stacked receiver functions at MAW

S-wave models and the Vp/Vs ratio determined by the averaged one for the best 1,000 models in GA inversion (broken traces), compared with the observed mean (upper solid trace) and +/-1 standard error bounding (lower two solid traces) of the stacked receiver functions.

There were several noticeable later phases for all traces after the *P*-arrival. For example, large amplitudes were identified around 4-5 s, which were considered to be the directly converted *Ps* at the Moho discontinuity. Intra-crustal converted phases were identified around 1-2 s and 2.5-3.5 s, implying the mid-crustal velocity discontinuities. Later phases, after around 7 s, had a rather worse waveform fitting compared with the earlier phases, because of relatively poor signal-to-noise ratios for these later phases.

3. Results and discussion

In this section, we discussed the resultant shear velocity models for the individual FDSN station. Here, the averaged Vs models for the best 1000 misfits in GA inversion were mainly discussed (red lines in Figs. 7a, 7b and 7c).

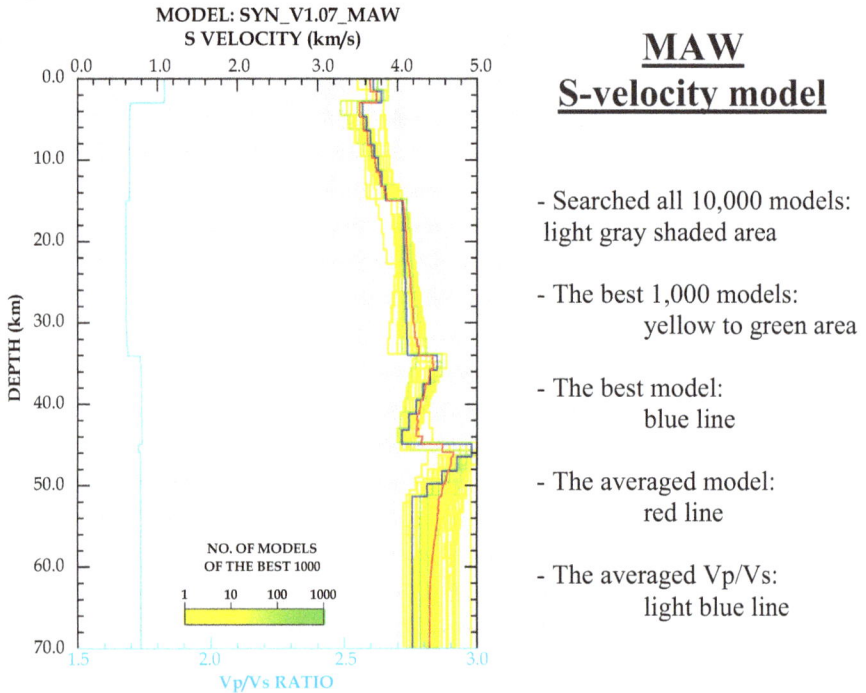

Fig. 7a. Seismic S-velocity models for MAW by GA inversion. For the S-wave velocity, all 10,000 models searched in GA inversion are shown as the light gray shaded area. The best 1,000 models are represented by the yellow to green area. The best model and the averaged model for the best 1,000 are shown by the blue line and the red line, respectively. For the Vp/Vs ratio, the light blue solid line corresponds to the averaged model

The inverted velocity model beneath MAW had a very sharp Moho discontinuity at a depth of 44 km (Fig. 7a). A discontinuity between the middle and the lower crust were recognized at 34 km depth. In Mac. Robertson Land, where MAW is located (Fig. 1), late-Proterozoic metamorphic events generated the granulite facies rocks in upper part of the crust (Tingey, 1982; Sheraton et al., 1987). The sharp and fairly deeper Moho around 44 km depth might have a relationship with the metamorphism of the Rayner Complex besides the Archaean Napier Complex. The intrusion of charnockites around MAW was an evidence of the compression tectonic setting in the Proterozoic mobile belt (Young and Ellis, 1991). Depletions of heavy rare-earth elements in the low-Ti charnockites suggested that garnet was a residual phase in partial melting, which required high pressures and an over-thickened crust. The deeper crustal thickness obtained from GA inversion at MAW appeared to be correspond with a signature of the crustal root what now have been remained as the deepened architecture comparing with the adjacent areas in Enderby Land.

The resultant velocity model around DRV (Fig. 7b, left) indicated a fairly sharp Moho at depths of 28 km. A high-velocity zone appeared in the upper and the lower crustal depths. A relatively lower velocity zone was obtained at depths around 20 km, which lied between the above two high-velocity zones. The velocities of the topmost mantle had lower values of about 4.2 km/s. In Adelie Land, where DRV is belonging (Fig. 1), a metamorphic event occurred in early-Proterozoic age (Bellair and Delbos, 1962). A rather sharp Moho and fluctuations of the crustal velocities might had been developed during the metamorphic event of the Adelie Land. In addition, high velocity zones in the upper crust together with a low-velocity discontinuity in the middle crust might be related to the early-Proterozoic tectonothermal activities.

A sharp Moho discontinuity was determined approximately at 40 km beneath SYO (Fig. 7b, right), in the Lüzow-Holm Bay region. The Moho depth was consistent with that obtained from previous large scale deep refraction / wide-angle reflection surveys around the region (Ikami and Ito, 1986; Yoshii et al., 2004). Velocity jumps were identified at 12 km and 20 km depths, which corresponded to the discontinuities between the upper-middle crust and middle-lower crust, respectively. The latter discontinuity between the middle and the lower crust significantly coincided with the depths by the recent refraction / wide-angle reflection study around the SYO (Miyamachi et al., 2003). High velocity zones in the upper crust were presumably corresponding to the granulite facies metamorphic rocks appeared in surface geology. The obtained velocities in the upper part of the crust were consistent with the velocities of granulite facies rocks found from high-pressure laboratory measurements (Christensen and Mooney, 1995; Kitamura et al., 2001). The considerable crustal evolution models to explain the velocity variations within the crust might be related to the compressional stress during the early-Paleozoic metamorphism in the Lützow-Holm Bay region (Hiroi et al., 1991; Shiraishi et al., 1994).

As for the Antarctic Peninsular, very broad Moho discontinuity was found around 36 km depths beneath PMSA (Fig. 7c, left). Several previously conducted refraction / wide-angle reflection experiments had revealed a complicated Moho topography around the region (Sroda et al., 1997; Grad et al., 2002). They determined the thickness of the crust in the range of 36-42 km at coastal area of the Peninsula, in contrast, decreased to about 25-28 km toward the Pacific Ocean. The dipping Moho obtained from our results supported a possibility of the transition zone between the oceanic and continental crust in the Antarctic Peninsula.

S-velocity model : DRV, SYO

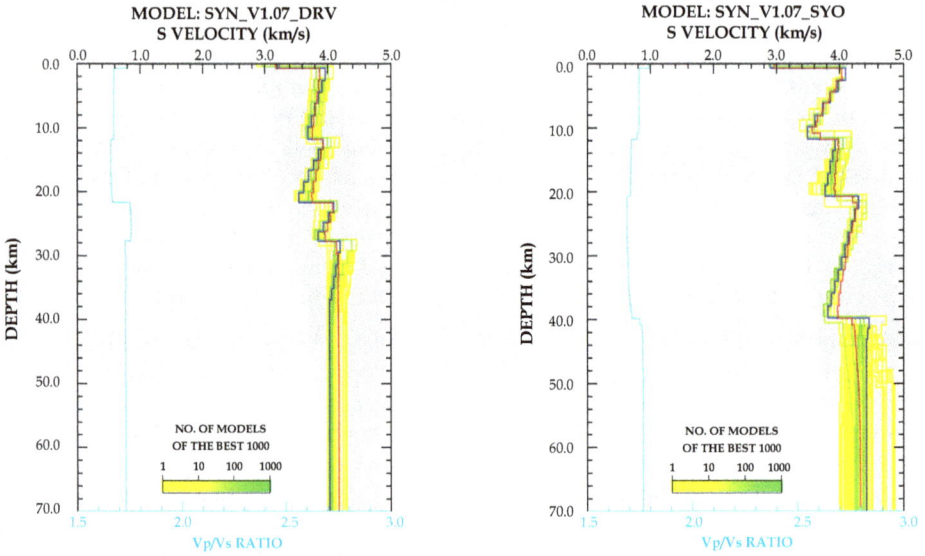

Fig. 7b. Seismic S velocity models for DRV and SYO by GA inversion

S-velocity model : PMSA, VNDA

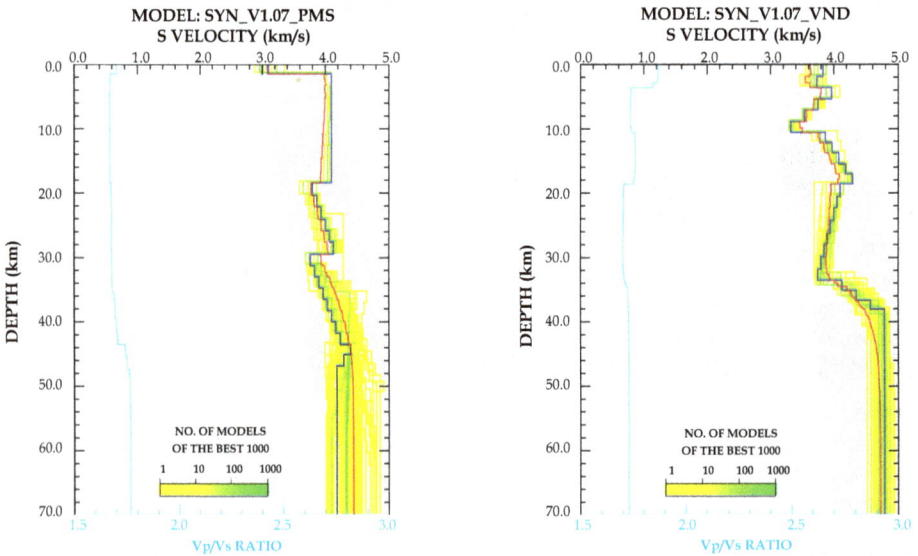

Fig. 7c. Seismic S velocity models for PMSA and VNDA by GA inversion

Since the GA inversion applied for this study had assumed a uniformed structure composed of the flat-lying layers, the dipping structure cannot be directly identified. However, the obtained crust-mantle boundary was considered to reflect the averaged structure for the complicated regime in the vicinity of the Moho discontinuity.

Broadening low velocity zones around 30 km depths and transitional Moho with few km widths at VNDA (Fig. 7c, right) might be involved in the uplift mechanism on the West Antarctic Rift System (WARS) nearby the Trans Antarctic Mountains (TAM) (Smith and Drewry, 1984; Stern and ten Brink, 1989; Ten Brink et al., 1997). Around station VNDA in the Terra Nova Bay region, the Moho depth was already estimated from Ps converted phases of the receiver functions by temporary seismic array data (Di Bona et al., 1997). They pointed out a thinned crust with thickness drastically varied from 17 to 29 km, which implied a transitional zone between East and West Antarctica, which crossing the WARS. The other seismic refraction data indicated the same regime of the Moho depths involving the crustal rift system at TAM (Vedova et al., 1997). Wiens et al. (2003; 2006) conducted the TransAntarctic Mountains SEISmic experiment (TAMSEIS) around the region and obtained a detailed distribution of the crustal thickness by receiver function analyses (Lawrence et al., 2006). Their results also indicated relatively shallow Moho depths together with low velocity zones beneath VNDA.

Figure 8 demonstrated a comparison of the Moho depths by three different methods of seismic refraction studies, gravity-based estimates (after Von Frese et al., 1999), together with receiver function GA inversion by this study. In spite of the existence of small differences in estimating the Moho depths between three methods, it might be mentioned that a principal difference between the East and West Antarctica, as well as the Antarctic Peninsula, was remarkably identified. In order to establish a crustal model of the whole regions in Antarctica, available broadband waveform data of the other seismic stations, such as SPA (90.0°S), CSY (66.3°S, 110.5°E), SBA (77.8°S, 166.8°E) and the other temporary stations should be compiled for comparison.

During the International Polar Year (IPY 2007-2008), a major geo-science program had been conducted such as the 'Antarctica`s GAmburtsev Province / GAmburtsev Mountain SEISmic experiment (AGAP/GAMSEIS)' (Wiens, 2007). The AGAP/GAMSEIS was an internationally coordinated deployment with few tens of broadband seismographs over the huge area of continental ice sheet in East Antarctica. The investigations on the high plateau inside the ice covered continent could surely provide detail information on the crustal thickness and mantle structure (Hansen et al., 2010). In contrast, the 'Polar Earth Observing Network (POLENET; http://www.polenet.org/)' was another major contribution to the IPY by establishing a geophysical network mostly weighted on West Antarctica.

The accumulated seismic data during the IPY will be utilized to clarify the heterogeneous structure of the crust and upper mantle, as well as the Earth's deep interior, including the features such as the Core-Mantle-Boundary (CMB) and the lowermost mantle layer (D" zone). The broadband seismic arrays in the Antarctic at IPY and beyond have been conducting a significant contribution to FDSN as viewed from high southern latitude. Mapping of the crustal velocities beneath the whole Antarctic continent could firmly address for the advance in interpreting the difference of structure and tectonics in various terrains of Gondwana super-continent.

Moho depth in Antarctica

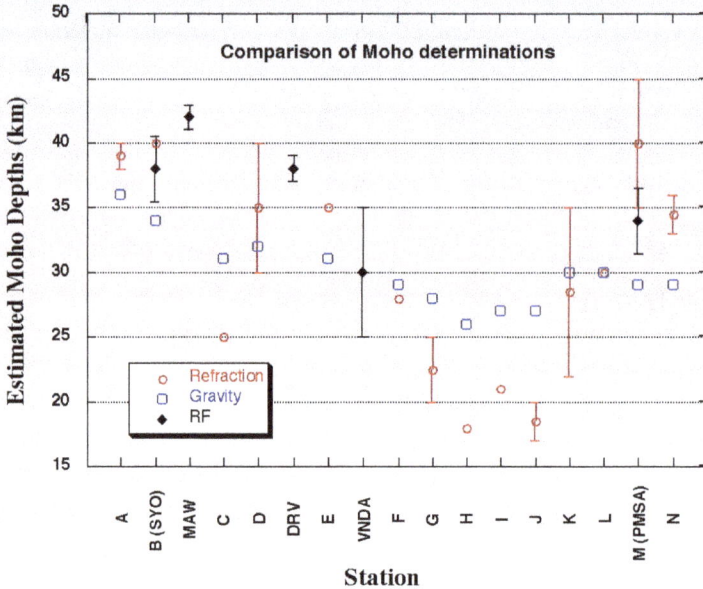

Fig. 8. Comparison of Moho depths determined from seismic refraction studies (open dots with points in center) and gravity-based estimates (open squares) (Von Frese et al., 1999), together with receiver function inversion study for permanent seismic stations (Solid diamonds; SYO, MAW, DRV, PMSA and VNDA) by this study. Alphabet numerals are location in Antarctica raveled after Von Frese et al., (1999), the same representation as in Fig. 1

4. Conclusions

In this chapter, seismic shear velocity models of the crust and the uppermost mantle were investigated by teleseismic receiver functions beneath the FDSN stations in Antarctica. In order to eliminate the starting model dependency, a non-linear GA was adopted in time domain inversion of the receiver functions. The shear velocity model beneath MAW represented a sharp Moho boundary at 44 km depth. A fairly sharp Moho was identified around 28 km and 40 km depths beneath DRV and SYO, respectively. Shear velocity variations in the crust for these stations might have a relationship with the lithologic variations of metamorphic rocks in the shallow crustal depths. Broadening low-velocity zones around 30 km depths with transitional crust-mantle boundary were identified at VNDA; which might be involved in the WARS associated with elevation of TAM. Moreover, a broad crust-mantle transition was determined around PMSA, in Antarctic Peninsula. These variations in shear velocities within the crust presumably reflected the tectonic history of each terrain where these permanent stations are belonging.

5. Acknowledgement

The authors express sincerely thankfulness to the data management centers for all the FDSN related stations in Antarctica and collaborators who were involved in data acquisition of the broadband seismic waveforms. We would like to acknowledge IRIS/Data Management System (DMS) for their grateful assistance to get PMSA and VNDA data accessible. Australian Government (AG) and the Australian Seismological Center of the Australian Geological Survey Organization (AGSO) took a grateful effort to make MAW data accessible. They also wish to acknowledge to the Ecole et Observatoire de Physique du Globe for their useful information on GEOSCOPE's DRV data accessibility. Data of SYO were available from Polar Data Center of NIPR.

We would like to express sincere thanks to Dr. C. Müller of the Alfred Wegener Institute for Polar and Marine Research, Germany, for his critical reading of the manuscript and giving useful opinions for modification. They would like to express their sincere thanks to Dr. A. Reading of Tasmania University, Prof. D. A. Wiens of Washington University in St. Luis, and Dr. M. Hoffmann in the Alfred Wegener Institute, Prof. Hiroshi Takenaka of Kyushu University, Dr. Genchi Toyokuni of NIPR for their useful discussions for analytical methods of receiver functions and regional tectonics of Antarctica.

The authors would like to express their sincere thankfulness for reviewers and the publisher for their useful comments and supports in publication management for the special issue on "Seismic Waves". The production of this paper was supported by an NIPR publication subsidy.

6. References

Ammon, C. J., Randall, G. E. & Zandt, G. (1990). On Nonuniqueness of Receiver Function Inversion, *J. Geophys. Res.* , Vol. 95, pp. 15303-15318.

Bellair, P. & Delbos, L. (1962). Age absolu de la derniere granitisation en Terre Adelie, *C. R. Ac. Sc. Paris,* Vol. 254, pp. 1465-1466.

Black, L., Harley, S. L., Sun, S. S. & McCulloch, M. T. (1987). The Rayner complex of East Antarctica: complex isotopic systematics within a Proterozoic mobile belt, *Jour. Metamor. Geol.*, Vol. 5, pp. 1-26.

Di Bona, M., Amato, A., Azzara, R., Cimini, G. B., Colombo, D. & Pondrelli, S. (1997). Constraints on the lithospheric structure beneath the Terra Nova Bay area from teleseismic P to S conversions, *In: C. A. Ricci. (ed.) The Antarctic Region: Geological Evolution and Processes, Siena, TERAPUB*, pp. 1087-1093.

Butler, R. & Anderson, K. (2008). Global Seismographic Network (GSN), *IRIS Annual Report,* pp. 6-7.

Christensen, N. I. & Mooney, W. D. (1995). Seismic velocity structure and composition of the continental crust: A global view, *J. Geophys. Res.*, Vol. 100, pp. 9761-9788.

Danesi, S. & Morelli, A. (2001). Structure of the upper mantle under the Antarctic Plate from surface wave tomography, Geophys. Res. Lett., Vol. 28, pp. 4395-4398.

Ellis, D. J. (1987). Origin and evolution of granulites in normal and thickened crusts., *Geology*, Vol. 15, pp. 167-170.

Evison, F.F., Ingham, C.E., Orr, R.H., & le Fort, J.H. (1960). Thickness of the Earth's crust in Antarctica and the surrounding oceans, *Geophys. J. R. Astron. Soc.*, Vol. 3, pp. 289-306.

Lawrence, J. F., Wiens, D. A., Nyblade. A.A., Anandakrishan, S., Shore, P. J. & Voigt, D., (2006). Upper mantle thermal variations beneath the Transantarctic Mountains inferred from teleseismic S-wave attenuation, Geophys. Res. Lett., Vol. 33, L03303, doi:10.1029/2005GL024516.

Grad, M., Guterch, A., Janik, T. & Sroda, P. (2002). Seismic characteristics of the crust in the transition zone from the Pacific Ocean to the northern Antarctic peninsula, West Antarctica, *Antarctica at the close of a millennium, Royal Society of New Zealand Bulletin*, Vol. 35, pp. 493-498.

Hansen, S. E., Nyblade, A. A., Heeszel, D. S., Wiens, D. A., Shore, P. & Kanao, M. (2010). Crustal Structure of the Gamburtsev Mountains, East Antarctica, from S-wave Receiver Functions and Rayleigh Wave Phase Velocities, *Earth Planet. Sci. Lett.*, 10.1016/j.epsl.2010.10.022.

Hiroi, Y., Shiraishi, K. & Motoyoshi, Y. (1991). Late Proterozoic paired metamorphic complexes in East Antarctica, with special reference to the tectonic significance of ultramafic rocks, *Geological Evolution of Anatrcica, M. R. A. Thomson et al. (eds.)*, *Cambridge, Cambridge Univ. Press*, pp. 83-87.

Ikami, A. & Ito, K. (1986). Crustal structure in the Mizuho Plateau, East Antarctica, by a Two-Dimensional Ray Approximation, *Jour. Geod.*, Vol. 6, pp. 271-283.

Kanao, M. & Akamatsu, J. (1995). Shear Wave Q Structure for the Lithosphere in the Lützow-Holm Bay Region, East Antarctica,Polar Geosci., Vol. 8, pp. 1-14.

Kanao, M. (1997). Variations in the crust structure of the Lüzow-Holm Bay region, East Antarctica using shear wave velocity, *Tectonophysics,* Vol. 270, pp. 43-72.

Kanao, M., Kubo, A., Shibutani, T., Negishi, H. & Tono, Y. (2002). Crustal structure around the Antarctic margin by teleseismic receiver function analyses, *In. Antarctica at the close of a millennium, Royal Society of New Zealand Bulletin,* Vol. 35, pp. 485-491.

Kind, R., Kosarev, G. L. & Petersen, N. V. (1995). Receiver functions at the stations of the German Regional Seismic Network (GRSN), *Geophys. J. Int.,* Vol. 121, pp. 191-202.

Kitamura, K., Ishikawa, M., Arima, M. & Shiraishi, K. (2001). Laboratory measurements of P-wave velocity of granulites from Lützow-Holm Complex, East Antarctica: Preliminary report, *Polar Geosci.,* Vol. 14, pp. 180-194.

Kobayashi, R. & Zhao, D. (2004). Rayleigh-wave group velocity distribution in the Antarctic region, *Phys. Earth Planet. Inter.,* Vol. 141, pp. 167-181.

Kovach, R.L. & Press, F. (1961). Surface wave dispersion and crustal strucutre in Antarctica and the surrounding oceans, *Ann. Geofis.,* Vol. 14, pp. 211-224.

Langston, C. A. (1979). Structure under Mount Rainier, Washington, inferred from teleseismic body waves, *J. Geophys. Res.,* Vol. 84, pp. 4749-4762.

Miyamachi, H., Toda, S., Matsushima, T., Takada, M., Watanabe, A., Yamashita, M. & Kanao, M. (2003). A refraction and wide-angle reflection seismic exploration in JARE-43 on the Mizuho Plateau, East Antarctica, *Polar Geosci.,* Vol. 16, pp. 1-21.

Owens, T. J., Zandt, G. & Taylor, S. R. (1984). Seismic Evidence for an Ancient Rift Beneath the Cumberland Plateau, Tennessee: A Detailed Analysis of Broadband Teleseismic P Waveforms, *J. Geophys. Res.,* Vol. 89, pp. 7783-7795.

Ritzwoller, M. H., Shapino, N. M., Levshin, A. L. & Leahy, G. M. (2001). Crustal and upper
    mantle structure beneath Antarctica and surrounding oceans, *J. Geophys. Res.*, Vol.
    106, pp. 30645-30670.
Roult, G., Rouland, D. & Montagner, J. P. (1994). Antarctia II: Upper-mantle structure from
    velocities and anisotropy, *Phys. Earth Planet. Inter.*, Vol. 84, pp. 33-57.
Sambridge, M. & Drijkoningen, G. (1992). Genetic algorithms in seismic waveform
    inversion, *Geophys. J. Int.*, Vol. 109, pp. 323-342.
Sheraton, J. W., Tingey, R. J., Black, L. P., Offe, L. A. & Ellis, D. J. (1987). Geology of Enderby
    Land and western Kemp Land, Antarctica., *Bulletin, Bureau of Mineral Resources,
    Australia*, Vol. 223, pp. 51.
Shibutani, T., Sambridge, M. & Kennett, B. L. (1996). Genetic algorithm inversion for
    receiver functions with application to crust and uppermost mantle structure
    beneath Eastern Australia, *Geophys. Res. Lett.*, Vol. 23, pp. 1829-1832.
Shiraishi, K., Ellis, D. J., Hiroi, Y., Fanning, C. M., Motoyoshi, Y. & Nakai, Y. (1994).
    Cambrian orogenic belt in East Antarctica and Sri Lanka: Implications for
    Gondwana assembly, J. Geology, Vol. 102, pp. 47-65.
Smith, A. G. & Drewry, D. J. (1984). Delayed phase change due to hot asthenosphere causes
    Transantarctic uplift ?, *Nature*, Vol. 309, pp. 536-538.
Sroda, P., Grad, M. & Guterch, A. (1997). Seismic Models of the Earth's Crustal Structure
    between the South Pacific and the Antarctic Peninsula, *The Antarctic Region:
    Geological Evolution and Processes, Siena*, pp. 685-689.
Stern, T. A. & ten Brink, U. S. (1989). Flexural uplift of the Transantarctic Mountains,
    J. Goephys. Res., Vol. 94, pp. 10,315-10,330.
Ten Brink , U. S., Hackney, R. I., Bannister, S., Stern, T. A. & Makovsky, Y. (1997). Uplift of
    Transantarctic Mountains and the bedrock beneath the East Antarctic ice sheet,
    J. Goephys. Res., Vol. 102, pp. 27,603-27,622.
Tingey, R. J. (1982). The geological evolution of the Prince Charles Mountains - an Antarctic
    Archaean cratonic block, *Antarctic Geoscience, C. Craddock, C. et al. (Eds.), Univ.
    Wisconsin Press, Madison*, pp. 455-464.
Tsuboi, S. (1995). POSEIDON (PACIFIC21), *IRIS Newsletter*, Vol. 1, pp. 8-9.
Vedova, B. D., Pellis, G., Trey, H., Zhang, J., Cooper, A. K., Marris, J. & the ACRUP Working
    Group (1997). Crustal structure of the Transantarcic Mountains, Western Ross Sea,
    *In: C. A. Ricci. (ed.) The Antarctic Region: Geological Evolution and Processes, Siena,
    TERAPUB*, pp. 609-618.
Von Frese, R. B., Tan, L., Kim, J. W. & Bentley, C. R. (1999). Antarctic crustal modeling from
    the spectral correlation of free-air gravity anomalies with the terran, *J. Geophys. Res.
    *, Vol. 104, pp. 25275-25296.
Wiens D A, Anandakrishnan S, Nyblade A, Fisher J L, Pozgay S, Shore P J, Voigt D (2003):
    Preliminary results from the Trans-Antarctic Mountains Seismic Experiment
    (TAMSEIS). IX Intern. Sympo. Antarc. Earth Sci. Programme and Abstracts: 340
    (September 8-12, Potsudam, Germany).
Wiens, D. A., Anandakrishnan, S. & Nyblade A. A. (2006). Remote detection and monitoring
    of glacial slip from Whillans Ice Streeam using seismic rayleigh waves recorded by
    the TAMSEIS array, *EOS Trans. AGU*, vol. 87, pp. 52.

Wiens, D. A. (2007). Broadband Seismology in Antarctica: Recent Progress and plans for the International Polar Year, *Proceedings of International Symposium –Asian Collaboration in IPY 2007-2008-*, March 1, Tokyo, Japan, pp. 21-24.

Yoshii, K., Ito, K., Miyamachi, H. & Kanao, M. (2004). Crustal structure derived from refractions and wide-angle reflections in the Mizuho Plateau, East Antarctica, *Polar Geosci.*, Vol. 17, pp. 112-138.

Young, D. N. & Ellis, D. J. (1991). The intrusive Mawson charnockites: evidence for a compressional plate margin setting of the Proterozoic mobile belt, *In. Geological Evolution of Antarctica, Thomson, M. R. A. et al. (Eds.), Cambridge, Cambridge Univ. Press*, pp. 25-31.

2.5-D Time-Domain Finite-Difference Modelling of Teleseismic Body Waves

Hiroshi Takenaka[1] and Taro Okamoto[2]
[1]*Kyushu University*
[2]*Tokyo Institute of Technology*
Japan

1. Introduction

Teleseismic-waveform analysis is one of the most effective approaches for study of the earthquake source process. It is also useful for investigation of the subsurface structures since the teleseismic seismograms have much information on the structures beneath the stations as well as on those near source and around the paths between the source and the stations.

Analysing the teleseismic waveforms, we often calculate the synthetic seismograms. For complex structures such as subduction zones, however, it may be difficult to calculate accurate synthetic waveforms because of strong lateral heterogeneity. The laterally varying features such as steep sea-bottom topography and thick sedimentary layers can have a large effect even on long-period teleseismic body waveforms. For example, we can expect the large-amplitude later phases as the result of the structural effect, which cannot be predicted by the flat-layered model structure usually assumed in teleseismic-waveform analysis.

A full treatment of such effect requires a three-dimensional (3-D) model of the structure and a 3-D calculation for the wavefield, which requires 3-D numerical techniques such as the 3-D finite-difference or finite-element method. Recent advances in high performance computers have already brought full 3-D elastic modelling for seismic wave propagation within reach. Even a single CPU computer could now be used for full 3-D numerical simulations by exploitation of a single or multi-GPU (Graphics Processing Units) computing (e.g., Okamoto et al., 2010). However full 3-D modelling of large-scale seismic wave propagation is still computationally expensive due to its requirements for large memory and a large number of fast processors, and would be too costly even on parallel hardwares for solutions of large-sized problems in routine-like real data analyses because of many case computations. Nevertheless, in order to provide a quantitative analysis of real seismic records from complex regions such as subduction zones, we need to be able to calculate the 3-D wavefields.

An economical approach to modelling of seismic wave propagation which includes many important aspects of the propagation process is to examine the three-dimensional response of a model where the material parameters vary two-dimensionally. Such a configuration in which a 3-D field is calculated for a 2-D medium is sometimes called *two-and-a-half-dimensional* (2.5-D) problem (e.g., Eskola & Hongisto, 1981). As a compromise between realism and computational efficiency, 2.5-D methods for calculating 3-D wavefields in 2-D varying structures have been developed. Bleistein (1986) developed the ray-theoretical implications of

2.5-D modelling for acoustic problems. Luco et al. (1990) proposed a formulation for a 2.5-D indirect boundary method using Green's functions for a harmonic moving point force in order to obtain the 3-D response of an infinitely long canyon, in a layered half-space, for plane elastic waves impinging at an arbitrary angle with respect to the axis of the canyon. Pedersen et al. (1994) also presented a 2.5-D indirect boundary element method based on moving Green's functions to study 3-D scattering of plane elastic waves by 2-D topographies. Takenaka et al. (1996) have developed the 2.5-D discrete wavenumber–boundary integral equation method, coupled with a Green's function decomposition into P and S wave contributions, to consider the problem of the interaction of the seismic wavefield excited by a point source with 2-D irregular topography. Randall (1991) developed a 2.5-D velocity-stress finite-difference technique in time domain to calculate waveforms for multipole excitation of azimuthally nonsymmetric boreholes and formations. Okamoto (1994) also presented a 2.5-D finite-difference time-domain method, coupled with the reciprocal principle, to simulate the teleseismic records of a subduction earthquake. Furumura & Takenaka (1996) have developed an efficient 2.5-D formulation for the pseudospectral time-domain method for point source excitation and have applied this approach successfully to modelling the waveforms recorded in a refraction survey. Such 2.5-D methods can calculate 3-D wavefields without huge computer memory requirements, since they require storage only slightly larger than those of the corresponding 2-D calculations.

In this article we consider a 2.5-D elastodynamic equation in the time domain for obliquely incident plane waves as a means of modelling teleseismic wavefields for media with a 2-D variation in structure. For a 2-D medium, applying a spatial Fourier transform to the 3-D time-domain elastodynamic equation in the medium-constant direction along which the material parameters are constant, we get equations in the mixed coordinate-wavenumber domain. These can be solved as independent sets of 2-D equations for a set of wavenumbers. Okamoto (1994) solved the equations for each wavenumber by the staggered-grid finite-difference time-domain method and then applied an inverse Fourier transform over wavenumber (i.e. wavenumber summation) in order to obtain theoretical seismograms in the spatial domain. His time-domain approach solves the source-free elastodynamic equation in the time domain and needs to perform a large number of 2-D calculations. On the other hand, frequency-domain methods, such as the indirect boundary methods mentioned above, require only one 2-D calculation for solving plane-wave incidence problems; they do not require wavenumber summation because in 2.5-D plane-wave incidence problems the waveslowness (medium-constant directional component) is invariant, and so at each frequency the wavenumber (medium-constant directional component) is constant and equal to that of the incident wave. It may be related to the fact that arbitrary phase shift can easily be operated in the frequency domain while in the time domain the time shift operation is more difficult. Takenaka & Kennett (1996a) proposed a 2.5-D "time-domain" elastodynamic equation for plane-wave incidence, which does not require wavenumber summation. Takenaka & Okamoto (1997) then applied the staggered-grid finite-difference technique to this new 2.5-D equation for teleseismic body-waveform synthesis. It requires computation time only similar to the corresponding 2-D ones, and could reduce the computation time by nearly three order as compared to Okamoto (1994)'s method.

In the following sections of this article we describe the 2.5-D time-domain elastodynamic equation for plane-wave incidence and a staggered-grid finite-difference scheme for solving the equation, which do not require wavenumber summation, by following Takenaka &

Kennett (1996a;b) and Takenaka & Okamoto (1997). We then show two subjects of applications done by our group: one is an example of application to source-side structures, teleseismic waveform synthesis for source inversion; the other is an example of application to receiver-side structures, modelling for receiver function analysis.

2. 2.5-D elastodynamic equation for a plane-wave incidence

We first use the physical properties of the wavefield to derive a 2.5-D elastodynamic equation in the time domain for the situation of an incident plane wave. Throughout this article we employ a Cartesian coordinate system $[x, y, z]$, where the x and y are the horizontal coordinates and z is the vertical one.

For an isotropic linear elastic medium, the source-free 3-D elastodynamic equation in the time domain is given by

$$\rho \partial_{tt} u = \partial_x \tau_{xx} + \partial_y \tau_{xy} + \partial_z \tau_{zx},$$
$$\rho \partial_{tt} v = \partial_x \tau_{xy} + \partial_y \tau_{yy} + \partial_z \tau_{yz}, \qquad (1)$$
$$\rho \partial_{tt} w = \partial_x \tau_{zx} + \partial_y \tau_{yz} + \partial_z \tau_{zz},$$

where $\rho = \rho(x, y, z)$ is the density, $[u, v, w] = \boldsymbol{u} = [u, v, w](x, y, z, t)$ are the displacements at a point (x, y, z) at time t, and the stress components are $\tau_{rs} = \tau_{rs}(x, y, z, t), (r, s = x, y, z)$. We have used a contracted notation for derivatives $\partial_{tt} \equiv \partial^2 / \partial t^2$, and $\partial_r \equiv \partial / \partial r, (r = x, y, z)$. The stress and displacement components are related by the 3-D Hooke's law through the Lamé constants $\lambda = \lambda(x, y, z)$ and $\mu = \mu(x, y, z)$ as follows:

$$\tau_{xx} = (\lambda + 2\mu)\partial_x u + \lambda(\partial_y v + \partial_z w), \quad \tau_{yy} = (\lambda + 2\mu)\partial_y v + \lambda(\partial_z w + \partial_x u),$$

$$\tau_{zz} = (\lambda + 2\mu)\partial_z w + \lambda(\partial_x u + \partial_y v), \quad \tau_{yz} = \mu(\partial_z v + \partial_y w), \qquad (2)$$

$$\tau_{zx} = \mu(\partial_x w + \partial_z u), \qquad \tau_{xy} = \mu(\partial_y u + \partial_x v).$$

Numerical modelling schemes such as the finite-difference and pseudospectral methods can compute directly discretised versions of equations (1) and (2), where the bounded computational domains are usually represented by grids.

We now derive a 2.5-D equation of motion for a 3-D wavefield in a 2-D medium which is constant with respect to one coordinate and varies with the other two coordinates. We will assume the medium is constant in the y-direction throughout the rest of this article, so that the material properties take the form

$$\lambda = \lambda(x, z), \quad \mu = \mu(x, z), \quad \rho = \rho(x, z). \qquad (3)$$

Furthermore we assume the medium includes a homogeneous half-space underlying the 2-D heterogeneous region whose top may be bounded by a free surface.

Consider an upgoing plane wave with horizontal slowness $[p_x, p_y]$, which passes a point $[x_0, y_0, z_0]$ in the homogeneous half-space at a time $t = t_0$. The 3-D wavefield at arbitrary time and position in the medium including the 2-D heterogeneous region has the characteristic of repeating itself with a certain time delay for different observers along the medium-constant axis (i.e., y-axis). For instance, the wavefield in the vertical plane $y = y_0$ at the time $t = t_0$ must be identical to that in the vertical plane $y = 0$ at the time $t = t_0 - p_y y_0$. This means

$$\boldsymbol{u}(x, y, x, t) = \boldsymbol{u}(x, 0, z, t - p_y y), \quad \tau_{rs}(x, y, x, t) = \tau_{rs}(x, 0, z, t - p_y y), \quad (r, s = x, y, z). \quad (4)$$

If the structure is invariant in both of the horizontal (x- and y-) directions so that the material properties depends only on the vertical (z-) direction (i.e., 1-D heterogeneous medium), equation (4) might reduce to

$$u(x, y, x, t) = u(0, 0, z, t - p_x x - p_y y), \quad \tau_{rs}(x, y, x, t) = \tau_{rs}(0, 0, z, t - p_x x - p_y y), \quad (r, s = x, y, z).$$
(5)

Note that this is 'Snell's law' for plane-wave propagation in a 1-D heterogeneous medium. Equation (4) is thus a 2.5-D *version* of the 'Snell's law' which is also mentioned below. Equation (5) could be used for modelling three-component seismic plane waves in vertically heterogeneous media in the time domain (JafarGandomi & Takenaka, 2007; 2010; Tanaka & Takenaka, 2005).

Let us be back to the 2.5-D problem. From relations (4), the derivatives of the displacement and the stress with respect to y can be expressed as

$$\partial_y u(x, y, x, t) = -p_y \partial_t u(x, y, z, t), \quad \partial_y \tau_{rs}(x, y, x, t) = -p_y \partial_t \tau_{rs}(x, y, z, t),$$
(6)

where $\partial_t \equiv \partial/\partial t$. The equivalent relations for the stress in (4) and (6) can be derived directly from those for the displacement in (4) and (6) through equations (2) and (3).

Substituting (6) into (1) and (2) we obtain the equation of motion

$$\rho \partial_{tt} u = \partial_x \tau_{xx} - p_y \partial_t \tau_{xy} + \partial_z \tau_{zx},$$

$$\rho \partial_{tt} v = \partial_x \tau_{xy} - p_y \partial_t \tau_{yy} + \partial_z \tau_{yz},$$
(7)

$$\rho \partial_{tt} w = \partial_x \tau_{zx} - p_y \partial_t \tau_{yz} + \partial_z \tau_{zz},$$

and the stress representations

$$\tau_{xx} = (\lambda + 2\mu)\partial_x u + \lambda(-p_y \partial_t v + \partial_z w), \quad \tau_{yy} = -(\lambda + 2\mu)p_y \partial_t v + \lambda(\partial_z w + \partial_x u),$$

$$\tau_{zz} = (\lambda + 2\mu)\partial_z w + \lambda(\partial_x u - p_y \partial_t v), \quad \tau_{yz} = \mu(\partial_z v - p_y \partial_t w),$$
(8)

$$\tau_{zx} = \mu(\partial_x w + \partial_z u), \quad \tau_{xy} = \mu(-p_y \partial_t u + \partial_x v).$$

This set of equations represents the 2.5-D elastodynamic response of a medium in the absence of source. Note that all the variables in this set of equations are real-valued. When we solve equations (7) and (8), we can set $y = y_0$, so that these equations are reduced to 2-D ones. Once equations (7) and (8) have been solved for $y = y_0$, we can deduce the wavefield at any y from that at $y = y_0$ by shifting the time origin by $p_y(y - y_0)$ (see equation (4)).

We next give an alternative derivation of these time-domain 2.5-D equations (7) and (8), from the 2.5-D equations in the frequency-wavenumber domain, and recover the characteristic of the 2.5-D wavefield, equation (4), in the process of deriving these equations. Fourier-transforming the 3-D equations (1) and (2) with respect to t and y, we obtain the following source-free 2.5-D elastodynamic equation in the frequency-wavenumber domain:

$$-\omega^2 \rho \tilde{u} = \partial_x \tilde{\tau}_{xx} - ik_y \tilde{\tau}_{xy} + \partial_z \tilde{\tau}_{zx},$$

$$-\omega^2 \rho \tilde{v} = \partial_x \tilde{\tau}_{xy} - ik_y \tilde{\tau}_{yy} + \partial_z \tilde{\tau}_{yz},$$
(9)

$$-\omega^2 \rho \tilde{w} = \partial_x \tilde{\tau}_{zx} - ik_y \tilde{\tau}_{yz} + \partial_z \tilde{\tau}_{zz},$$

and stress-displacement relations:

$$\tilde{\tau}_{xx} = (\lambda + 2\mu)\partial_x\tilde{u} + \lambda(-ik_y\tilde{v} + \partial_z\tilde{w}), \quad \tilde{\tau}_{yy} = (\lambda + 2\mu)(-ik_y)\tilde{v} + \lambda(\partial_z\tilde{w} + \partial_x\tilde{u}),$$

$$\tilde{\tau}_{zz} = (\lambda + 2\mu)\partial_z\tilde{w} + \lambda(\partial_x\tilde{u} - ik_y\tilde{v}), \quad \tilde{\tau}_{yz} = \mu(\partial_z\tilde{v} - ik_y\tilde{w}), \tag{10}$$

$$\tilde{\tau}_{zx} = \mu(\partial_x\tilde{w} + \partial_z\tilde{u}), \quad \tilde{\tau}_{xy} = \mu(-ik_y\tilde{u} + \partial_x\tilde{v}),$$

where we have used the y-invariance of the medium, i.e. equation (3), and have used the notation

$$\tilde{g}(x, k_y, z, \omega) = \frac{1}{2\pi} \int_{-\infty}^{\infty} dy\, e^{+ik_y y} \int_{-\infty}^{\infty} dt\, e^{-i\omega t} g(x, y, z, t), \tag{11}$$

for the transform to the frequency-wavenumber domain. For a fixed value of the wavenumber k_y, equations (9) and (10) depend on only two space coordinates, i.e. x and z. For each value of k_y, these equations can therefore be solved as independent 2-D equations. The invariance of the medium in the y direction means that there is no coupling between different k_y components. Whereas for full 3-D problems there would be coupling between different k_y wavenumber components. For 2.5-D problems with an incident plane wave, we need to consider only one k_y for each ω, which is that of the incident plane wave.

The inverse transform of the double Fourier transform (11) is

$$g(x, y, z, t) = \frac{1}{2\pi} \int_{-\infty}^{\infty} d\omega\, e^{+i\omega t} \int_{-\infty}^{\infty} dk_y\, e^{-ik_y y} \tilde{g}(x, k_y, z, \omega). \tag{12}$$

Changing the order of the integrations, and inserting the following relation between the wavenumber k_y and the slowness p_y:

$$k_y = \omega p_y, \tag{13}$$

equation (12) leads to

$$g(x, y, z, t) = \frac{1}{2\pi} \int_{-\infty}^{\infty} dp_y \int_{-\infty}^{\infty} d\omega\, e^{i\omega(t - p_y y)} |\omega| \tilde{g}(x, \omega p_y, z, \omega)$$

$$= \int_{-\infty}^{\infty} dp_y\, \hat{g}(x, p_y, z, t - p_y y), \tag{14}$$

where $\hat{g}$ in the time-slowness domain has been defined as

$$\hat{g}(x, p_y, z, t) \equiv \frac{1}{2\pi} \int_{-\infty}^{\infty} d\omega\, e^{i\omega t} |\omega| \tilde{g}(x, \omega p_y, z, \omega). \tag{15}$$

We then find

$$\frac{1}{2\pi} \int_{-\infty}^{\infty} d\omega\, e^{i\omega t} e^{-ik_y y} |\omega| \tilde{u}(x, k_y, z, \omega) = \hat{u}(x, p_y, z, t - p_y y), \tag{16}$$

and

$$\frac{1}{2\pi} \int_{-\infty}^{\infty} d\omega e^{i\omega t} e^{-ik_y y} ik_y |\omega| \tilde{u}(x, k_y, z, \omega) = p_y \partial_t \hat{u}(x, p_y, z, t - p_y y). \tag{17}$$

For an incident plane wave in a 2.5-D situation, the horizontal wavenumber k_y of all wavefields is constant for each ω, and equal to that of the incident plane wave. Further from (13) we require p_y to be invariant and equal to the y-component of the slowness of the incident plane wave, which represents 'Snell's law' for 2.5-D problems as mentioned above. Thus, when the slowness of the incident wave is p_{y0}, $\hat{u}(x, p_y, z, t - p_y y)$ can be represented as

$$\hat{u}(x, p_y, z, t - p_y y) = \hat{u}(x, p_{y0}, z, t - p_{y0} y)\, \delta(p_y - p_{y0}), \tag{18}$$

where $\delta(x)$ is the Dirac delta function.   Applying the inverse transform (12) to the displacement in the frequency-wavenumber domain $\tilde{u}(x, k_y, z, \omega)$, and using equations (14), (16) and (18), we obtain

$$u(x, y, z, t) = \hat{u}(x, p_{y0}, z, t - p_{y0} y). \tag{19}$$

Then,

$$\partial_t u(x, y, z, t) = \partial_t \hat{u}(x, p_{y0}, z, t - p_{y0} y). \tag{20}$$

We can recover (4) from (19) by appropriate substitutions: setting y to 0 we obtain an equation at time t and then making the particular choice $t - p_{y0} y$ gives

$$u(x, 0, z, t - p_{y0} y) = \hat{u}(x, p_{y0}, z, t - p_{y0} y) = u(x, y, z, t). \tag{21}$$

In a similar way, we can obtain the corresponding equation for the stress. Applying the partial Fourier inversions (16) and (17) to (9) and (10), and using equations (19) and (20), we recover the earlier forms (7) and (8).

In equations (7) and (8) the time derivatives appear on both sides of these equations, which may be inconvenient for direct discretisation with the finite-difference method. Here instead of direct use of equations (7) and (8), we employ the 2.5-D equation in terms of velocity-stress that is well suited to the use of the staggered-grid finite-difference technique. After some manipulation of (7) and (8) (Takenaka & Kennett, 1996b), we get the following velocity-stress formulation of the 2.5-D elastodynamic equation:

$$\partial_t \dot{u} = -p_y \rho_\beta^{-1} \mu \partial_x \dot{v} + \rho_\beta^{-1}(\partial_x \tau_{xx} + \partial_z \tau_{zx}),$$

$$\partial_t \dot{v} = -p_y \rho_\alpha^{-1} \lambda(\partial_x \dot{u} + \partial_z \dot{w}) + \rho_\alpha^{-1}(\partial_x \tau_{xy} + \partial_z \tau_{yz}),$$

$$\partial_t \dot{w} = -p_y \rho_\beta^{-1} \mu \partial_z \dot{v} + \rho_\beta^{-1}(\partial_x \tau_{zx} + \partial_z \tau_{zz}),$$

$$\partial_t \tau_{xx} = \nu \partial_x \dot{u} + \eta \partial_z \dot{w} - p_y \rho_\alpha^{-1} \lambda(\partial_x \tau_{xy} + \partial_z \tau_{yz}),$$

$$\partial_t \tau_{yy} = \rho_\alpha^{-1} \rho \lambda(\partial_x \dot{u} + \partial_z \dot{w}) - p_y \rho_\alpha^{-1}(\lambda + 2\mu)(\partial_x \tau_{xy} + \partial_z \tau_{yz}), \tag{22}$$

$$\partial_t \tau_{zz} = \eta \partial_x \dot{u} + \nu \partial_z \dot{w} - p_y \rho_\alpha^{-1} \lambda(\partial_x \tau_{xy} + \partial_z \tau_{yz}),$$

$$\partial_t \tau_{yz} = \rho_\beta^{-1} \rho \mu \partial_z \dot{v} - p_y \rho_\beta^{-1} \mu(\partial_x \tau_{zx} + \partial_z \tau_{zz}),$$

$$\partial_t \tau_{zx} = \mu(\partial_x \dot{w} + \partial_z \dot{u}),$$

$$\partial_t \tau_{xy} = \rho_\beta^{-1} \rho \mu \partial_x \dot{v} - p_y \rho_\beta^{-1} \mu(\partial_x \tau_{xx} + \partial_z \tau_{zx}),$$

where $[\dot{u}, \dot{v}, \dot{w}] = [\dot{u}, \dot{v}, \dot{w}](x, y, z, t)$ is the particle velocity, and

$$\rho_\alpha \equiv \rho - p_y^2(\lambda + 2\mu) = \rho(1 - \alpha^2 p_y^2),$$

$$\rho_\beta \equiv \rho - p_y^2 \mu = \rho(1 - \beta^2 p_y^2), \tag{23}$$

$$\nu \equiv (\lambda + 2\mu) + p_y^2 \rho_\alpha^{-1} \lambda^2, \quad \eta \equiv \lambda + p_y^2 \rho_\alpha^{-1} \lambda^2,$$

with P-wave velocity α and S-wave velocity β. When we solve (22), we can set $y = 0$ as well as the case of (7) and (8) .

3. Finite-difference scheme

We use a staggered-grid finite-difference scheme (e.g., Hayashida et al., 1999; Levander, 1988; Virieux, 1986), which is stable for any values of Poisson's ratio, making it ideal for modelling marine problems. The finite-difference grid is staggered in time and two-dimensional space (x-z plane) as shown in Fig. 1. The y-components of particle velocity $\dot{v}$ locates at the same grid points as the normal stresses both in time and space. We should note the y coordinate is not discretised but continuous. Letting $x = i\Delta x$ or $x = (i \pm 1/2)\Delta x$, $z = j\Delta z$ or $z = (j \pm 1/2)\Delta z$, and $t = l\Delta t$ or $t = (l \pm 1/2)\Delta t$; Δx and Δz are the grid spacings in the x- and z-direction, respectively, and Δt is the time step, and using Levander's notation (Levander, 1988), the difference equations for (22) are, for example,

$$
\begin{aligned}
D_t^+ \dot{u}(i,j+1/2,l) = {}& \rho_\beta^{-1}(i,j+1/2) \times \\
& [-p_y \mu(i,j+1/2) D_x^- \dot{v}(i+1/2,j+1/2,l+1/2) \\
& + D_x^- \tau_{xx}(i+1/2,j+1/2,l+1/2) + D_z^+ \tau_{zx}(i,j,l+1/2)],
\end{aligned}
$$

$$
\begin{aligned}
D_t^+ \tau_{xx}(i+1/2,j+1/2,l+1/2) = {}& \nu(i+1/2,j+1/2) D_x^+ \dot{u}(i,j+1/2,l+1) \\
& + \eta(i+1/2,j+1/2) D_z^+ \dot{w}(i+1/2,j,l+1) \\
& - p_y \rho_\alpha^{-1}(i+1/2,j+1/2)\lambda(i+1/2,j+1/2) \times \\
& [D_x^+ \tau_{xy}(i,j+1/2,l+1) + D_z^+ \tau_{yz}(i+1/2,j,l+1)],
\end{aligned}
\tag{24}
$$

where D_t^+ is forward difference operator in time, and $D_x^\pm$ and $D_z^\pm$ are the forward- or backward-difference operators in space, with sign chosen to center the difference operator about the quantity being updated. For example, in case of a second-order accurate in time and fourth-order accurate in space scheme we used,

$$
D_t^+ \dot{u}(i,j+1/2,l) = \frac{1}{\Delta t}[\dot{u}(i,j+1/2,l+1) - \dot{u}(i,j+1/2,l)],
$$

$$
\begin{aligned}
D_x^- \dot{v}(i+1/2,j+1/2,l+1/2) = {}& \frac{1}{\Delta x}\{c_1[\dot{v}(i+1/2,j+1/2,l+1/2) - \dot{v}(i-1/2,j+1/2,l+1/2)] \\
& - c_2[\dot{v}(i+3/2,j+1/2,l+1/2) - \dot{v}(i-3/2,j+1/2,l+1/2)]\},
\end{aligned}
$$

$$
\begin{aligned}
D_z^+ \tau_{zx}(i,j,l+1/2) = {}& \frac{1}{\Delta z}\{c_1[\tau_{zx}(i,j+1,l+1/2) - \tau_{zx}(i,j,l+1/2)] \\
& - c_2[\tau_{zx}(i,j+2,l+1/2) - \tau_{zx}(i,j-1,l+1/2)]\},
\end{aligned}
\tag{25}
$$

where $c_1 = 9/8$, $c_2 = 1/24$.

The fourth-order spatial finite-difference scheme usually needs more than five grid points per wavelength (Levander, 1988). The finite-difference region (computational domain) is divided into two parts: a *model zone* for the upper part and an *initial wave zone* for the lower part. The model zone fully includes the target structural model and may be heterogeneous. The initial wave zone locates under the model zone and should be homogeneous. It is needed for the incident wave, and an upgoing plane wave is input as the initial condition within it. The initial wave zone should have no velocity contrast at the interface contacting the model zone to prevent artificial reflections and conversions. The size of the computational domain is set sufficiently large so that artificial noises, such as noises from the both ends of the input plane wave, can be ignored. As Okamoto (1994) we here select the time step as 62 % of the maximum allowed time step by the stability condition for the 2-D $P - SV$ finite-difference scheme of a second-order accurate in time and fourth-order accurate in space (Levander, 1988).

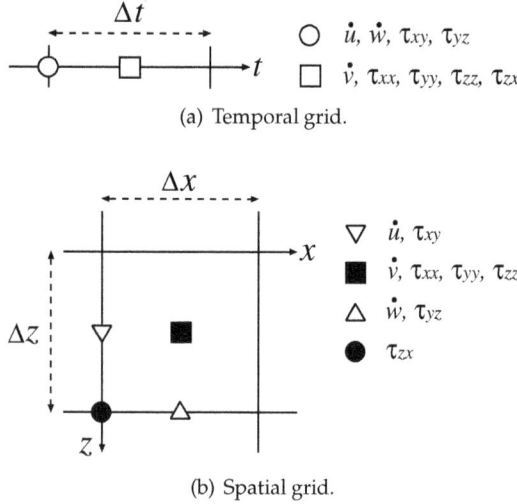

(a) Temporal grid.

(b) Spatial grid.

Fig. 1. Discretisation of (a) time and (b) space.

4. Teleseismic waveform synthesis for source inversion

Seismic displacement due to a point source of earthquake may be expressed as

$$u_n(x, t) = M_{pq}(t) * \partial_q G_{n;p}(x, t; \xi, 0), \tag{26}$$

where $M_{pq}(t)$ is moment tensor with time varying components, operation $*$ is convolution with respect to time t, and $G_{n;p}(x, t; \xi, \tau)$ is displacement Green's tensor representing the nth component of elastic displacement at a receiver position x and time t caused by a unit point force in the p-direction at a source position ξ and time τ (e.g., Aki & Richards, 2002). We have used the convention of summation over repeated suffices. In source inversions derivative of the GreenAfsÂ tensor $\partial_q G_{n;p}(x, t; \xi, 0)$ is empirically called just "Green's function". We here follow this custom.

Applying the spatial reciprocity:

$$G_{n;p}(x, t; \xi, 0) = G_{p;n}(\xi, t; x, 0), \tag{27}$$

equation (26) reduces to

$$u_n(x, t) = M_{pq}(t) * \partial_q G_{p;n}(\xi, t; x, 0)$$
$$= M_{pq}(t) * E_{pq;n}(\xi, t; x, 0), \tag{28}$$

where $E_{pq;n}(\xi, t; x, 0)$ is the strain tensor at the source position corresponding to $G_{p;n}(\xi, t; x, 0)$. This equation shows as follows. The reciprocity of the elastodynamic theory is exploited to calculate the Green's functions. The displacement at receiver due to the moment tensor at source can be calculated by solving the reciprocal problem; the reciprocal relation is applied to the response displacement at the source position due to a unit body force acting at the receiver position. In the teleseismic problem, the virtual force is applied at infinity (i.e., station location). Since we are only concerned with the teleseismic body waves, the wavefield due to this virtual force can be approximated by a plane wave. The response of the near-source structure to an incident plane wave is then calculated and converted to the far-field displacement. This approach using the reciprocity for teleseisimic body wave synthesis was employed by Bouchon (1976) where a simple method was presented, which combines the reciprocity theorem and the flat layer theory (Thomson-Haskell matrix formulation: Haskell, 1953; Thomson, 1950) to yield teleseismic body wave radiation from seismic sources embedded in the horizontally layered crustal models.

For the conversion to the far-field displacement, based on the assumption of ray propagation in the mantle, the response of the near-source structure mentioned above is multiplied by the geometrical spreading effect within the mantle upon the amplitude of the body wave and convolution with the source time function, with atenuation operator for the path in the mantle, with the response at the surface of the receiver crust for the incident impulsive teleseismic body wave, and with the instrumental response, where the actually employed response of the receiver crust often accounts for only the free surface effect at the receiver or is calculated from a very simple crustal model (this issue may be related to the subject of the next subsection). This method has been extensively used to calculate teleseismic waveforms for studies on earthquake source processes (e.g., Miyamura & Sasatani, 1986). Even now it is one of the most popular methods for calculation of teleseismic Green's functions in routine-like source inversions (e.g., Kikuchi & Kanamori, 1991; Nakamura et al., 2009).

The reciprocal formulation has the following advantages: for a single station the Green's functions for many point sources at different positions can be obtained simultaneously in a single numerical calculation, which facilitates the waveform analysis to find the best position for the earthquake source location without repeating the time-consuming numerical calculations. Although Bouchon (1976) actually treated calculation for horizontally layered structures, he mentioned in the paper that the reciprocal approach is also applicable to the case of a source in a layered structure with irregular interfaces. Wiens (1987; 1989) and Yoshida (1992) applied this approach to planar-dipping structures including the sea floor by using a ray-theoretical technique developed by Langston (1977). Furthermore Okamoto & Miyatake (1989) and Okamoto (1993) extended it to arbitrary 2-D heterogeneous structures by using the finite-difference method in time domain. However, their calculation is based on the 2-D elastodynamic equation, and is limited to two dimenions, so that the available stations are restricted to those located in the direction perpendicular to the medium-constant axis (y-axis in the previous sections). This restriction in the azimuthal coverage makes it difficult to examine the source process in detail.

In order to get teleseismic synthetics for arbitrary azimuth Okamoto (1994) proposed a method for calculating the 2.5-D telseismic body waves, which solves the time-domain version of 3-D elastodynamic equation in the mixed coordinate (x and z)-wavenumber (k_y) domain (9) and (10) for each of a number of discretised wavenumber k_y with the finite-difference time-domain scheme and perform an inverse Fourier transform over wavenumber k_y (i.e. wavenumber summation) to obtain the synthetic seismograms in the spatial domain. His method requires the computation time more than hundreds times of the corresponding 2-D computations.

Takenaka et al. (1997) presented a method without wavenumber summation so that 2.5-D teleseismic synthetics requires only computation time similar to the corresponding 2-D ones. This method uses the 2.5-D elastodynamic equation for a plane-wave incidence (22) and its finite-difference time-domain scheme proposed by Takenaka & Okamoto (1997) which was described in the previous section. It has been employed for calculating teleseismic Green's functions in several source inversion studies (e.g., Okamoto & Takenaka, 2009a;b) and for modelling the teleseismic waveforms including a near-source scattering inside a subducted plate (Kaneshima et al., 2007). We here show two results for source inversion among them as examples.

Figure 2 shows comparison of the teleseismic Green's function waveforms from the 2-D model with those from 1-D (flat-layered) models for a source inversion of the 17-July-2006 Java tsunami earthquake (M_W7.8 by Okamoto & Takenaka, 2009a, USGS PDE: 08:19:26.6, 9.284°S, 107.419°E, depth 20 km, http://earthquake.usgs.gov/research/data). The material properties of the assumed Java trench model (Fig. 2(a)) are allowed to vary with respect to the trench-perpendicular axis, while they are assumed to be invariant with respect to the trench-parallel axis. This model was constructed from the results of seismic surveys in the nearby area (Kopp et al., 2002) and global reference models (Dziewonski & Anderson, 1981; Kennett & Engdahl, 1991; Laske et al., 2001). In Fig. 2(a) the point sources for Green's function computations are located along the dip of the main-shock fault plane. The along-dip interval of source points is 8.0 km for the section from S1 to S8 and 8.1 km for the section from S8 to S15. The rigidity for sources S1-S7 is 16.3 GPa and for sources S8-S15, 38.6 GPa. The best double couple of the Global CMT solution (http://www.globalcmt.org) shown in Fig. 2(b) is employed for each Green's function computation. Following the standard 1-D teleseismic wave computations, mantle attenuation is incorporated by choosing t^* = 1.0 for P-waves and 4.0 for SH-waves, while anelastic attenuation is not included in the finite-difference computations that evaluate near-source response. The 1-D Green's functions were computed by the method of Kroeger & Geller (1983). The comparison between waveforms of the 2.5-D and flat-layered Green's functions (Fig. 2(c)) clearly illustrates the large effect of the heterogeneous structure on the body waves. The waveforms of 2.5-D Green's functions have prolonged, large amplitude later phases. They appear irrespective of station azimuth, and are not reproduced by 1-D model. At the oceanic trench regions large effect of laterally heterogeneous structure is expected to appear on the teleseismic body waveforms: thick water layer, dipping ocean bottom, and thick sediments near the source distort ray paths to teleseismic stations and often cause large later phases on the teleseismic body waveforms. This effect must be evaluated carefully before a detailed source process analysis is applied to real earthquake records.

Okamoto & Takenaka (2009b) studied strong effect of near-source structure on teleseismic body waveforms from two well-recorded aftershocks of the 2006 Java tsunami earthquake. Figure 3 shows the results of one of the two events: M_W6.1, 2006/07/17 15:45:59.8, 9.420°S 108.319°E. They applied a "waveform relocation technique" which combines a waveform inversion of source parameters with a grid search procedure to correct possible systematic bias in hypocentral parameters. In the waveform inversion 2.5-D teleseismic Green's functions are used. The grid spacing for grid search is 2 km horizontally and 1 km vertically. In Fig. 3(b), the 2.5-D synthetic seismograms are compared with the observation records. In most of the stations, peaks and troughs in the observed later arrivals are well reproduced by the synthetics. The best position (Fig. 3(a), (c)) and the mechanism (Fig. 3(b)) of the obtained point

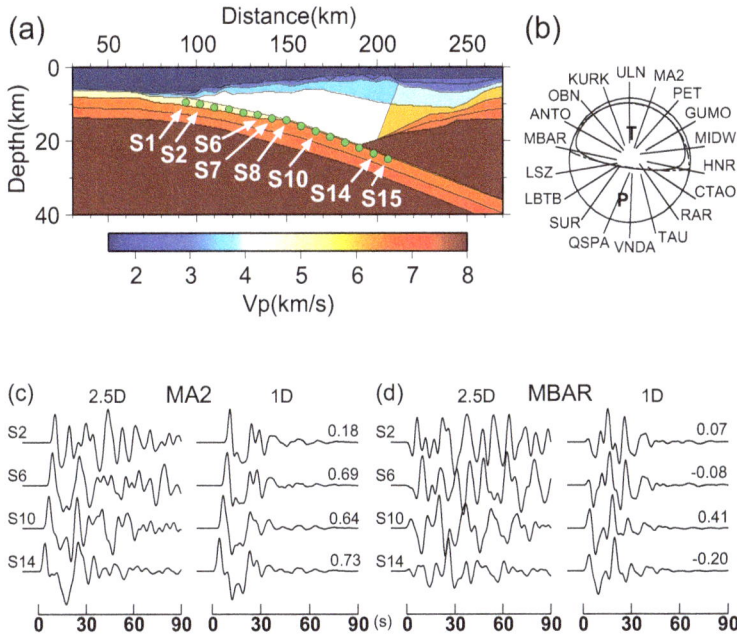

Fig. 2. (a) 2-D model of the Java trench. *P*-wave velocity is shown in colour scale (*S*-wave velocity is zero in the ocean). Green circles denote point source positions along the dip for Green's function computations. (b) Global CMT solution of the main shock and the teleseismic station coverage. (c) Examples of synthetic *P*-waveforms (Green's functions) for station MA2. 2.5D denotes those computed for the 2-D model of the near-source structure, and 1D denotes those for the flat-layered model. Attached indexes (S2-S14) denote the source positions. Numbers attached to the 1-D waveforms denote cross-correlation coefficients between 1-D and corresponding 2.5-D Green's functions for a dulation from 0 to 90 s. The 1-D model consists of a standard crust (Kennett & Engdahl, 1991) additionally overlain by a 3-km-thick ocean and 2-km-thick sediment. (d) Same as (c) but for station MBAR. (Reproduced from Okamoto & Takenaka, 2009a).

source are close to those of the Global CMT. The residual contour distribution in Fig. 3(a) and the RMS error plots in Fig. 3(c) indicate that the source location could be well constrained both vertically and horizontally. Use of the 2.5-D modelling makes it possible to obtain improved source parameters at the trench regions where only teleseismic data are available.

5. Modelling for receiver function analysis

Receiver function analysis is one of the effective and popular methods for study of the crust and upper mantle structures using teleseismic waveform data (e.g., Ammon, 1991; Cassidy, 1992; Dugda et al., 2005; Farra & Vinnik, 2000; Kanao & Shibutani, 2011; Langston, 1979; Owens et al., 1988; Saita et al., 2002; Suetsugu et al., 2004; Zhu & Kanamori, 2000). In the analysis it is often necessary to calculate synthetic waveforms for the structure models. For this purpose horizontally layered structure models have been assumed, because the response

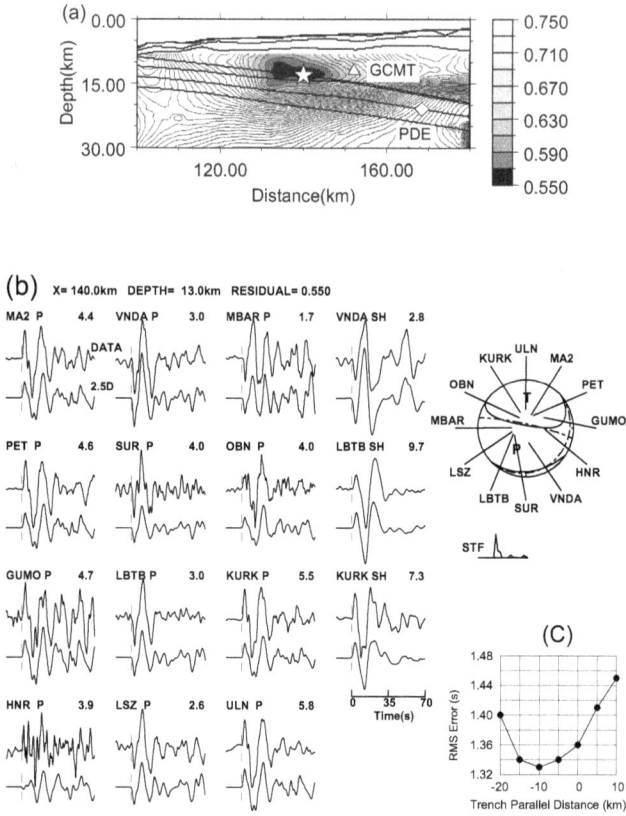

Fig. 3. (a) Open star indicates the best point source position of the 17-July-2006 event (M_W6.1). Locations of Global CMT (triangle) and PDE (diamond) are also projected. The contour lines denote the residual distribution of the grid search relocation by the waveform inversion. The contour interval is 0.02. (b) Observed (top) and 2.5-D synthetic (bottom) waveforms. Attached number denotes the maximum amplitude of the observed waveform in μm. Also plotted are the source time function (STF) and the focal mechanism. A time window of 70 s after the onset (indicated by vertical lines) is used for the inversion. The estimated moment tensor components in unit of 10^{17} Nm are: $M_{rr} = -1.81$, $M_{\theta\theta} = 4.63$, $M_{\phi\phi} = -2.81$, $M_{r\theta} = 11.1$, $M_{r\phi} = -2.04$, $M_{\theta\phi} = 0.67$, which yield a scalar moment of 1.19×10^{18} Nm (M_W6.0). (c) RMS error in travel time analysis plotted versus the distance with respect to trench-parallel axis (positive toward N116°E with the origin placed on the cross section through the PDE epicenter). Most of the travel times listed in USGS NEIC Monthly Earthquake Data Report are used. (Reproduced from "Effect of near-source trench structure on teleseismic body waveforms: an application of a 2.5D FDM to the Java trench" by T. Okamoto & H. Takenaka, in *Advances in Geosciences*, Vol. 13 (Solid Earth), Ed. Kenji Satake, Copyright (C) 2009 by World Scientific Publishing.)

of such a simple structure model to teleseismic P wave (plane wave) can be calculated easily and accurately by a semi-analytical method such as the Thomson-Haskell matrix method (Haskell, 1953; Thomson, 1950). However, for structures with strong lateral heterogeneity such as subdunction zones, it is often difficult to consider horizontally layered media for modelling the teleseismic body waves that propagate through such complex structures. Full 3-D modelling of seismic wave propagation is still computationally intensive. Takenaka & Okamoto (1997) used the 2.5-D finite-difference method described in the previous sections to simulate teleseismic seismograms at ocean-bottom stations for assessing the effect of sea-bottom topography. Ando et al. (2003) applied this approach to a profile across a realistic model of subduction zone structure to simulate the effect of a subducting slab on the receiver function waveforms observed at subduction zone, and Takenaka et al. (2004) then clearly demonstrated the azimuthal dependence of the slab-converted phases in the receiver functions. We here illustrate their results.

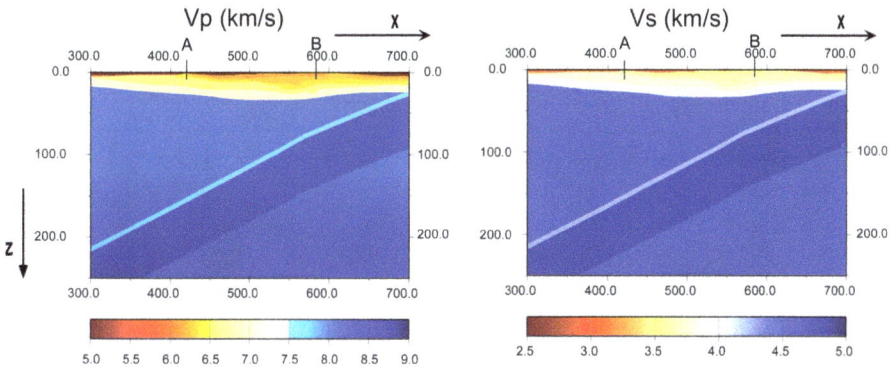

Fig. 4. P- and S-wave velocity profiles of a subduction zone model. The slab consists of two layers: the upper layer of 7 km thickness corresponding to the oceanic crust has velocities of 6 % lower relative to the mantle of the ak135 model (Kennett et al., 1995), while the lower layer corresponding to the oceanic slab mantle has velocities of 5 % higher relative to the ak135 mantle.

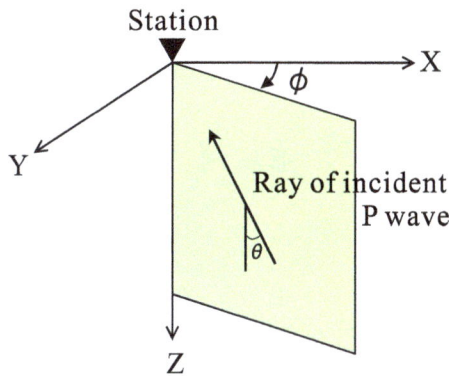

Fig. 5. Geometrical definitions. θ is the incident angle of a plane P wave, and ϕ is the backazimuth.

Figure 4 shows a cross-section of a subduction zone model which is an analogue to that of Tohoku, Japan. The shown area is the target where all the seismic phases for the receiver functions are modelled. The actual computational domain is set sufficiently larger so that artificial reflection noises from the bottom and both sides of the domain do not contaminate the synthetic seismograms. The synthetic seismograms of teleseismic P wave at the ground surface between A and B in Fig. 4 were calculated for events of epicentral distance $80°$ with backazimuths of $0°$ to $180°$ in the interval of $22.5°$. The definition of the backazimuth (ϕ) is indicated in Fig. 5. The epicentral distance $80°$ corresponds to the P-wave incident angle of around $17°$ at the surface. Three-components of the synthetic seismograms for the teleseismic events of backazimuths of $0°$ and $90°$ are shown in Fig. 6. The signal of the incident wave (source wavelet) was assumed to be a simple Gaussian pulse. The radial and transverse receiver functions for each backazimuth event can be calculated from the synthetic seismograms (Fig. 7).

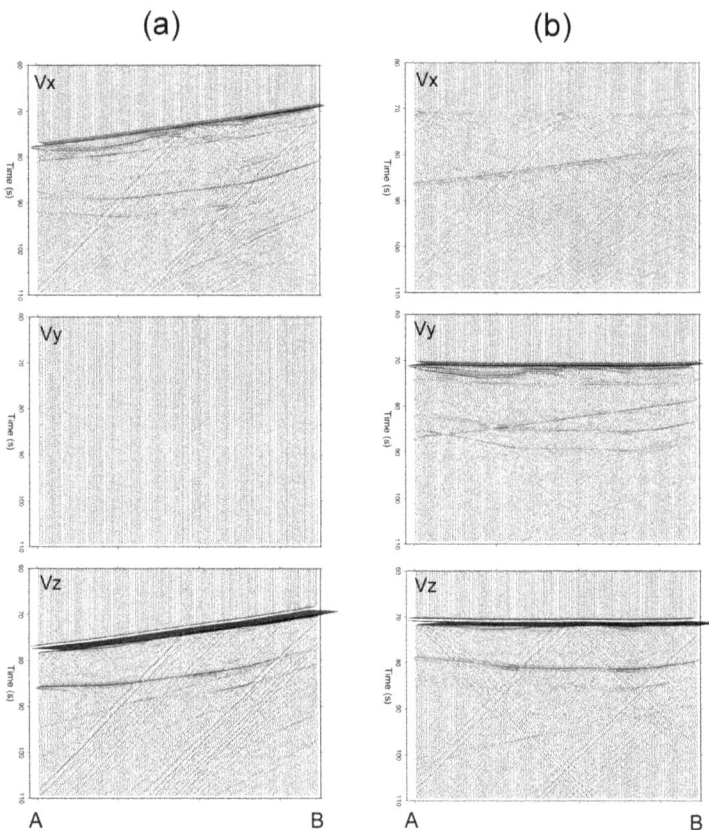

Fig. 6. Examples of three-component waveforms for events of (a) $\phi = 0°$ and (b) $\phi = 90°$. V_x, V_y, and V_z are the x-, y-, and z-components, respectively.

Figure 8(a) displays circular plots of the radial receiver functions over backazimuth of events for three stations. The receiver functions for backazimuths of more than $180°$ can be obtained from the synthetic seismograms for backazimuths of less than $180°$ through symmetrical

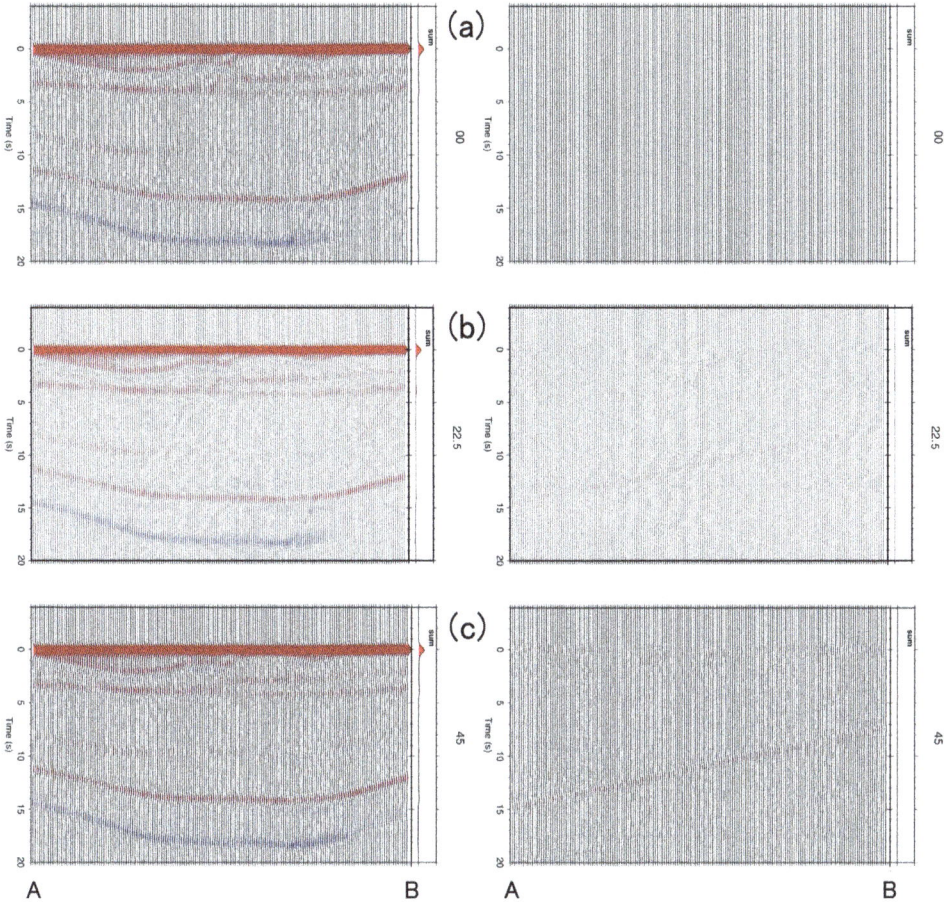

Fig. 7. Receiver function profiles between A and B at the surface (see Fig. 4) for nine teleseismic events of (a) $\phi = 0°$, (b) $\phi = 22.5°$, (c) $\phi = 45°$, (d) $\phi = 67.5°$, (e) $\phi = 90°$, (f) $\phi = 112.5°$, (g) $\phi = 135°$, (h) $\phi = 157.5°$, and (i) $\phi = 180°$. Left column: radial receiver functions. Right column: transverse receiver functions. Red and blue colours are positive and negative amplitudes, respectively. The Gaussian filter $G(\omega) = \exp[-\omega^2/(4a^2)]$ of $a = 2.5$ has been applied to all receiver functions.

properties of seismic wavefields with respect to the azimuth of the incident plane wave. Figure 8(b) shows linear plots of the radial and transverse receiver functions for the left station among the three stations shown in Fig. 8(a). In both components of the receiver functions the slab Ps phases generated at dipping interfaces are clearly seen as convex arrival patterns with the latest arrival around 17 s at backazimuth of 180°. The slab-converted phases exhibit the amplitude and arrival-time variations as a function of the backazimuth: the arrival is the latest for the backazimuth (ϕ) which is equal to the dip direction of the slab ($\phi_0 = 180°$); and the amplitude variation shows a pattern with the backazimuth like $\cos[(\phi - \phi_0)/2]$ for the radial receiver function, and $\cos(\phi - \phi_0)$ for the transverse one, respectively.

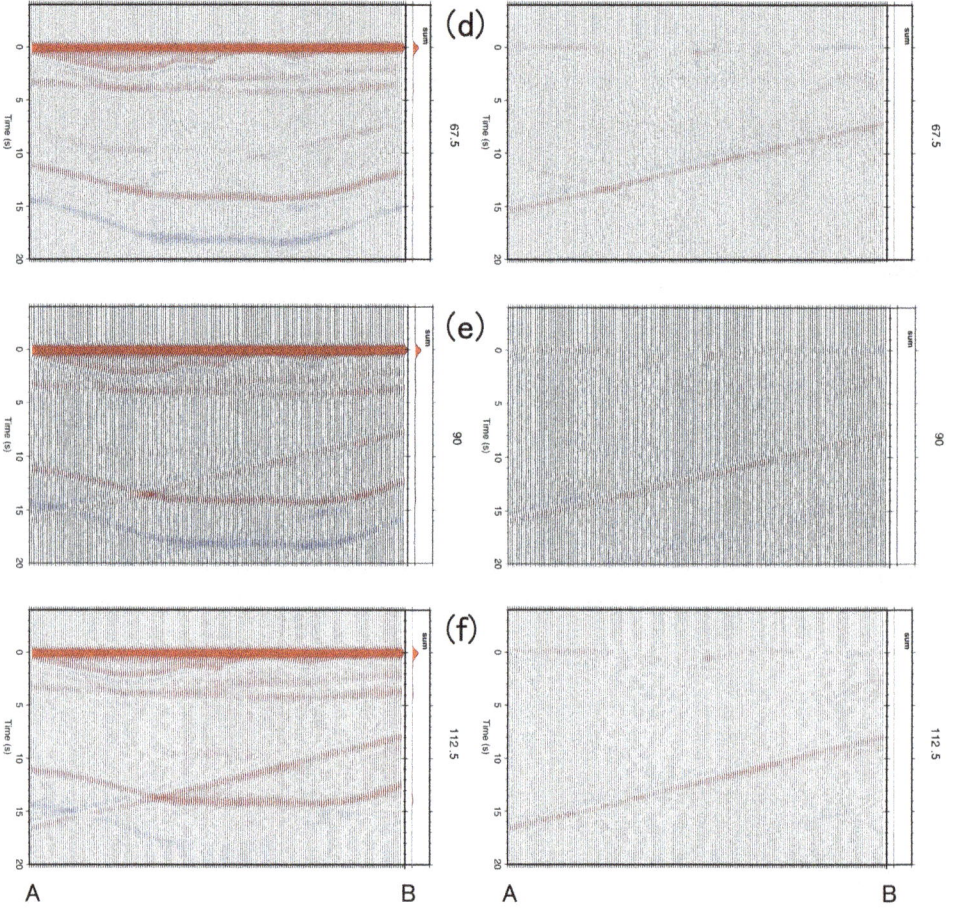

Fig. 7. (**Continued.**)

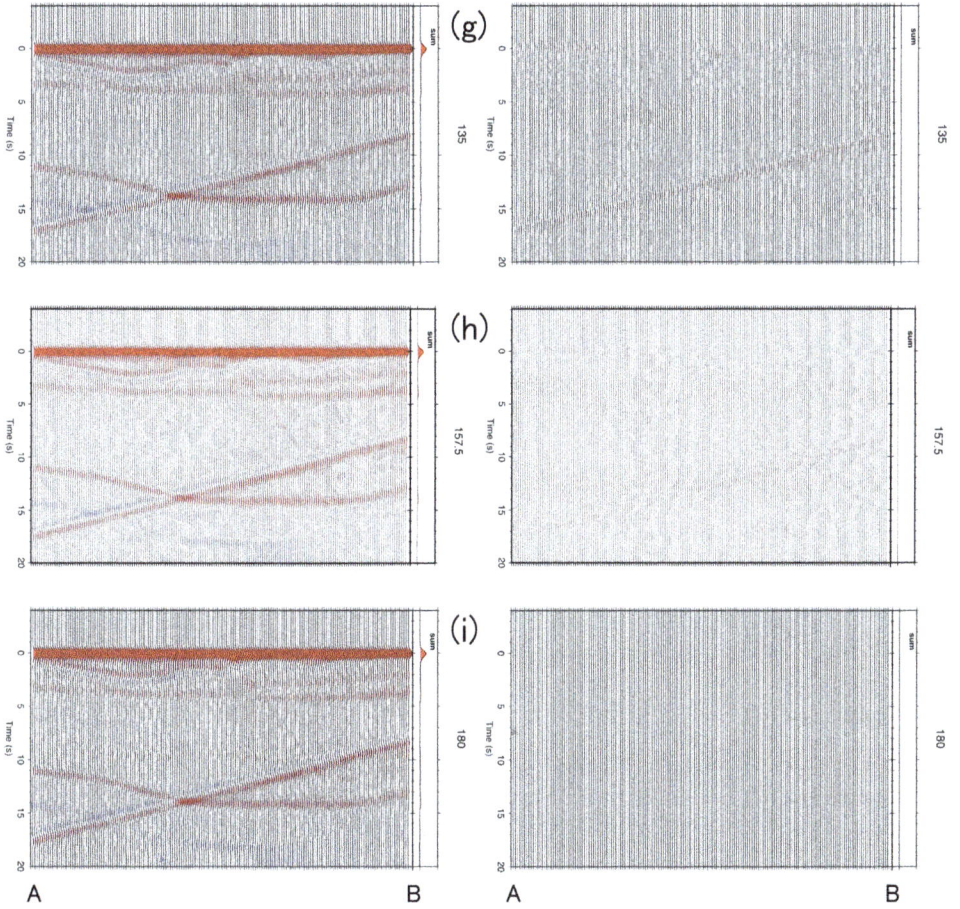

Fig. 7. **(Continued.)**

(a)

(b)

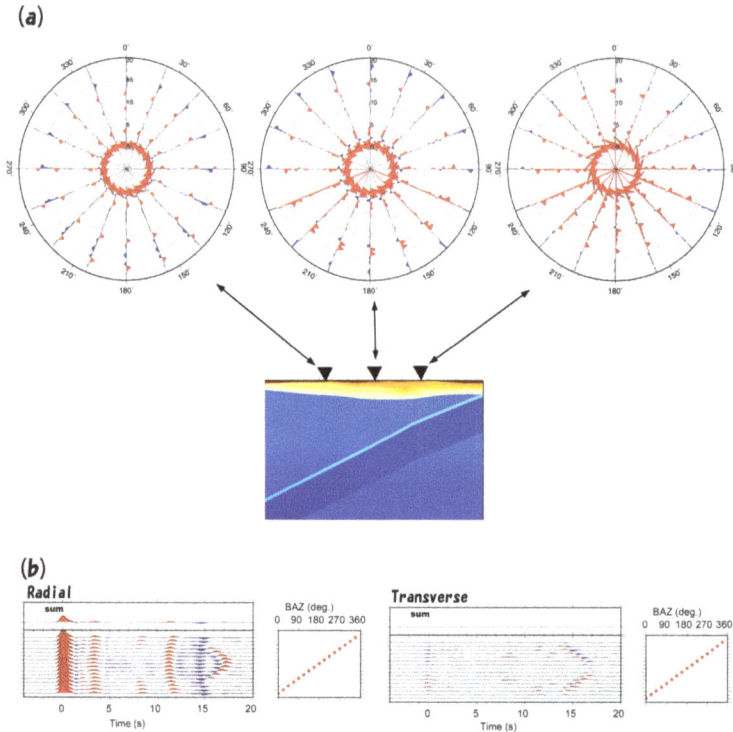

Fig. 8. Receiver functions relative to backazimuth. (a) Circular plots at three stations. Delay time is plotted from −5 to 20 s from the center of the plot outward. (b) Linear plots at the left station among the three stations. Left column: radial receiver functions and their backazimuth plots, where the backazimuth of each trace is plotted at the same level as that of the trace. Right column: transverse receiver functions and their backazimuth plots.

6. Conclusion

Two-and-half-dimensional approach in the time domain has been considered for calculating three-dimensional teleseismic body wave propagation in two-dimensionally varying medium. It is an economical approach for calculating 3-D wavefields in real problems, and requires storage only slightly larger than, and computation time only slightly longer than those of the corresponding 2-D calculation. A finite-difference scheme solving the 2.5-D elastodynamic equation for 3-D seismic response of a 2-D structure model due to an obliquely incident plane wave has been described. The modelling of such seismic wavefields for a 2.5-D situation with an incident plane wave is of considerable practical importance. For instance, this approach can be applied to modelling of teleseismic body waves observed on complex crustal structures or radiated from shallow earthquakes occurring in subduction zones, where the laterally heterogeneous media can have large effects on the waveforms. We showed some numerical examples which include modelling of teleseismic body waves for the earthquake source analysis and the receiver function analysis. The method described here is very efficient, so that we expect it could be used for waveform inversions for routine-like source retrievals and subsurface structure reconstructions from teleseismic seismograms in the near future, which need many forward modelling computations.

7. Acknowledgment

We thank Prof. Masaki Kanao for inviting us to submit this article as a chapter of this book. We also thank Dr. Takumi Murakoshi, Mr. Toshihiko Ando, and Dr. Genti Toyokuni for helping to make figures on receiver function analysis.

8. References

Aki, K. & Richards, P. (2002). *Quantitative Seismology (second ed.)*, University Science Books, Sausalito, California.

Ammon, C. J. (1991). The isolation of receiver effects from teleseismic *P* waveform, *Bull. Seism. Soc. Amer.* 81, 2504-2510.

Ando, T., Takenaka, H., Okamoto, T. & Murakoshi, T (2003). Modeling of receiver functions for laterally heterogeneous structures using a 2.5D FDM, *American Geophysical Union 2003 Fall Meeting, Eos Trans. AGU* 84 (46), S32A-0840, San Francisco, California, December 2003.

Bleistein, N. (1986). Two-and-one-half dimensional in-plane wave propagation, *Geophys. Prospect.* 34, 686–703.

Bouchon, M. (1976). Teleseismic body wave radiation from a seismic source in a layered medium, *Geophys. J. R. Astr. Soc.* 47, 515–530.

Cassidy, J. F. (1992). Numerical experiments in broadband receiver function analysis, *Bull. Seism. Soc. Am.* 82(3), 1453–1474.

Dugda, M. T., Nyblade, A. A., Julia, J., Langston, C. A., Ammon, C. J. & Simiyu, S. (2005). Crustal structure in EthiopiaA@and Kenya from receiver function analysis: Implications for rift development in eastern Africa, *J. Geophys. Res.*, 110, B01303, doi:10.1029/2004JB003065.

Dziewonski, A. M. & Anderson, D. L. (1981). Preliminary reference Earth model, *Phys. Earth Planet. Inter.* 25, 297–356.

Eskola, L. & Hongisto, H. (1981). The solution of the stationary electric field strength and potential of a point current source in a $2\frac{1}{2}$-dimensional environment, *Geophys. Prospect.* 29, 260–273.

Farra, V. & Vinnik, L. (2000). Upper mantle stratification by *P* and *S* receiver functions, *Geophys. J. Int.* 14(3), 699–712.

Furumura, T. & Takenaka, H. (1996). 2.5-D modelling of elastic waves using the pseudospectral method, *Geophys. J. Int.* 124, 820–832.

Haskell, N. A. (1953). The dispersion of surface waves in multilayered media, *Bull. Seism. Soc. Am.* 43, 17–34.

Hayashida, T., H. Takenaka, H. & Okamoto, T. (1999). Development of 2D and 3D codes of the velocity-stress staggered-grid finite-difference method for modeling seismic wave propagation, *Sci. Repts., Dept. Earth & Planet. Sci., Kyushu Univ.* 20(3), 99–110 (in Japanese with English abstract).

JafarGandomi, A. & Takenaka, H. (2007). Efficient FDTD algorithm for plane-wave simulation for vertically heterogeneous attenuative media, *Geophysics* 72(4), H43–H45.

JafarGandomi, A. & Takenaka, H. (2010). Three-component 1D viscoelastic FDM for plane-wave incidence, *Advances in Geosciences, Volume 20: Solid Earth (SE)*, edited by Kenji Satake, World Scientific Publishing Company (ISBN-10: 981-283-817-1; ISBN-13: 978-981-283-817-9), Singapore, 299–312.

Kanao, M. & Shibutani, T. (2011). Shear wave velocity models beneath Antarctic margins inverted by genetic algorithm for teleseismic receiver functions, this book.

Kaneshima, S., T. Okamoto, T. & Takenaka, H. (2007). Evidence for a metastable olivine wedge inside the subducted Mariana slab, *Earth and Planetary Science Letters* 258(1-2), 219–227.

Kennett, B. L. N. & Engdahl, E. R. (1991). Traveltimes for global earthquake location and phase identification, *Geophys. J. Int.* 105, 429–465.

Kennett, B. L. N., Engdahl, E. R. & Buland, R. (1995). Constraints on seismic velocities in the Earth from travel times. *Geophys. J. Int.* 122, 108–124.

Kikuchi, M. & Kanamori, H. (1991). Inversion of complex body waves-III, *Bull. Seism. Soc. Am.* 81(6), 2335–2350.

Kopp, H., Klaeschen, D., Flueh, E. R. & Bialas, J. (2002). Crustal structure of the Java margin from seismic wide-angle and multichannel reflection data, *J. Geophys. Res.* 107, doi:10.1029/2000JB000095.

Kroeger, G. C. & Geller, R. J. (1983). An efficient method for computing synthetic reflections for plane layered models, *EOS Trans. AGU* 64, 772.

Langston, C. A. (1977). The effect of the planar dipping structure on source and receiver responses for constant ray parameter, *Bull. Seism. Soc. Am.* 67, 1029–1050.

Langston, C. A. (1979). Structure under Mount Rainier, Washington, inferred from teleseismic body waves, *J. Geophys. Res.* 84, 4749–4762.

Laske, G., Masters, G. & Reif, C. (2001). CRUST 2.0: A new global crustal model at 2×2 degrees, http://mahi.ucsd.edu/Gabi/rem.html.

Levander, A.R. (1988). Fourth-order finite-difference *P-SV* seismograms, *Geophysics* 53, 1425–1436.

Luco, J.E., Wong, H.L. & De Barros, F.C.P. (1990). Three-dimensional response of a cylindrical canyon in a layered half-space, *Earthquake Eng. Struct. Dyn.* 19, 799–817.

Miyamura, J. & Sasatani, T. (1986). Accurate determination of source depths and focal mechanisms of shallow earthquakes occurring at the junction between the Kurile and the Japan trenches *Journal of the Faculty of Science, Hokkaido University. Series 7, Geophysics* 8(1), 37–63. http://hdl.handle.net/2115/8752

Nakamura, T., Tsuboi, S., Kaneda, Y. & Yamanaka, Y. (2009). Rupture process of the 2008 Wenchuan, China earthquake inferred from teleseismic waveform inversion and forward modeling of broadband seismic waves, *Tectonophysics* 491(1–4), 72–84, doi 10.1016/j.tecto.2009.09.020.

Okamoto, T. (1993). Effects of sedimentary structure and bathymetry near the source on teleseismic *P* waveforms from shallow subduction zone earthquakes, *Geophys. J. Int.* 112, 471–480.

Okamoto, T. (1994). Teleseismic synthetics obtained from 3-D calculations in 2-D media, *Geophys. J. Int.* 118, 613–622.

Okamoto, T. & Miyatake, T. (1989). Effects of near source seafloor topography on long-period teleseismic *P* waveforms, *Geophys. Res. Lett.* 16, 1309–1312.

Okamoto, T. & Takenaka, H. (2009a). Waveform inversion for slip distribution of the 2006 Java tsunami earthquake by using 2.5D finite-difference Green's function, *Earth Planets*

Space 61(5), e17–e20.
http://www.terrapub.co.jp/journals/EPS/pdf/2009e/6105e017.pdf
Okamoto, T. & Takenaka, H. (2009b). Effect of near-source trench structure on teleseismic body waveforms: an application of a 2.5D FDM to the Java Trench, *Advances in Geosciences, Volume 13: Solid Earth (SE)*, edited by Kenji Satake, World Scientific Publishing Company, Singapore (ISBN-10: 981-283-617-9; ISBN-13: 978-981-283-617-5), 215–229.

Okamoto, T., Takenaka, H., Nakamura, T. & Aoki, T. (2010). Accelerating large-scale simulation of seismic wave propagation by multi-GPUs and three-dimensional domain decomposition, *Earth Planets Space* 62(12), 939–942.
http://www.terrapub.co.jp/journals/EPS/abstract/6212/62120939.html

Owens, T. J., Crosson, R. S. & Hendrickson, M. A. (1988). Constraints on the subduction geometry beneath western Washington from broadband teleseismic waveform modeling, *Bull. Seism. Soc. Am.* 78(3), 1319–1334.

Pedersen, H.A., Sánchez-Sesma, F.J. & Campillo M. (1994). Three-dimensional scattering by two-dimensional topographies, *Bull. seism. Soc. Am.* 84, 1169–1183.

Randall, C.J. (1991). Multipole acoustic waveforms in nonaxisymmetric boreholes and formations, *J. Acoust Soc. Am.* 90, 1620–1631.

Saita, T., Suetsugu, D., Ohtaki, T., Takenaka, H., Kanjo, K. & Purwana, I. (2002). Transition zone thickness beneath Indonesia as inferred using the receiver function method for data from the JISNET regional broadband seismic network, *Geophys. Res. Lett.* 29(7), doi:10.1029/2001GL013629.

Suetsugu, D., Saita, T., Takenaka, H. & Niu, F. (2004). Thickness of the mantle transition zone beneath the South Pacific as inferred from analyses of ScS reverberated and Ps converted waves, *Phys. Earth Planet. Inter.* 146, 35–46, doi:10.1016/j.pepi.2003.06.008.

Takenaka, H. & B.L.N. Kennett (1996a). A 2.5-D time-domain elastodynamic equation for plane-wave incidence, *Geophys. J. Int.* 125, F5–F9.

Takenaka, H. & B.L.N. Kennett (1996b). A 2.5-D time-domain elastodynamic equation for a general anisotropic medium, *Geophys. J. Int.* 127, F1–F4.

Takenaka, H., Kennett, B.L.N., & Fujiwara, H. (1996). Effect of 2-D topography on the 3-D seismic wavefield using a 2.5-D discrete wavenumber - boundary integral equation method, *Geophys. J. Int.* 124, 741–755.

Takenaka, H. & Okamoto, T. (1997). Teleseismic waveform synthesis for ocean-bottom stations using a new, very effective 2.5-D finite difference technique, *Proceedings of International Workshop on Scientific Use of Submarine Cables*, pp. 23–26, Okinawa, Japan, February 1997.

Takenaka, H., Okamoto, T. & Kennett, B. L. N. (1997). A very effective 2.5-D method for teleseismic body-wave synthesis, *Programme and Abstracts, The Seismological Society of Japan 1997, No.2*, B65 (in Japanese), Hirosaki, Japan, September 1997.

Takenaka, H., Ando, T. & Okamoto, T. (2004). Investigation on azimuthal dependence of slab-converted phases observed in receiver functions by 2.5-D numerical simulations, *Abstracts 2004 Japan Earth and Planetary Science Joint Meeting*, I021-002, Chiba, May 2004.

Tanaka, H. & Takenaka, H. (2005). An efficient FDTD solution for plane-wave response of vertically heterogeneous media, *Zisin (Journal of the Seismological Society of Japan)* 57(3), 343–354 (in Japanese with English abstract).

`http://www.journalarchive.jst.go.jp/english/jnltop_en.php?`
`cdjournal=zisin1948`

Thomson, W. T. (1950). Transmission of elastic waves through the stratified solid medium, *J. Appl. Phys.* 21, 89–93.

Virieux, J. (1986). *P-SV* wave propagation in heterogeneous media: Velocity-stress finite-difference method, *Geophysics* 51, 889–901.

Wiens, D. A. (1987). Effects of near source bathymetry on teleseismic *P* waveforms, *Geophys. Res. Lett.* 14, 761–764.

Wiens, D. A. (1989). Bathymetric effects on body waveforms from shallow subduction zone earthquakes and application to seismic processes in the Kurile trench, *J. Geophys. Res.* 94, 2955–2972.

Yoshida, S. (1992). Waveform inversion for rupture process using a non-flat seafloor model: application to 1986 Andreanof islands and 1985 Chile earthquakes, *Tectonophysics* 211, 45–59.

Zhu, L. & Kanamori, H. (2000). Moho depth variation in southern California from teleseismic receiver functions, *J. Geophys. Res.* 105, 2969–2980.

Permissions

The contributors of this book come from diverse backgrounds, making this book a truly international effort. This book will bring forth new frontiers with its revolutionizing research information and detailed analysis of the nascent developments around the world.

We would like to thank Masaki Kanao, for lending his expertise to make the book truly unique. He has played a crucial role in the development of this book. Without his invaluable contribution this book wouldn't have been possible. He has made vital efforts to compile up to date information on the varied aspects of this subject to make this book a valuable addition to the collection of many professionals and students.

This book was conceptualized with the vision of imparting up-to-date information and advanced data in this field. To ensure the same, a matchless editorial board was set up. Every individual on the board went through rigorous rounds of assessment to prove their worth. After which they invested a large part of their time researching and compiling the most relevant data for our readers. Conferences and sessions were held from time to time between the editorial board and the contributing authors to present the data in the most comprehensible form. The editorial team has worked tirelessly to provide valuable and valid information to help people across the globe.

Every chapter published in this book has been scrutinized by our experts. Their significance has been extensively debated. The topics covered herein carry significant findings which will fuel the growth of the discipline. They may even be implemented as practical applications or may be referred to as a beginning point for another development. Chapters in this book were first published by InTech; hereby published with permission under the Creative Commons Attribution License or equivalent.

The editorial board has been involved in producing this book since its inception. They have spent rigorous hours researching and exploring the diverse topics which have resulted in the successful publishing of this book. They have passed on their knowledge of decades through this book. To expedite this challenging task, the publisher supported the team at every step. A small team of assistant editors was also appointed to further simplify the editing procedure and attain best results for the readers.

Our editorial team has been hand-picked from every corner of the world. Their multi-ethnicity adds dynamic inputs to the discussions which result in innovative outcomes. These outcomes are then further discussed with the researchers and contributors who give their valuable feedback and opinion regarding the same. The feedback is then collaborated with the researches and they are edited in a comprehensive manner to aid the understanding of the subject.

Apart from the editorial board, the designing team has also invested a significant amount of their time in understanding the subject and creating the most relevant covers. They scrutinized every image to scout for the most suitable representation of the subject and create an appropriate cover for the book.

The publishing team has been involved in this book since its early stages. They were actively engaged in every process, be it collecting the data, connecting with the contributors or procuring relevant information. The team has been an ardent support to the editorial, designing and production team. Their endless efforts to recruit the best for this project, has resulted in the accomplishment of this book. They are a veteran in the field of academics and their pool of knowledge is as vast as their experience in printing. Their expertise and guidance has proved useful at every step. Their uncompromising quality standards have made this book an exceptional effort. Their encouragement from time to time has been an inspiration for everyone.

The publisher and the editorial board hope that this book will prove to be a valuable piece of knowledge for researchers, students, practitioners and scholars across the globe.

List of Contributors

Masaki Kanao
National Institute of Polar Research, Tokyo
Japan

Alessia Maggi
Institut de Physique du Globe de Strasbourg, CNRS and University of Strasbourg, France

Yoshiaki Ishihara
National Astronomical Observatory, National Institutes of Natural Sciences, Iwate, Japan

Masa-yuki Yamamoto
Kochi University of Technology, Kochi, Japan

Kazunari Nawa
National Institute of Advanced Industrial Science and Technology, Tsukuba, Japan

Akira Yamada
Geodynamics Research Center, Ehime University, Ehime, Japan

Terry Wilson
Ohio State University, USA

Tetsuto Himeno and Genchi Toyokuni
National Institute of Polar Research, Tokyo, Japan

Seiji Tsuboi and Yoko Tono
Japan Agency for Marine-Earth Science and Technology, Yokohama, Japan

Kent Anderson
Incorporated Research Institutions for Seismology, Washington, DC, USA

Akihiro Takeuchi
1Earthquake Prediction Research Center, Institute of Oceanic Research and Development, Tokai University 3-20-1 Orido, Shimizu-ku, Shizuoka 424-8610,

Kan Okubo
Division of Information and Communications Systems Engineering, Tokyo Metropolitan University 6-6 Asahigaoka, Hino 191-0065, Japan

Nobunao Takeuchi
Research Center for Predictions of Earthquakes and Volcanic Eruptions, Graduate School of Sciences, Tohoku University, 6-6 Aza-aoba, Aramaki, Aoba-ku, Sendai 980-8578, Japan

Antonio Lira and Jorge A. Heraud
Pontificia Universidad Católica del Perú, Peru

Sri Atmaja P. Rosyidi
Universitas Muhammadiyah Yogyakarta, Indonesia

Mohd. Raihan Taha
Universiti Kebangsaan Malaysia, Malaysia

Genti Toyokuni and Masaki Kanao
National Institute of Polar Research, Japan

Hiroshi Takenaka
Department of Earth and Planetary Sciences, Faculty of Sciences, Kyushu University, Japan

Snezana Gjorgji Stamatovska
'Ss. Cyril and Methodius' University-Institute of Earthquake Engineering and Engineering
Seismology (UKIM-IZIIS), Skopje, Republic of Macedonia

Hervé Chauris and Daniela Donno
Centre de Géosciences, Mines Paristech, UMR Sisyphe 7619, France

Louis De Barros and Gareth S. O'Brien
School of Geological Sciences, University College Dublin, Ireland

Bastien Dupuy, Jean Virieux and Stéphane Garambois
ISTerre, CNRS - Université J. Fourier, Grenoble, France

Yoshio Murai
Hokkaido University, Japan

Jun Matsushima
The University of Tokyo, Japan

Behzad Alaei
Rocksource ASA, Norway

Jean Virieux
ISTerre, Université Joseph Fourier, Grenoble, France
Romain Brossier, Emmanuel Chaljub, Olivier Coutant and Stéphane Garambois, France

Vincent Etienne
GeoAzur, Centre National de la Recherche Scientifique, Institut de Recherche
pour le développement, France
Diego Mercerat, Vincent Prieux, Stéphane Operto and Alessandra Ribodetti, France

Victor Cruz-Atienza
Instituto de Geofisica, Departamento de Sismologia, Universidad Nacional
Autónoma de México, Mexico
Josué Tago, Mexico

Masaki Kanao
National Institute of Polar Research, Research Organization of Information and Systems,
Tokyo, Japan

Takuo Shibutani
Disaster Prevention Research Institute, Kyoto University, Gokasho, Kyoto, Japan

Hiroshi Takenaka
Kyushu University, Japan

Taro Okamoto
Tokyo Institute of Technology, Japan